# Advancing Environmental Research Through Applied GIS and Remote Sensing

Jamal Al Karkouri
*Ibn Tofail University, Morocco*

Adil Moumane
*Ibn Tofail University, Morocco*

Abdessamad Elmotawakkil
*Ibn Tofail University, Morocco*

Mouhcine Batchi
*Ibn Tofail University, Morocco*

IGI Global
Scientific Publishing
Publishing Tomorrow's Research Today

Vice President of Editorial Melissa Wagner
Director of Acquisitions Mikaela Felty
Director of Book Development Jocelynn Hessler
Production Manager Mike Brehm
Cover Design Jose Rosado

Published in the United States of America by
IGI Global Scientific Publishing
701 East Chocolate Avenue
Hershey, PA, 17033, USA
Tel: 717-533-8845 | Fax: 717-533-7115
Website: https://www.igi-global.com E-mail: cust@igi-global.com



Library of Congress Cataloging-in-Publication Data

Names: Al Karkouri, Jamal, 1961- editor | Moumane, Adil, 1984- editor | Elmotawakkil, Abdessamad, 1992- editor | Batchi, Mouhcine, 1981- editor
Title: Advancing environmental research through Applied GIS and Remote Sensing / edited by Jamal Al Karkouri, Adil Moumane, Abdessamad Elmotawakkil, Mouhcine Batchi.
Other titles: Advancing environmental research through applied Geographic information systems & Remote Sensing
Description: Hershey, PA : IGI Global Scientific Publishing, [2026] | Includes bibliographical references and index. | Summary: "This book showcases how GIS and remote sensing technologies are applied to address climate, water, land, and ecosystem challenges. Featuring methodological advances and case studies, it highlights innovative geospatial approaches for environmental monitoring, resilience building, and sustainable resource management"-- Provided by publisher.
Identifiers: LCCN 2025040074 (print) | LCCN 2025040075 (ebook) | ISBN 9798337366081 h/c | ISBN 9798337366098 s/c | ISBN 9798337366104 ebook
Subjects: LCSH: Environmental sciences--Geographic information systems | Environmental sciences--Remote sensing | Geographic information systems | Remote sensing
Classification: LCC G70.212 .A4265 2026 (print) | LCC G70.212 (ebook) | DDC 910.285--dc23/eng/20260202
LC record available at https://lccn.loc.gov/2025040074
LC ebook record available at https://lccn.loc.gov/2025040075
British Cataloguing in Publication Data
A Cataloguing in Publication record for this book is available from the British Library.

# Table of Contents

**Preface** .......... xvii

**Acknowledgment** .......... xx

**Chapter 1**
Flood Risk and Inundation Mapping in a Changing Climate: Advances in GIS, Remote Sensing, and Artificial Intelligence Methods .......... 1
*Mehmet Ali Celik, Igdir University, Turkey*
*Ahmed Elbeltagi, Mansoura University, Egypt*
*Rajat Mishra, G.B. Pant University of Agriculture and Technology, India*
*Shekhar Singh, G.B. Pant University of Agriculture and Technology, India*
*Dinesh Kumar Vishwakarma, G.B. Pant University of Agriculture and Technology, India*
*Kusum Pandey, G.B. Pant National Institute of Himalayan Environment, India*
*Muhammed Ernur Akiner, Akdeniz University, Turkey*

**Chapter 2**
Shrinkage of the Lake Dayet Er-Roumi, Morocco: A Decadal Remote Sensing Analysis .......... 35
*Adil Moumane, Ibn Tofail University, Morocco*

**Chapter 3**
Flood Risk Assement Using AHP, Frequency Ratio, Logistic Regression, and Random Forest in Naraipur Municipality .......... 63
*Rubi Chaulagain, Westernal Regional Campus, Nepal*
*Rekha Paudel, Westernal Regional Campus, Nepal*
*Fatima Ezzahra El Ghazali, Geosciences, Georesources, and Environment, Semlalia Laboratory, Department of Geology, Faculty of Sciences Semlalia, Cadi Ayyad University, Morocco*
*Adil Moumane, Ibn Tofail University, Morocco*

**Chapter 4**
Land Use and Environmental Dynamics of the Moroccan Coastal Strip 1984–2025: A Google Earth Engine Approach for Monitoring Vegetation, Water, and Urban Transformation - Case Study ........ 91
*Tarik Bahouq, Ibn Tofail University, Morocco*
*Amina Moumane, Faculté des Sciences Humaines et Sociales, Departement de Geographie, Ibn Tofail University, Morocco*
*Nadia Touhami, Faculté des Sciences Humaines et Sociales, Departement de Geographie, Ibn Tofail University, Morocco*

**Chapter 5**
Harnessing Precision Agriculture and Artificial Intelligence in Sustainable Farming ........ 119
*Muhammad Safdar, Agricultural Remote Sensing Lab, University of Agriculture, Faisalabad, Pakistan*
*Wasiq Farooq, Agricultural Remote Sensing Lab, University of Agriculture, Faisalabad, Pakistan*
*Muhammad Sajid Mehmood, School of Tourism and Planning, Pingdingshan University, Pingdingshan, China*
*Abdul Rauf, Department of Irrigation and Drainage, University of Agriculture, Faisalabad, Pakistan*
*Amina Rashid, Department of Agronomy, University of Agriculture, Faisalabad, Pakistan*
*Kashif Mehmood, Department of Irrigation and Drainage, University of Agriculture, Faisalabad, Pakistan*
*Muntaha Munir, Institute of Botany, University of the Punjab, Lahore, Pakistan*
*Nalain E. Muhammad, Department of Irrigation and Drainage, University of Agriculture, Faisalabad, Pakistan*
*Hafiz Muhammad Bilawal Akram, Department of Agronomy, University of Agriculture, Faisalabad, Pakistan*
*Hafiz Muhammad Mohsin Raza, Department of Agricultural Sciences, Allama Iqbal Open University, Islamabad, Pakistan*
*Aisha Nazir, Institute of Botany, University of the Punjab, Lahore, Pakistan*
*Hafiz Muhammad Awais, Agricultural Remote Sensing Lab, University of Agriculture, Faisalabad, Pakistan*

**Chapter 6**
Applications of Artificial Intelligence and Remote Sensing in Water Resources Management ........ 147
*Wasiq Farooq, Agricultural Remote Sensing Lab, University of Agriculture, Faisalabad, Pakistan*
*Muhammad Safdar, Agricultural Remote Sensing Lab, University of Agriculture, Faisalabad, Pakistan*
*Hamza Anjum, Department of Computer Science, Ghulam Ishaq Khan Institute of Engineering Sciences and Technology, Khyber Pakhtunkhwa, Pakistan*
*Muhammad Adnan Shahid, Agricultural Remote Sensing Lab, University of Agriculture, Faisalabad, Pakistan*
*Muhammad Sajid Mehmood, School of Tourism and Planning, Pingdingshan University, Pingdingshan, China*
*Amina Rashid, Department of Agronomy, University of Agriculture, Faisalabad, Pakistan*
*Abdul Rauf, Department of Irrigation and Drainage, University of Agriculture, Faisalabad, Pakistan*
*Fahd Rasul, Department of Agronomy, University of Agriculture, Faisalabad, Pakistan*
*Imran Shauket, Department of Structures and Environmental Engineering, University of Agriculture, Faisalabad, Pakistan*
*Naeem Saddique, Department of Irrigation and Drainage, University of Agriculture, Faisalabad, Pakistan*
*Hafiz Muhammad Bilawal Akram, Department of Agronomy, University of Agriculture, Faisalabad, Pakistan*
*Muntaha Munir, Institute of Botany, University of the Punjab, Lahore, Pakistan*

**Chapter 7**
Machine Learning and Remote Sensing for Soil Moisture Prediction: A Smart Agriculture Approach in Morocco ........ 175
*Abdessamad Elmotawakkil, Department of Computer Science, Ibn Tofaïl University, Kenitra, Morocco*
*Saad Jaldi, Faculty of Humanities and Social Sciences, Ibn Tofail University, Kenitra, Morocco*
*Mohammed Bouhassane, Faculty of Sciences, Ibn Tofail University, Kenitra, Morocco*
*Adil Moumane, Faculty of Sciences, Ibn Tofail University, Kenitra, Morocco*
*Nourddine Enneya, Faculty of Sciences, Ibn Tofail University, Kenitra, Morocco*

**Chapter 8**
Integration of Machine Learning and AI With Geospatial Analysis ................ 201
*Oladapo Muyiwa Oladoja, Abiola Ajimobi Technical University, Ibadan, Nigeria*
*Latifat Adebisi Abimbola, Abiola Ajimobi Technical University, Ibadan, Nigeria*
*Elizabeth Ayodele Oladejo, Abiola Ajimobi Technical University, Ibadan, Nigeria*

**Chapter 9**
Contribution of DSAS and GIS Tools in Shoreline Evolution: The Case of Mehdia ........................................................................................................ 229
*Mohamed Drissi, Ibn Tofail University, Morocco*
*Mouhcine Batchi, Department of Geography, Ibn Tofail University, Morocco*
*Said Laaribya, Department of Geography, Ibn Tofail University, Morocco*

**Chapter 10**
Geo-Spatial Perspectives on Urban Sprawl and Environmental Sustainability 251
*Sitesh Kumar Singh, College of Engineering, National University of Science and Technology, Muscat, Oman*
*Jaishri Gothania, Lingaya's Vidyapeeth, Faridabad, India*

**Chapter 11**
Quantum Fields of Vision UAV-Driven Rice Growth Stage Mapping With Quantum-Inspired Algorithms ........................................................................ 277
*R. Deepa, SRM Institute of Science and Technology, India*
*S. Pushpalatha, Saveetha School of Engineering, India*
*B. Yasotha, SRM Institute of Science and Technology*
*R. Swathy, SRM Institute of Science and Technology, India*
*K. L. Shoba, SRM Institute of Science and Technology, India*
*P. Thilakavathy, Vels Institute of Science, Technology, and Advanced Studies, India*

**Chapter 12**
Graph-Theoretic and Geospatial Intelligence Framework for Optimizing Smart Grid Structures Using Hybrid Drones ........ 301
*R. Deepa, SRM Institute of Science and Technology, India*
*A. Bennet Prabhu, Congruent Solutions, India*
*G. Niranjana, SRM Institute of Science and Technology, India*

**Compilation of References** ........ 321

**About the Contributors** ........ 371

**Index** ........ 379

# Detailed Table of Contents

**Preface** .......... xvii

**Acknowledgment** .......... xx

**Chapter 1**
Flood Risk and Inundation Mapping in a Changing Climate: Advances in GIS, Remote Sensing, and Artificial Intelligence Methods .......... 1
*Mehmet Ali Celik, Igdir University, Turkey*
*Ahmed Elbeltagi, Mansoura University, Egypt*
*Rajat Mishra, G.B. Pant University of Agriculture and Technology, India*
*Shekhar Singh, G.B. Pant University of Agriculture and Technology, India*
*Dinesh Kumar Vishwakarma, G.B. Pant University of Agriculture and Technology, India*
*Kusum Pandey, G.B. Pant National Institute of Himalayan Environment, India*
*Muhammed Ernur Akiner, Akdeniz University, Turkey*

Floods are among the most recurrent and destructive natural hazards worldwide, posing significant threats to human life, infrastructure, and ecosystems. They lead to widespread socioeconomic and environmental consequences, including the displacement of populations, disruption of livelihoods, contamination of water supplies, and degradation of natural resources. In recent decades, the impacts of climate change, rapid urbanization, and anthropogenic alterations to natural landscapes such as deforestation, land-use change, and the construction of impervious surfaces have intensified both the frequency and magnitude of flooding events. These evolving environmental and societal conditions have created unprecedented challenges for flood risk assessment, prediction, and mitigation, necessitating the development of more sophisticated and comprehensive approaches to understanding and managing flood hazards.

**Chapter 2**
Shrinkage of the Lake Dayet Er-Roumi, Morocco: A Decadal Remote Sensing Analysis .......... 35
*Adil Moumane, Ibn Tofail University, Morocco*

Freshwater lakes in semi-arid regions are critically vulnerable to climate change and anthropogenic pressures. Lake Dayet Er-Roumi, a permanent natural lake in Morocco recognized as a Site of Biological and Ecological Interest (SIBE), exemplifies such threats. This study quantifies the spatiotemporal dynamics of its water surface area from 2015 to 2025. Using Google Earth Engine, we analyzed Landsat 8/9 TOA imagery. The NDWI and MNDWI were applied to derive water masks. Seasonal water areas were computed by averaging cloud-filtered (5%) images per year. The lake underwent a 50% surface area reduction over the decade. Critical declines occurred in 2016–2017, 2019–2020, and most severely in 2023–2024 (-0.137 km$^2$), with recoveries too brief to reverse the trend.

**Chapter 3**
Flood Risk Assement Using AHP, Frequency Ratio, Logistic Regression, and Random Forest in Naraipur Municipality ........................................................ 63
*Rubi Chaulagain, Westernal Regional Campus, Nepal*
*Rekha Paudel, Westernal Regional Campus, Nepal*
*Fatima Ezzahra El Ghazali, Geosciences, Georesources, and Environment, Semlalia Laboratory, Department of Geology, Faculty of Sciences Semlalia, Cadi Ayyad University, Morocco*
*Adil Moumane, Ibn Tofail University, Morocco*

Floods, a natural hazard characterized by the overflow of water from its normal riverbed onto dry land, are a recurrent problem. The study area we choose is Narainapur Municipality. This study aims to develop a detailed flood risk assessment using the Analytic Hierarchy Process, Frequency Ratio, Logistic Regression,and Random Forest models. A flood inventory map was created, with 102 flooded and 95 non-flooded points Flood conditioning factors include slope, NDVI, soil type, distance from streams and roads, rainfall, TWI, aspect, and curvature. The Logistic Regression model (AUC = 0.944) achieved the highest AUC value, indicating superior predictive accuracy compared to other models. The Random Forest model (AUC = 0.936) also performed well, followed by the Frequency Ratio model (AUC = 0.855) and the AHP model (AUC = 0.822). These results highlight the effectiveness of machine learning-based models in our study area.The findings of this study provide valuable insights for flood risk management, offering a scientific basis for better planning and decision-making in vulnerable regions.

**Chapter 4**
Land Use and Environmental Dynamics of the Moroccan Coastal Strip 1984–2025: A Google Earth Engine Approach for Monitoring Vegetation, Water, and Urban Transformation - Case Study ........ 91
*Tarik Bahouq, Ibn Tofail University, Morocco*
*Amina Moumane, Faculté des Sciences Humaines et Sociales, Departement de Geographie, Ibn Tofail University, Morocco*
*Nadia Touhami, Faculté des Sciences Humaines et Sociales, Departement de Geographie, Ibn Tofail University, Morocco*

This study examines land use, vegetation cover, and water body changes in the Moulay Bousselham area of Kenitra Province, focusing on the impact of agricultural activities on the Merja Zerga wetland. Using multi-source satellite imagery (Landsat and Sentinel) processed in Google Earth Engine, vegetation dynamics were tracked from 2014 to 2024, and water surface changes from 1984 to 2024. Results reveal a steady annual increase of about 9.5 $km^2$ in vegetation cover, driven by intensive farming and modern irrigation that boosted productivity and mitigated desertification. Conversely, the water surface of Lake Merja Zerga declined by 0.0352 $km^2$ per year, threatening aquatic habitats, water quality, and local climate regulation. Land use analysis highlights urban expansion, a shift to intensive production systems, and forest degradation. These findings stress the need for sustainable land management and wetland conservation to balance development with ecosystem protection.

**Chapter 5**
Harnessing Precision Agriculture and Artificial Intelligence in Sustainable Farming ........ 119
*Muhammad Safdar, Agricultural Remote Sensing Lab, University of Agriculture, Faisalabad, Pakistan*
*Wasiq Farooq, Agricultural Remote Sensing Lab, University of Agriculture, Faisalabad, Pakistan*
*Muhammad Sajid Mehmood, School of Tourism and Planning, Pingdingshan University, Pingdingshan, China*
*Abdul Rauf, Department of Irrigation and Drainage, University of Agriculture, Faisalabad, Pakistan*
*Amina Rashid, Department of Agronomy, University of Agriculture, Faisalabad, Pakistan*
*Kashif Mehmood, Department of Irrigation and Drainage, University of Agriculture, Faisalabad, Pakistan*
*Muntaha Munir, Institute of Botany, University of the Punjab, Lahore, Pakistan*
*Nalain E. Muhammad, Department of Irrigation and Drainage, University of Agriculture, Faisalabad, Pakistan*
*Hafiz Muhammad Bilawal Akram, Department of Agronomy, University of Agriculture, Faisalabad, Pakistan*
*Hafiz Muhammad Mohsin Raza, Department of Agricultural Sciences, Allama Iqbal Open University, Islamabad, Pakistan*
*Aisha Nazir, Institute of Botany, University of the Punjab, Lahore, Pakistan*
*Hafiz Muhammad Awais, Agricultural Remote Sensing Lab, University of Agriculture, Faisalabad, Pakistan*

The chapter explores the role of Artificial Intelligence (AI) in revolutionizing precision agriculture, focusing on data-driven, adaptive, and efficient practices. It traces the evolution of AI-integrated digital farming systems, including GNSS, drones, sensors, and IoT technologies. Key AI applications include crop classification, yield forecasting, pest detection, irrigation scheduling, and nutrient management. The chapter also discusses technologies like UAVs, hyperspectral imaging, edge computing, and cloud-based platforms for improved farm-level decision-making. The chapter discusses the potential of AI in large-scale and smallholder farming systems but also highlights challenges like heterogeneous data, economic barriers, limited digital literacy, and low adoption among smallholder farmers. The chapter discusses future opportunities in AI-integrated robotics, blockchain-based supply chains, and big data analytics for climate-resilient agriculture and calls for inclusive policy reforms, capacity-building initiatives, and collaborative innovation for equitable scale.

**Chapter 6**
Applications of Artificial Intelligence and Remote Sensing in Water Resources Management ........................................................................................ 147
*Wasiq Farooq, Agricultural Remote Sensing Lab, University of Agriculture, Faisalabad, Pakistan*
*Muhammad Safdar, Agricultural Remote Sensing Lab, University of Agriculture, Faisalabad, Pakistan*
*Hamza Anjum, Department of Computer Science, Ghulam Ishaq Khan Institute of Engineering Sciences and Technology, Khyber Pakhtunkhwa, Pakistan*
*Muhammad Adnan Shahid, Agricultural Remote Sensing Lab, University of Agriculture, Faisalabad, Pakistan*
*Muhammad Sajid Mehmood, School of Tourism and Planning, Pingdingshan University, Pingdingshan, China*
*Amina Rashid, Department of Agronomy, University of Agriculture, Faisalabad, Pakistan*
*Abdul Rauf, Department of Irrigation and Drainage, University of Agriculture, Faisalabad, Pakistan*
*Fahd Rasul, Department of Agronomy, University of Agriculture, Faisalabad, Pakistan*
*Imran Shauket, Department of Structures and Environmental Engineering, University of Agriculture, Faisalabad, Pakistan*
*Naeem Saddique, Department of Irrigation and Drainage, University of Agriculture, Faisalabad, Pakistan*
*Hafiz Muhammad Bilawal Akram, Department of Agronomy, University of Agriculture, Faisalabad, Pakistan*
*Muntaha Munir, Institute of Botany, University of the Punjab, Lahore, Pakistan*

The chapter discusses the need for improved water resource management due to climate change, population growth, and freshwater issues. The potential of artificial intelligence, including machine learning, deep learning, and hybrid models, in reshaping water management systems. The chapter explores AI-powered solutions for improved water assessment practices, including rainfall-runoff modeling, groundwater prediction, drought/flood forecasting, streamflow forecasting, aquifer recharge analysis, and hydrological anomaly detection. The integration of AI with remote sensing and geographic information systems is particularly important for enabling spatial-temporal watershed monitoring. Pakistan's AI deployments in early warning systems and water governance have made significant progress, but challenges like data scarcity, model overfitting, computational complexity, and skill gaps persist. The chapter discusses the need for scalable, transferable, and policy-supported AI solutions for water management in both developed and developing regions, emphasizing the adoption of explainable AI.

**Chapter 7**
Machine Learning and Remote Sensing for Soil Moisture Prediction: A Smart Agriculture Approach in Morocco ........................................................ 175
*Abdessamad Elmotawakkil, Department of Computer Science, Ibn Tofaïl University, Kenitra, Morocco*
*Saad Jaldi, Faculty of Humanities and Social Sciences, Ibn Tofail University, Kenitra, Morocco*
*Mohammed Bouhassane, Faculty of Sciences, Ibn Tofail University, Kenitra, Morocco*
*Adil Moumane, Faculty of Sciences, Ibn Tofail University, Kenitra, Morocco*
*Nourddine Enneya, Faculty of Sciences, Ibn Tofail University, Kenitra, Morocco*

Artificial intelligence (AI) plays a crucial role in soil moisture prediction, essential for sustainable agriculture in arid regions where water scarcity and climate variability threaten crop yields. This study introduces an AI-driven framework to forecast soil moisture across five sites in Morocco's Draa Valley (Agdz, Tagounite, Tamegroute, Tansikht, Zagora). A dataset (2003–2024) was built by integrating historical climate records with remote sensing indicators using Google Earth Engine (GEE). Six models Random Forest (RF), XGBoost, CatBoost, k-Nearest Neighbors (KNN), Long Short-Term Memory (LSTM), and Temporal Convolutional Networks (TCN) were assessed with RMSE, NSE, MSE, and MAPE. Results showed that tree-based models clearly outperformed deep learning, with RF, XGBoost, and CatBoost achieving RMSE of 2.89%–9.11% and NSE > 0.965. Findings highlight the potential of AI-based soil moisture prediction to enhance irrigation scheduling, optimize water allocation, and support climate-resilient farming, offering a scalable solution for precision agriculture.

**Chapter 8**
Integration of Machine Learning and AI With Geospatial Analysis ................ 201
*Oladapo Muyiwa Oladoja, Abiola Ajimobi Technical University, Ibadan, Nigeria*
*Latifat Adebisi Abimbola, Abiola Ajimobi Technical University, Ibadan, Nigeria*
*Elizabeth Ayodele Oladejo, Abiola Ajimobi Technical University, Ibadan, Nigeria*

The aim of the chapter is to provide a more comprehensive vision of how machine learning (ML) and artificial intelligence (AI) technologies can be used together with geospatial analysis to help solve modern environmental problems. This study shows that spatial data, remote sensing, and powerful computational techniques can be used to transform environmental monitoring, prediction, and decision-making processes using the power of ML/AI-enhanced geospatial tools. Demonstrated via hands-on case studies based on wind speed data of Lagos, Nigeria, it was shown how several ML algorithms, such as Random Forest (RF), Naïve Bayes (NB), XGBoost (XGB), Linear Regression (LR) and Neural Networks (NN), can be utilised to recognise and predict patterns in the environment. The models were validated using Root Mean Square Errors (RMSE) and Mean Absolute Error (MAE).

**Chapter 9**
Contribution of DSAS and GIS Tools in Shoreline Evolution: The Case of Mehdia ........................................................................................................ 229
*Mohamed Drissi, Ibn Tofail University, Morocco*
*Mouhcine Batchi, Department of Geography, Ibn Tofail University, Morocco*
*Said Laaribya, Department of Geography, Ibn Tofail University, Morocco*

The present study manipulates satellite imagery (Landsat TM+ and ETM+) and GIS tools to analyze the changes in the Mehdia coastline between 1990 and 2023. We estimated shoreline change rates, particularly the End Point Rate (EPR) index, using a diachronic approach and the DSAS 5.0 model in ArcGIS. Results show that the coastline has significantly retreated, particularly between 1990 and 2005 and 2005 and 2023, with erosion concentrated in the central and northern sectors. Across the estuary, the overall trend shows an average retreat of –3.25 m/year. In addition to human actions including the construction of dams and dike systems, urbanization, tourism, and dredging, this vulnerability is also linked to natural forces like storms and climate change.By affecting sediment dynamics, these factors have accelerated beach loss, particularly in the north. The findings highlight the critical need for sustained monitoring and resilient coastal management to prevent present erosion and anticipate future environmental challenges.

**Chapter 10**
Geo-Spatial Perspectives on Urban Sprawl and Environmental Sustainability 251
*Sitesh Kumar Singh, College of Engineering, National University of Science and Technology, Muscat, Oman*
*Jaishri Gothania, Lingaya's Vidyapeeth, Faridabad, India*

Urban sprawl has a great question mark over the issues of environmental sustainability, infrastructure, and urban resilience. This chapter explains how geospatial technologies (remote sensing, GIS, and spatial modeling) can be used to monitor and manage developments in cities. It discusses the ecological, social, and infrastructural effects of sprawl, taking Muscat, Oman, as an illustrative example, and outlines development-planning concepts of compact cities, smart growth, and eco-friendly modes of transport. Cities can optimize planning needs between development and ecosystem conservation through the incorporation of spatial intelligence in governance and policy frameworks. The chapter highlights the planning and participatory strategies as suitable in ensuring a sustainable, resilient, and livable urban environment in accordance with global sustainability objectives.

**Chapter 11**
Quantum Fields of Vision UAV-Driven Rice Growth Stage Mapping With Quantum-Inspired Algorithms ........ 277
*R. Deepa, SRM Institute of Science and Technology, India*
*S. Pushpalatha, Saveetha School of Engineering, India*
*B. Yasotha, SRM Institute of Science and Technology*
*R. Swathy, SRM Institute of Science and Technology, India*
*K. L. Shoba, SRM Institute of Science and Technology, India*
*P. Thilakavathy, Vels Institute of Science, Technology, and Advanced Studies, India*

Food safety and precision farming need accurate plot-scale rice yield predictions, so we developed a method combining UAV-derived vegetation indices (VIs) with brightness, greenness and moisture data from tasseled cap transformation (TCT). Eight nitrogen gradients of rice were used during the booting and heading stages to obtain ground truth and six-band UAV imagery. We propose a hybrid quantum learning model that uses Bi-LSTM for extracting temporal features and quantum circuits for quantum feature processing. These enhanced features are combined with Bi-LSTM outputs into an XGBoost regressor. Our Quantum BiLSTM + XGBoost approach outperformed traditional models by 7-10%, achieving ~94% accuracy.

**Chapter 12**
Graph-Theoretic and Geospatial Intelligence Framework for Optimizing Smart Grid Structures Using Hybrid Drones ........ 301
*R. Deepa, SRM Institute of Science and Technology, India*
*A. Bennet Prabhu, Congruent Solutions, India*
*G. Niranjana, SRM Institute of Science and Technology, India*

This paper proposes G-O-HA, a graph theory-based framework integrating heuristic algorithms and hybrid drones to optimize smart grid performance. Smart grids are modelled as dynamic graphs, with Particle Swarm Optimization (PSO) identifying efficient routing paths and enhancing energy distribution. Hybrid drones equipped with multispectral sensors and LiDAR conduct aerial inspections and real-time fault localization. Geospatial data from drones continuously updates the optimization model, enabling adaptive maintenance and network reconfiguration. Applied to real-world scenarios, G-O-HA reduces fault recovery time by 35% and improves communication efficiency and throughput. The study demonstrates how combining graph theory, heuristic optimization, and drone intelligence can build resilient, responsive power systems.

**Compilation of References** ........................................................................... 321

**About the Contributors** ................................................................................ 371

**Index** ............................................................................................................. 379

# Preface

The accelerating pace of environmental change across the globe—manifested through land degradation, biodiversity loss, water scarcity, and intensifying climatic extremes—demands research that is both innovative and actionable. In this rapidly evolving context, *Geographic Information Systems (GIS)*, *Remote Sensing (RS)*, and *Artificial Intelligence (AI)* have become indispensable tools for observing landscapes across scales, analyzing dynamic environmental processes, and transforming spatial data into meaningful insights for sustainable decision-making.

The present volume, *Advances in GIS, Remote Sensing, and Artificial Intelligence Methods: Flood Risk and Inundation Mapping in a Changing Climate*, brings together a collection of studies that highlight the transformative potential of spatial technologies in addressing the world's most pressing environmental challenges. Each chapter represents a distinct yet interconnected effort to bridge science, technology, and practice in environmental monitoring, risk management, and sustainable resource planning.

## CHAPTER OVERVIEW

The opening chapter, *"Flood Risk and Inundation Mapping in a Changing Climate,"* sets the scientific foundation by exploring the complex interplay between climatic variability, hydrological hazards, and flood vulnerability. Building upon this theme, *"Shrinkage of the Lake Dayet Er-Roumi, Morocco: A Decadal Remote Sensing Analysis"* presents a compelling case study on the alarming ecological transformation of inland water bodies revealed through long-term satellite observations.

The next contribution, *"Flood Risk Assessment Using AHP, Frequency Ratio, Logistic Regression, and Random Forest in Naraipur Municipality,"* showcases the integration of multi-model analytical frameworks to enhance the accuracy of flood hazard mapping. Expanding the geographical and methodological scope, *"Land Use and Environmental Dynamics of the Moroccan Coastal Strip (1984–2025): A*

*Google Earth Engine Approach for Monitoring Vegetation, Water, and Urban Transformation"* demonstrates how cloud-based geospatial analysis can effectively track decades of environmental change.

Chapters five through eight explore the convergence of artificial intelligence, precision agriculture, and geospatial science. *"Harnessing Precision Agriculture and Artificial Intelligence in Sustainable Farming"* illustrates how AI-driven analytics can optimize crop management and resource efficiency. *"Applications of Artificial Intelligence and Remote Sensing in Water Resources Management"* examines advanced modeling techniques for monitoring and conserving water systems under increasing stress. *"Machine Learning and Remote Sensing for Soil Moisture Prediction: A Smart Agriculture Approach in Morocco"* and *"Integration of Machine Learning and AI with Geospatial Analysis"* further demonstrate how hybrid modeling approaches are redefining agricultural intelligence and environmental monitoring.

Coastal and urban environments receive dedicated attention in *"Contribution of DSAS and GIS Tools in Shoreline Evolution – The Case of Mehdia,"* which applies Digital Shoreline Analysis Systems to quantify coastal retreat, and *"Geo-Spatial Perspectives on Urban Sprawl and Environmental Sustainability,"* which investigates how urban expansion reshapes ecological and socio-economic systems.

The final chapters reflect the emerging frontiers of geo-intelligent research. *"Quantum Fields of Vision: UAV-Driven Rice Growth Stage Mapping with Quantum-Inspired Algorithms"* explores the fusion of quantum computing concepts and drone-based sensing for precision crop monitoring, while *"Graph-Theoretic and Geospatial Intelligence Framework for Optimizing Smart Grid Structures Using Hybrid Drones"* extends geospatial analytics to the domain of renewable energy and infrastructure optimization.

Together, these chapters exemplify the technological sophistication and interdisciplinary integration that define modern environmental research. By merging geospatial analytics with socio-economic, climatic, and hydrological datasets, the contributing authors advance holistic and predictive models of environmental change—offering decision-makers new tools for resilience planning and adaptive governance.

Ultimately, this book seeks to inspire researchers, practitioners, and policymakers to harness spatial intelligence for sustainability and climate adaptation. In an era defined by uncertainty and interdependence, the ability to observe, model, and manage Earth's systems spatially is not merely a scientific advantage—it is a moral and practical imperative for securing a sustainable future.

## AUDIENCE AND IMPACT

This volume is designed for a wide and diverse readership:

- **Researchers and students** seeking to understand applied GIS, RS, and AI across varied ecological and geographic contexts.
- **Practitioners and agencies** engaged in land, water, hazard, and agricultural management who require robust spatial evidence for informed decision-making.
- **Policy makers and planners** navigating the challenges of uncertainty and environmental change through data-driven strategies.

By presenting validated methodologies, innovative case studies, and policy-relevant insights, this book aims to empower its audience to apply geospatial science toward sustainable development and resilience building.

**Jamal Al Karkouri**
*Ibn Tofail University, Kenitra, Morocco*

**Adil Moumane**
*Ibn Tofail University, Kenitra, Morocco*

**Abdessamad Elmotawakkil**
*Ibn Tofail University, Kenitra, Morocco*

**Mouhcine Batchi**
*Ibn Tofail University, Kenitra, Morocco*

# Acknowledgment

We extend our deepest gratitude to all contributing authors for their intellectual rigor, creativity, and dedication. Our appreciation also goes to the institutions, reviewers, and collaborators whose support and constructive feedback shaped the quality of this volume. Finally, we acknowledge the landscapes and communities represented in these studies—their challenges, resilience, and adaptive spirit continue to inspire the advancement of applied environmental research.

# Chapter 1
# Flood Risk and Inundation Mapping in a Changing Climate:
## Advances in GIS, Remote Sensing, and Artificial Intelligence Methods

**Mehmet Ali Celik**
https://orcid.org/0000-0002-7729-6650
*Igdir University, Turkey*

**Ahmed Elbeltagi**
*Mansoura University, Egypt*

**Rajat Mishra**
*G.B. Pant University of Agriculture and Technology, India*

**Shekhar Singh**
*G.B. Pant University of Agriculture and Technology, India*

**Dinesh Kumar Vishwakarma**
*G.B. Pant University of Agriculture and Technology, India*

**Kusum Pandey**
https://orcid.org/0009-0005-4888-6650
*G.B. Pant National Institute of Himalayan Environment, India*

**Muhammed Ernur Akiner**
https://orcid.org/0000-0002-5192-2473
*Akdeniz University, Turkey*

## ABSTRACT

*Floods are among the most recurrent and destructive natural hazards worldwide, posing significant threats to human life, infrastructure, and ecosystems. They lead to widespread socioeconomic and environmental consequences, including the displacement of populations, disruption of livelihoods, contamination of water supplies, and degradation of natural resources. In recent decades, the impacts of climate change, rapid urbanization, and anthropogenic alterations to natural landscapes such as*

DOI: 10.4018/979-8-3373-6608-1.ch001

*deforestation, land-use change, and the construction of impervious surfaces have intensified both the frequency and magnitude of flooding events. These evolving environmental and societal conditions have created unprecedented challenges for flood risk assessment, prediction, and mitigation, necessitating the development of more sophisticated and comprehensive approaches to understanding and managing flood hazards.*

## INTRODUCTION

Floods are the most disturbing natural phenomena, termed disasters in terms of frequency, financial cost, and, most significantly, the impact on the population and interruption of socioeconomic activity (UNDRR, 2021). Flooding is the inundation of an area caused by an unexpected rise in water, dam failure, or severe rainfall duration and intensity, putting the lives and property of those in the affected area at risk (Ward et al., 2020). According to the European Environment Agency (EEA, 2010), floods and wind-water storms were Europe's most expensive natural disasters from 1998 to 2009, with total losses of almost EUR 52 billion for floods and EUR 44 billion for storms. One thousand one hundred twenty-six people died as a result of 213 flood disasters, while 729 people died as a result of 155 storms, by 2009 (EEA, 2010). The extreme events with return periods exceeding 100 years are projected to double in frequency in Europe by 2035, according to collaborative efforts of EURO-CORDEX climate projections using the RCP8.5 scenario, with a further increase through the end of the century (Kotlarski et al., 2014).

Between 1985 and 2009, numerous flooding events occurred in Europe to chronicle the spatiotemporal role of inundation under weather and geographical changes. In recent years, several authors have employed empirical studies, including field surveys, GIS, remote sensing technologies, and various statistical methodologies, to address the complexities of flood inundation (Cvetković et al., 2024). Different combinations of inundation data and DEM can build valuable flood extent maps within a GIS environment. Herein, remote sensing, GIS, geological data, fuzzy logic, and statistical analyses of peak discharge and rainfall intensities have been combined to form reliable flood hazard maps (Hamidifar et al., 2024). Likewise, using optimized weighting methods, GIS combined with multi-criteria analysis techniques was employed to assess flood susceptibility, exposure, and risk accurately. Key variables such as DEMs, land use, precipitation, geological features, and catchment geometry have been utilized to construct detailed flood risk maps.

The problem with satellite pictures produced by optical sensors is that bad weather or very dense established vegetation cover might affect their reception. Synthetic aperture radar (SAR), a newly created technology, can be critical to the extent of flooded areas and assess temporal variations thereon, generating probabilistic inundation maps (Bagja et al., 2025; Schumann et al., 2009). The integrated flood inundation mapping (FIM) workflow is seen in Figure 1.

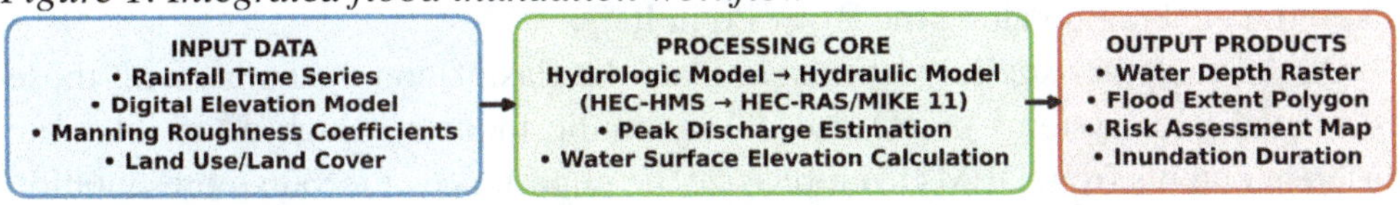


*Figure 1. Integrated flood inundation workflow*

Numerical models designed to simulate extreme flood events, such as FLO and ORSADEM, have been employed to estimate flood depths and delineate inundated areas (Teng et al., 2017). Hydrodynamic models are frequently used to produce accurate flood inundation maps (Horritt and Bates, 2002). The tools used for modeling complexity include raster-based AutoRoute, Height Above the Nearest Drainage (HAND), and two-dimensional hydrodynamic HEC-RAS, and have been applied in different scenarios (Follum et al., 2016). The studies in the United States and Australia suggest that raster-based models can generate flood extents comparable to those produced by hydraulic models (Inman & Lyons, 2020; Nobre et al., 2011). The estimates returned by AutoRoute, when coupled with the raster outputs, more closely resemble those of the 2D HEC-RAS model (Follum et al., 2016; Wing et al., 2018).

The uncertainties in hydraulic flood modeling have been theoretically detailed through extensive research. Among these is the primary factor: Manning's roughness coefficient, which regards the resistance to flow (Pappenberger, 2005). Drawn down with a lower value, its emphasis in bed values can aid fast propagation of the flood waves, thereby decreasing the inundation area (Hall et al., 2005). Evidence of attempts at calibrating this coefficient exists and encompasses analysis of paleo-stage indicators such as scars on tree trunks, produced by floating debris during past floods (Sigafoos, 1964).

An extended Kalman filter was applied to eight flash flood events in a French catchment to assimilate measured water depths and refine boundary conditions in the 1D hydraulic MASCARET model to improve the flood modeling accuracy approach. Dimitriadis et al. used Monte Carlo simulation methods to address uncertainty issues linked to different hydraulic model types (1D and quasi-2D), and the sensitivity of flood extents to flow release rate, terrain gradients, Manning coefficient, and grid resolution variations. In Vorarlberg, Austria, Winter et al. (2020) studied the sources of uncertainty arising from threshold parameter choices in models, selections of frequency distribution, geometric representation, water level estimation, and probabilistic risk assessment frameworks. Domeneghetti et al. (2013) addressed the uncertainties from upstream and downstream boundary conditions linked to rating curve estimates and dike failure mechanisms. They then used bivariate frequency analysis methods within a hybrid probabilistic-deterministic hydraulic model to produce flood maps with confidence intervals in a flood-prone, dike-protected region along the River Po in Italy.

One of the other studies went to simulate the flash flood dynamics with those associated with systems of 1D and 2D hydraulic models, which often integrate different outputs from DEMs to emphasize the importance of terrain representation (Neal et al., 2011). Although the coupling of 1D/2D hydraulic modeling is relatively a recent development in HEC-RAS software, a few investigations have been conducted into its use for enhancing reliability in flood mapping (Shustikova et al., 2019). Table 1 gives a brief comparative study of the hydraulic modeling tools applied to flood inundation studies.

*Table 1. Comprehensive comparison of hydraulic modeling software for flood inundation analysis.*

| Feature | HEC-RAS | MIKE 11 | FLO-2D | TUFLOW |
|---|---|---|---|---|
| Developer | US Army Corps of Engineers | DHI Group | FLO-2D Software Inc. | BMT Group |
| License Type | Free (Public Domain) | Commercial | Commercial | Commercial |
| 1D Capability | Excellent | Excellent | Limited | Good |
| 2D capability | Excellent (v5.0+) | Good (MIKE 21) | Excellent | Excellent |
| 1D–2D Coupling | Integrated | Separate modules | Integrated | Fully integrated |
| GIS Integration | Excellent (ArcGIS) | Good | Moderate | Excellent |
| User Interface | Graphical (RAS Mapper) | Graphical | Graphical | Graphical |
| Computational Efficiency | Good | Very Good | Moderate | Excellent |
| Steady Flow | Excellent | Excellent | Limited | Excellent |
| Unsteady Flow | Excellent | Excellent | Good | Excellent |
| Sediment Transport | Good | Excellent | Limited | Excellent |
| Water Quality | Basic | Excellent | No | Excellent |
| Technical Support | User community | Professional | Professional | Professional |
| Learning Curve | Moderate | Steep | Moderate | Steep |
| Documentation | Extensive | Excellent | Good | Excellent |

continued on following page

*Table 1. Continued*

| Feature | HEC-RAS | MIKE 11 | FLO-2D | TUFLOW |
|---|---|---|---|---|
| Global Usage | Very High | High | Moderate | High |
| Cost Category | Free | High Cost | Medium Cost | Premium |
| Strengths | Free and widely adopted; Strong user community; Good documentation; Regular updates. | Comprehensive features; Well-validated; Professional support; Multipurpose | Specialized for flood modeling; Easy to use; Good for mud/debris flow | High performance; Advanced features; Excellent for complex hydraulics |
| Limitations | Limited 3D Capability; Requires GIS preprocessing | Expensive license; Complex setup; Resource-intensive | Limited 1D capabilities; Less validation; Moderate support | Costly; Complex setup; Requires expertise |

Excellent: Full-featured, well-validated, industry standard. Very Good: Strong capabilities with minor limitations. Good: Adequate for most practical applications. Moderate/Basic: Functional yet with considerable limitations. Limited/No: So poorly functioning that complex projects are not suggested.

## Flood Conditioning Factors

Flood-prone areas must be precisely marked by considering an array of physical and environmental factors, which build or diminish a land's susceptibility to flooding (Tehrany et al., 2014). These flood-conditioning factors are of several kinds and are interconnected, from topographic, hydrological, geological, to anthropogenic (human-induced) ones (Tehrany et al., 2014). The topographic factors mainly affecting the water movement include elevation, slope, and curvature (Moore et al., 1991). Low-lying plains and flat terrains are more prone to inundation (Tehrany et al., 2014). Hydrological factors such as distance to streams or drainage networks, drainage density, and discharge of rivers are important since these factors measure the capacity of a land to carry excess water (Rahmati et al., 2015). Land Use/Land Cover (LULC) is a leading anthropogenic factor; impervious surfaces such as urban structures heighten runoff, whereas vegetated areas allow infiltration and aid in lowering flood peaks (Mir et al., 2025). Soil factors such as soil texture, type, and hydraulic conductivity control infiltration rate and retention capacity (Rawls et al., 1982). Geological features such as lithology and impermeable layers can significantly influence groundwater movement and surface runoff (Freeze and Cherry, 1979). Furthermore, other factors such as rainfall intensity and duration, geomorphological indices (e.g., Topographic Moisture Index (TWI)), vegetation indices (NDVI), and channel geometry are used as key variables in flood susceptibility models, allowing for precise mapping of potential flood-prone areas (Sørensen et al., 2006). Each of these factors influences flood formation mechanisms in different ways and to varying degrees, creating a complex system that must be evaluated in an integrated manner (IPCC, 2023).

## Impact and Vulnerability of Flood

Extreme flooding has the potential to change landscapes, damage people, and, as a result, property and livelihoods are lost (Kundzewicz et al., 2013). We must develop adequate flood prevention and adaptation methods since expected variations in rainfall and climate arrays might affect provincial flood frequency (Cutter et al., 2014). Enormous floods, such as those that wreaked havoc on the Mississippi and Ohio Rivers in spring 2011, produce extremely varied spatial patterns of erosion and deposition (Aalto et al., 2006). These concentrated pockets of change, or "hotspots," reveal underlying landscape vulnerabilities and offer an opportunity to examine their causes in the past to help guide future actions.

Flooding has the first and most immediately apparent effect of inundating areas usually free of water. Increased water depth causes saturation, thus reducing the soil's load-bearing capacity and damaging buildings and infrastructures through foundation failure and eventual damage to the structures themselves. Crops and vegetation made to stay underwater for longer usually end up dying due to oxygen deficiency (Kozlowski, 1997).

Infrastructure installation, including power lines, communication networks, pipelines, and transportation systems (roads, railways, and airports), would remain underwater and cease functioning or be terminated altogether (Jonkman, 2005). Floods cause substantial financial losses through irreversible damage to tangible assets, such as electrical equipment, vehicles, and furniture (Jonkman and Kelman, 2005).

Usually, a water depth of 0.8 m is regarded as a threshold beyond which damage increases, as this depth submerges objects on tables or countertops, sometimes termed by these authors as the "table effect" (Penning-Rowsell and Green, 2000). The spread of contaminants through floodwaters increases the risk of damaging hazardous facilities and materials, such as landfills, chemical plants, and power stations (Spencer, 2024).

Greater water depths threaten human and animal safety, especially sudden changes like steps or dips (Jonkman and Penning-Rowsell, 2008). Vulnerable groups, including infants, the elderly, and individuals with disabilities, face heightened risks (Alderman et al., 2012). The duration of flooding is equally important; longer durations would expose the crops to rotting due to anaerobic bacteria and may worsen damage to the foundation (Kozlowski, 1997). In contrast, rapid drainage can reduce harm by limiting sediment deposition.

Seawater intrusion during coastal floods leads to soil and groundwater reservoir salinization (Akiner & Lačnjevac, 2022; Werner and Simmons, 2009). Stream velocity represents another significant hazard. Even moderate flow speeds can dislodge and transport vehicles and equipment (Martínez-Gomariz et al., 2016). Velocities exceeding one meter per second, particularly when combined with specific depths, are enough to sweep away people. Sudden velocity surges such as those caused by hydraulic jumps near waterfalls or constrictions trap objects and individuals, complicating rescue efforts (Kvočka et al., 2016). Flash floods may reach speeds of 4-6 m/s, sufficient to move big debris. Eddies, a suction effect through the Venturi mechanism, and extra pressure from cavitation can also occur (Scott et al., 2001).

The energy fluid that flash floods may carry can erode riverbeds and banks, destabilizing the slopes and infrastructures on them (Knighton, 1998). Slumping landslides then come into play, damaging roads, hydraulic works, and constructions (Iverson, 2000). Sediment transport represents a further hazard. Materials like rocks, gravel, sand, and organic matter may travel suspended in water or along the bed, damaging structures and posing physical threats to humans and animals (Julien, 2010). These materials increase water density and viscosity, which enhances erosion and sediment transport in a self-reinforcing cycle.

High sediment loads can make the floodwaters dangerous, with debris or mudflows (Iverson, 1997). Floating debris such as plant materials, hail, or ice may block channels or culverts, leading to failures in bridges or sewer lines (Lamb et al., 2010). Sediments settling in low-energy areas obstruct transport routes, drainage, and utility infrastructure. Sedimentation affects fish stocks and damages appliances by clogging their filters, aside from hindering shellfishing operations (Walling and Fang, 2003).

Piping activities in riverbanks and shorelines are less frequent but equally damaging, where subsoil infiltration post-flooding removes fine particles to form subsurface channels that collapse (Foster et al., 2000). Flooding over clay or anhydrite layers also promotes land-form changes due to soil expansion. Karst systems may also reactivate, with floods removing obstructions or collapsing subterranean cavities, altering the underground landscape (Holtz, 1981).

## Environmental and Socioeconomic Benefits

Natural flooding regimes fuelled by rainfall and river waters provide natural maintenance in many floodplain habitats (Tockner and Stanford, 2002). Natural floods play a crucial role in this ecosystem. By transporting sediments and nutrients, this natural flooding phase aids in maintaining soil fertility in natural and farmed floodplain areas (Junk, 1999). On the other hand, understanding the frequency, magnitude, and changes anticipated due to flooding in a specific floodplain is critical because it allows for improved scheduling and timing of urgent and non-urgent renovation and mitigation actions (Merz et al., 2010).

## Hydrological Indicators for Assessing the Flood Inundation

Floods are among the deadliest natural calamities, claiming thousands of lives yearly (Dilley et al., 2005). Population increase, rapid urbanisation, rapid industrial development, and notably irrational exploitation of land, water resources, and forests all contribute to the magnitude of its destruction (Douglas and Lawson, 2000). From 1900 to 2007, flood events affected an estimated 3.05 billion people worldwide, with China alone accounting for approximately 1.67 billion of those instances (EM-DAT, 2009). Managing regional flood risks remains arduous, prompting a growing body of research into flood detection and monitoring, flood management, flood risk mitigation, and flood forecasting (Plate, 2002). This is an increase in flood frequency worldwide, and according to the IPCC's 2007 Fourth Assessment Report, extreme weather events will increase throughout the 21st century (IPCC, 2007; Reggiani and Weerts, 2008).

Human activities, including land use changes and encroachment of urbanization to dam building and gravel-sand mining, have significantly altered the flood regime (Paul and Meyer, 2001). Researchers have become more concerned with flood magnitude; the criterion often used for flood severity is almost always directly related to the destructive capability of a flood event (De Moel, 2015). A range of parameters measures the intensity of the flooding, peak discharge, and flood volume (Merz and Blöschl, 2003). However, these studies mainly address a single parameter, such as peak discharge or flood level, with no multiple integration.

Peak discharges, consequently, have been shown to have regional variations (Ouarda et al., 2001; Fitri et al., 2021). The existence of these insights, however, often created some trouble in defining flood intensity, considering its multi-dimensional nature and locality-based hydrological conditions (Di Baldassarre et al., 2009). From the engineers' and planners' perspective, estimating flood intensity is essential for designing a dam, spillway, highway, or urban drainage system.

The primary purpose of this study is to yield an enhanced understanding of flood intensity through simultaneously developing different indicators (Lee and Choi, 2018; Fu et al., 2023). This becomes crucial in river basins such as the Wujiang, where flood patterns have changed drastically. Historical chronicles between 1850 and 1940 register only four major floods, whereas five severe floods occurred in 1961, 1968, 1994, 2002, and 2006 (Zhou et al., 2002).

A more nuanced understanding of flood events can be gained by evaluating multiple indicators. Flooding events that share some characteristics could be clustered using various methods; cluster analysis is one of the most potent methods (Merz and Blöschl, 2003). Zhao et al. (2014) clustered flood severity by fuzzy logic for Chinese provinces during 1997 to 2008 flood events as an example of this method in action.

Various works have sought to group hydrologically similar events through clustering, and thus are of increased use in conducting flood risk planning and management.

Basic features of a flood include peak discharge (Q), peak water level (H), maximum small flood volume for 24 hours (V24), maximum flood volume for 72 hours (V72), and total flood volume (V). Total flood volume depends on discharge and duration, since many flood types with varying durations exist. Hence, when measuring the effects of floods, duration should be taken into account (Shaw, 1994).

The flooding volume for particular times, as Vn, (where n = 1, 2, 3,_t hours), is one of the significant planning factors. In the Wujiang River Basin, flood durations are centered within three days, so V24 and V72 are good flood power indicators. By concentrating on capturing the most critical volumes within these time periods, a better appraisal of flood severity can be made (Dullo et al., 2021; Gaál et al., 2015)

Therefore, the distribution of flood concentration should best be investigated using multi-indicator analysis, including Q, H, V24, V72, and V. This paradigm then confirms that decisions relating to flood risk and infrastructure planning can be appropriately made (Merz and Blöschl, 2003)

High coastal and riverine flood plains are at some risk of being hit by floods. Flood damages may occur because of any physical scenario where abnormally high waters may result in financial loss or could disrupt property (Pielke and Downton, 2000). Overbank streamflows and coastal storms cause significant damage (Ingle Smith, 1999). However, another common source of water damage is intermittent surface runoff from precipitation or spring thaws, which causes erosion and siltation, floods basements, undermines buildings and highways, and interferes with agriculture and other land uses as standing water (Montgomery, 2007). Overflow flows can cause damage in streams, along coasts, and improperly drained locations in any area of the country (Ingle Smith, 1999). Cloudbursts can occur in deserts and temperate regions alike. Local mountain rains cause all the flood damage when the floodwaters are facilitated downward into the valley plains. These rare surges in floodwaters and the loss of lives cause terrible floods in the parched western states (Costa, 1987). Part of the damage is attributable to land-use rules, which can help reduce flood damage wherever it occurs.

However, "drainage" issues and rare flood concerns in poorly defined valley plains pose unique challenges regarding hazard identification and standard formulation. Communities should ideally produce complete drainage plans that establish drainage design requirements. When a piece of ground is graded, a highway or railroad is built, a structure is built, or a swamp is drained, the normal flow of surface water is disrupted, posing a hazard to adjoining places. Most municipalities have not undertaken such complete investigations. Instead, some local ordinances only broadly direct developers to provide proper drainage for each subdivision and building site without specifying any particular evaluation criteria. Occasionally, municipal engineers will evaluate drainage suitability for a development case-by-case basis. In a few towns, however, full-fledged drainage ordinances do exist. Identifying the flood hazard zones of an alluvial valley plain can be challenging because of the scarcity of flood data and the variety in flood flow routes. These considerations further hinder the definition of development standards and the abruptness and unpredictability of the flows (Smith and Bedient, 1981).

## METHODOLOGY

### Numerical Methods

The numerical modeling technique is important for determining the dynamic behaviour of flooding and its sources and effects. The model and the types of simulation equations used determine the consistency of the modeling (Teng et al., 2017). As a result, this research was conducted to identify commonly used models for various flooding factors and compare frequently used 1D and 2D numerical models (Horritt and Bates, 2002).

The study focused on key parameters commonly employed in hydrological modeling and the models frequently used to estimate them. Loss parameters were accounted for in detail to foster understanding of individual processes occurring in the hydrological cycle. It was brought to attention that rainfall data are generally obtained from gauging stations and may sometimes be subjected to regionalization to maintain accuracy, especially in regions with a paucity of observational data.

Issues arising from the emptiness of data were also considered on some occasions. Some well-used methods were acknowledged for describing the losses: namely, the Priestley-Taylor model with CMOS modifications for evapotranspiration; the revised Gash model for rainfall interception; the modified Green-Ampt model for infiltration; and the adjusted SCS-CN method for surface runoff (Priestley and Taylor, 1972). These are most commonly accepted to improve the accuracy and applicability of hydrological simulation in varying environmentally based circumstances.

Fundamental equations for river modeling and the key factors influencing channel design and flow dynamics were reviewed. These influences emphasized vegetation within the channel, bed and bank materials composition, and stream power. The Large Eddy Simulation (LES) model was cited as a highly reliable and frequently implemented model, showing the effect of vegetation (Millar and Quick, 1998; Dhotre et al., 2009). First, the key points regarding the significant advantages and disadvantages of the one-dimensional and two-dimensional numerical models, most commonly used with HEC-RAS, MIKE 11, and FLO-2D, were given. HEC-RAS and MIKE 11 could simulate channel flooding accurately; however, HEC-RAS offers the advantage of being free software, making it more cost-effective. At the same time, MIKE 11 imports many types of data from HEC-RAS and thus may offer an opportunity to integrate analyses from two platforms for more efficient and accurate results (Horritt and Bates, 2002).

In contrast, the grid-based simulation in FLO-2D was superior to 1D modeling by considering the river channel and the adjacent floodplains, allowing for a more holistic spatial and temporal representation of floodwaters. Nevertheless, some limitations were pointed against elevation variability and vertical irregularities in grid cells, which are critical in modeling flood behavior over complex terrains (Teng et al., 2017).

These shortcomings raised the demand for 3D modeling. They proposed to develop a 3D flood model by incorporating the merits of the existing models while eliminating their drawbacks. This would therefore provide a much better evaluation of the flood-induced phenomena and effects, especially in areas with complex topography (Rong et al., 2020).

A comparative evaluation of hydraulic models used in flood simulations is provided in detail in Table 2.

*Table 2. Comparison of hydraulic models for flood modeling.*

| Model | Primary Applications | Target Users | Cost Range | Learning Time |
|---|---|---|---|---|
| HEC-RAS | Standard flood studies, Regulatory compliance, Academic research, Emergency planning | Beginners to experts, Government agencies, Consulting firms, and Researchers | Free | 2–4 weeks |
| MIKE 11 | Complex river systems, Water quality modeling, Research projects, Environmental impact | Experienced modelers, Research institutions, consultants, and government | High (10,000–50,000 USD) | 4–8 weeks |
| FLO-2D | Flash flood modeling/debris flow analysis, urban flooding, hazard mapping | Flood specialists, Emergency planners, Researchers, Insurance companies | Medium (5,000–20,000 USD) | 2–3 weeks |
| TUFLOW | Advanced 2D modeling, Coastal flooding, Complex hydraulics, Detailed flood risk | Expert modelers, specialized consultants, and research laboratory projects | Premium (20,000–100,000 USD) | 8–12 weeks |

## Machine Learning Method

Support vector machine (SVM): A sorting system and supervised model allowing a computer to classify and analyze data (Cortes and Vapnik, 1995). SVMs are popular because of their strong empirical performance (when equated to alternative models like artificial neural networks), ease of training, prevention of local minima, and generally appropriate mathematics for various dimensional data, and a compromise between complication and error (Cortes and Vapnik, 1995). Advanced machine learning approaches in flood modeling are compared in Figure 2.

*Figure 2. Advanced machine learning approaches in flood modeling.*

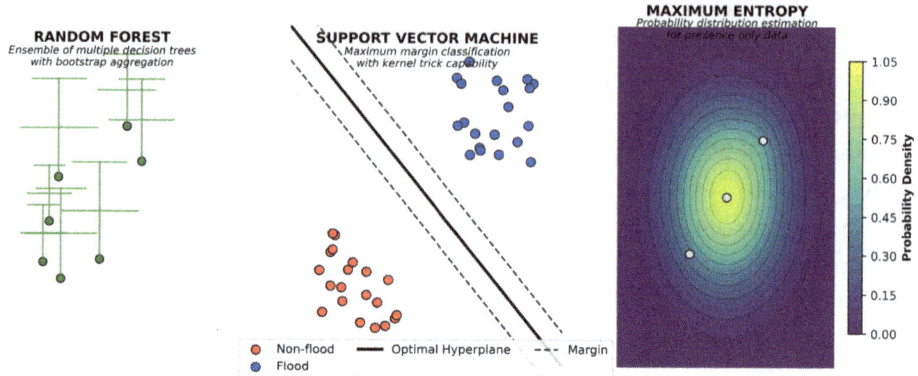


**Comparative Analysis of Machine Learning Methods for Flood Susceptibility Modeling**

| Method | Strengths | Limitations | Best For |
|---|---|---|---|
| Random Forest | Handles high dimensionality<br>Robust to outliers | Black box model<br>Memory intensive | Feature-rich datasets<br>Non-linear relationships |
| SVM | Effective in high dimensions<br>Strong theoretical basis | Sensitive to parameters<br>Slow for large datasets | Clear separation boundaries<br>Kernel applications |
| MaxEnt | Works with presence-only data<br>Probabilistic output | Sensitive to background points<br>Complex parameter tuning | Species distribution modeling<br>Limited data scenarios |

Random Forest (RF), introduced by Breiman (2001), is a widely adopted ensemble learning technique for classification and regression tasks. It constructs many decision trees, each built from a bootstrapped training data sample. On the other hand, prediction can be obtained by aggregating the outputs of all trees, primarily through majority voting in classification problems. It applies two principles: bagging and random feature selection. Among the trees, each is built from a different subset of data. For splitting at each node, a random set of predictor variables is selected to build the tree, which gives it greater robustness and generalization ability.

Phillips et al.(2006) developed Maxent, a machine learning algorithm sometimes implemented for modeling species distributions and other spatial phenomena. According to the maximum entropy principle, Maxent estimates the most uniform distribution of a species or phenomenon constrained by the environment. In the presence of sample data only, Maxent performs spatial inference of distributions by relating the known occurrence points to environmental variables, somewhat indirectly applying Bayes' theorem, thus becoming especially useful when other data are unavailable or unreliable.

Boosted Regression Trees (BRT) are within the gradient boosting family of algorithms that refine a predictive model by sequentially building regression trees so that the objective of the second tree is to reduce the errors of the first one (Elith et al., 2008). The effect was that more weight was given to those samples that already had difficult-to-predict responses, as the BRT thus incrementally improved the overall model performance. BRT is especially effective for capturing complex, nonlinear relationships between predictor and response variables by combining many weak learners.

Multivariate Adaptive Regression Splines (MARS) is a nonparametric regression introduced by Friedman (1991) to address the problem of modeling high-dimensional, nonlinear relationships. Traditional methods have fixed functional forms, but MARS is essentially free-form: it uses piecewise linear basis functions to identify and model interactions among variables, providing a very flexible and interpretable framework for spatial prediction.

The Generalized Linear Model (GLM) generalizes the classical linear regression in which the response variables are not normally distributed. GLM relates the expected value of the dependent variable to a linear combination of predictors through a prespecified link function. The GLM consists of three components: a random component specifying the probability distribution of the outcome variable, a systematic component specifying the linear combination of predictors, and a link function relating the expected value to the linear predictor (Zheng and Agresti, 2000).

In broader terms, a Generalized Additive Model (GAM) has been widely defined as a class of models where a semiparametric extension of GLM allows for more flexible dynamics in determining nonlinear relationships. So instead of assuming a fixed linear predictor, GAM considers smooth functions of its covariates, thus enabling it to accommodate complex, usually non-monotone, interactions between variables. The models are fitted using local scoring and backfitting algorithm variants (Murase et al., 2009).

Flood records of the events in 2017–2018 in Amol were used to prepare a flood inventory. Verified flood-prone areas were identified using municipal records and photographic documentation. Ninety-two flood-affected locations were labeled inundated (value = 1), while 60 sites with no observed flooding were labeled non-flooded (value = 0).

Drawing from previous studies, eight variables related to environment and geography connected with flood susceptibility were identified as predictors: precipitation, LULC, elevation, slope, CN, distance from river, distance from channel, and depth to groundwater. All these variables were rasterized with a 5-m pixel resolution and used as inputs for the spatial distribution modeling (SDM) through machine-learning methods implemented in the R environment.

## Hybrid Machine Learning Method

Several methods and approaches have been designed to evaluate flood susceptibility when integrating soft computing and data mining techniques (Chapi et al., 2017; Tien Bui et al., 2020). One set of approaches has mapped flood susceptibility using fuzzy multi-criteria analysis (Tehrany et al., 2014), TOPSIS (Ozturk and Batuk, 2011), mixed with Shannon entropy artificial neural networks, fuzzy logic, and decision trees (Kesavan et al., 2024; Senthil Kumar et al., 2013). While one may argue that each technique has its merits, they similarly endure drawbacks. ANNs, for example, can manage to extract information from incomplete or uncertain datasets (Agatonovic-Kustrin and Beresford, 2000). Nevertheless, the end users do not generally prefer them mainly due to their complexity and lack of interpretive power.

To compensate for the shortcomings, hybrid models were ushered in, combining the potentialities of different techniques; this approach handles spatial and numerical data better (Chapi et al., 2017). Typically, these hybrids merge more than one prediction method. In flood modeling, different MCDM techniques have been applied, including AHP (Massam, 1988; Darko et al., 2018), fuzzy multi-criteria analysis (Locatelli et al., 2008), weighted sum models (Miljković et al., 2017), and decision trees (Khosravi et al., 2018). Nonetheless, with every virtue comes a few demerits of these techniques. AHP incorporates a high degree of subjectivity due to its reliance on expert judgment (Saaty, 2008), while fuzzy multi-criteria methods are also prone to incorporation of biases of decision makers (Chen and Hwang, 1992); instability can be brought about in decision trees if minimal changes bring about significant differences in the structure of the tree (Breiman et al., 2017). To overcome these issues, a hybrid MCDM model that combines ANP and DEMATEL is used for complex decision-making problems (Si et al., 2018; Wang, 2012).

*Figure 3. Hybrid modeling framework: Integrating AHP. DEMATEL and Machine Learning for flood susceptibility assessment.*

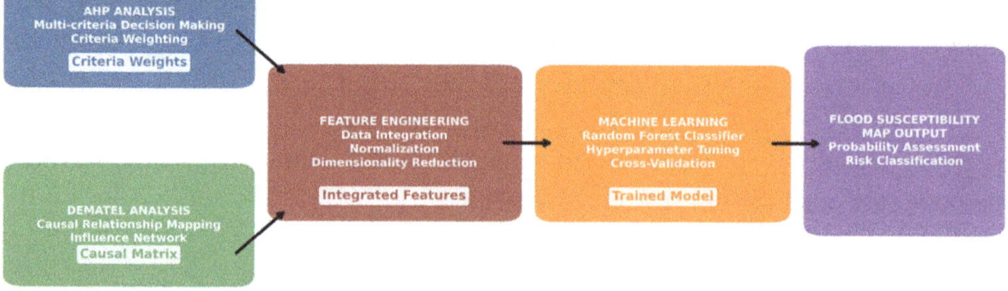


*Note: MCDM stands for multi-criteria decision-making. The structure of this hybrid modeling framework is illustrated in Figure 3.*

## Uncertainty in FIM

Uncertainty in FIM comes from uncertainties in estimating design flow, terrain elevation data, water surface elevations, and delineation methodologies for the flooded areas. Design flow is usually estimated with a method that aims at a 100-year return period, where: (I) actual on-site flood data is used for frequency analysis, (II) flood frequency analysis is conducted using data from nearby gauging stations lying on the same river, (III) regional flood frequency analysis is done using data from other areas that are hydrologically similar, and (IV) rainfall-runoff modeling is implemented through hydrologic simulations.

The approaches mentioned above can be exercised singly or in concert to estimate peak flows, with the probability distribution taken from within or externally. In all, the greater the number of ways to estimate, the more sources of uncertainty may find their way into peak flow calculations. Flow records determined from stage-discharge rating curves bear measurement uncertainties (Beven, 2012). Such uncertainties negatively affect the quality of input data in flood frequency analysis and hydrologic modeling. This uncertainty usually increases for high-return-period flows, as it is rare to have a rating curve prepared for such events (Di Baldassarre and Montanari, 2009). In the United States, Bulletin No. 17B of the Hydrology Subcommittee of the Interagency Advisory Committee on Water Data provides guidelines to estimate design flows (IACWD, 1982). The Log-Pearson Type III distribution is recommended, with parameter estimation done via historical weighting methods. However, more recent studies (Stedinger and Griffis, 2008; England et al., 2018) advocate that incorporating historical and paleoflood data might lead to more dependable flow estimates for selected return periods than in the Bulletin 17B method. This parameter estimation procedure brings additional uncertainty when applied to the Log-Pearson III distribution. Furthermore, all other choices-the one about which maxima to use: annual maxima or partial duration series; the other probability distribution; the alternative assumptions applied about climate variability, watershed change, and occurrence of zero-flow years; and methods of interpolating and handling missing data-all continue to influence the consistency and the credibility of a flood-frequency analysis. Then, there is yet another factor: two elements of data length, with the effects of data length preferably confounding; the shorter the series, the greater the uncertainty.

When data in other locations are used to estimate flows at ungauged sites, in addition to the transfer uncertainty, other methods create uncertainty in determining information about the flow. HEC-HMS (Nagarajan et al., 2022) and TR-20 (Endreny and Higgins, 2008) are the two popular hydrologic floodplain mapping models in the United States. Literature sets forth several sources of uncertainty in hydrologic modeling, such as model structure, parameter choices, or determining assumptions.

Singh (1997) identified four factors affecting streamflow hydrographs: antecedent conditions, storm precipitation dynamics, infiltration processes, and watershed characteristics. Further, it is necessary to realize that the output from hydrologic models is also affected by their assumptions, configurations, and modeling processes. Nevertheless, precipitation dynamics are the most powerful in determining streamflow forecast results among all these factors.

The perilous task of modeling the spatio-temporal variability of hydrologic phenomena such as precipitation, evapotranspiration, and infiltration has remained since immemorial times. Notwithstanding the evolution of distributed hydrologic models, inadequate spatial and temporal variability knowledge detracts from the models' predictive power. Accordingly, using spatially and temporally averaged values of hydrologic variables should compound uncertainties. This uncertainty can be dealt with in stochastic analyses by developing confidence and prediction intervals, among others. Recognizing such uncertainties from different sources adds value to modeling as a whole. An example for this is the estimation of design flows through the equations of the National Flood Frequency (NFF) (England 2011), for which prediction errors have been reported to range between 15% and 100% (Sauer et al., 1983). The above errors also depend on the coefficients and techniques of derivation that have been applied in the NFF models (Stuckey and Reed, 2000).

The formation of flood extent is sensitive to topographic data quality because of how high water hydrologic modeling, the extraction of hydraulic cross-sections, and the mapping of water surface elevations are performed. Terrain elevation uncertainty Azure FIM in three categories: modifications of discharge values estimated by hydrologic models (Brasington and Richards, 1998; Hancock, 2005), changes in water surface elevations from hydraulic models (Marks and Bates, 2000; Permana et al., 2022; Werner, 2001), and alterations to the inundation modelling.

Figure 4 illustrates the change in water surface elevations calculated using HEC-RAS from two digital elevation datasets - the 30-meter SRTM DEM and the 1-meter LIDAR DEM. Terrain data quality concepts include raw data accuracy, interpolation, methods to combine data, data formats (e.g., TIN vs. Raster/GRID), procedures for making hydrologically conditioned surfaces, and spatial resolution (grid cell size). Raw data sources include contour maps, LIDAR-derived elevation points, topographic survey points, cross-sections, and DEMs. Contour interval spacing, LIDAR posting density, and river cross-section intervals also contribute to the uncertainty associated with topographic data (Yilmaz et al., 2008). The influence of DEM resolution on the accuracy of FIM is also depicted in Figure 4.

*Figure 4. Impact of DEM resolution on flood inundation accuracy.*

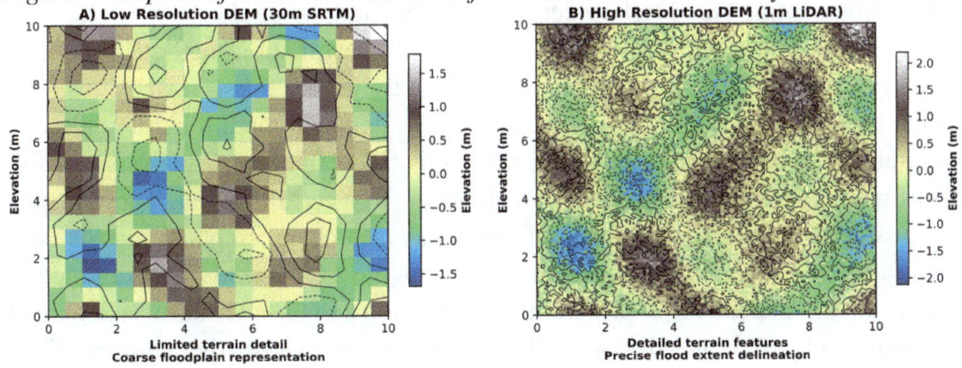


Hydraulic Model Flood inundation maps are also subject to uncertainty due to hydraulic modelling. Hydraulic models require geometric pictures, model constraints (Manning's n), and the demonstration of hydraulic structures like bridges and embankments. Apart from the variables previously cited concerning the hydrological model, the model application for simulating river hydrodynamics adds to any uncertainty in the process (Deodatis et al., 1996). In hydraulic models, channel geometry specification (Pappenberger et al., 2005) constitutes the most relevant factor (Pappenberger et al., 2005). Finite element mesh (1D model) or finite element mesh (2D/3D model). Changing WSE and extent estimation results when changing elevations from LIDAR DEM to 30 M coarse DEM. A careful geometric specification becomes critical in 2D and 3D models where elevation is defined at every mesh node, scattered along the channel and floodplains. Inadequately topographical data-assisted geometries can be more competitive than those in 1D and 2D models. Apart from the main channel and its disconnect from the floodplains, modelling line elements (breaklines) to confine the amount of inundation for different flow degrees is important in hydraulic modelling of high flows (Werner, 2004).

## DISCUSSION

The findings of this study reveal how complex FIM is and how its complexity develops from an interrelationship of many factors, such as hydrologic modeling, hydraulic modeling, quality of topographic data, and the use of newer computational techniques such as machine learning. It was further observed that uncertainties arise at each modeling step, influencing the prediction of flood extents. These uncertainties are particularly critical when estimating design flows, delineating inundation areas, and communicating flood risk to decision-makers and the public.

The results highlight that selecting design flow estimation methods is pivotal in determining flood risk outcomes. Data paucity, rating curve restrictions, and assumptions regarding climate variability and watershed conditions may hamper the application of different approaches, such as flood frequency analysis, rainfall runoff modeling, and regionalization techniques, to a certain extent. These results are consistent with the conclusions of previous researchers, emphasizing the concerns faced in extrapolating flows with a very high return period in the absence of reliable historical data. Also, probability distribution and the alternate use of annual maximum and partial duration series have been identified as further sources of uncertainties, pointing out the need to thoroughly assess these methods when applied for floodplain management.

The most influential factors affecting FIM accuracy were terrain representation quality and terrain data resolution. An inadequate topography description thus impacted hydraulic modeling, hydraulic extraction of cross-sections, and the delineation of water surface elevations. In agreement with previous research findings demonstrating that coarse DEMs usually tend to under-represent the complexity of floodplain hydraulics, resulting in an overestimation or underestimation of inundation extents, the above factors are at play. High-resolution datasets such as LiDAR ought to have helped significantly; these, however, are also prone to uncertainties due to posting density, interpolation algorithms used, and distance between cross-sections. Consequently, topographic data uncertainty propagates throughout the modeling chain, influencing discharge calculations and final inundation maps.

Hydraulic model selection and configuration were also significant contributors to overall uncertainty. It was shown that 1D, 2D, and coupled 1D/2D models vary in their ability to simulate flow dynamics and floodplain connectivity. While 2D and hybrid models were found to produce more realistic flood extents, they also require higher-quality input data and greater computational resources. Sensitivity to parameters such as Manning's n illuminates the need for rigorous calibration and validation algorithms. Structural features such as levees, bridges, and breaklines were required to constrain flow pathways and improve inundation predictions.

Integrating machine learning and hybrid modeling approaches enhanced the state of the art. In this practice, ensemble methods defeat single model methods for linking complex and nonlinear behavior between environmental variables and flood occurrence. These methods also allow one to integrate different data sources, including those from remote sensing and those variables derived from GIS, thus increasing the spatial-temporal accuracy of flood predictions. However, interpreting these models still poses a challenge, especially for end users in decision-making contexts. Future research should focus on rendering further transparency and developing frameworks that ease the current chasm between advanced analytics and applied flood management.

Finally, the study corroborated that uncertainties cannot be eradicated yet may be quantified and managed using rigorous modeling approaches. Confidence intervals, sensitivity analysis, and probabilistic frameworks were instrumental in establishing the uncertainty and reliability of FIM outputs. Now that these uncertainties have been communicated, planners and policymakers can weigh in further to choose flood mitigation methods, infrastructure designs, and modes of emergency preparedness. Overall, the findings stress the need for integrated multi-disciplinary approaches involving hydrologic, hydraulic, and data-driven methods to improve the accuracy and utility of flood inundation mapping in a changing climate.

## CONCLUSION

Flood hazard mapping was examined using various sources of information. They used various tools and models for the assessment and remote sensing, including ArcGIS, MIKE 11, ERDAS, HEC-RAS, and HEC-HMS. Flood inundation of low-lying flood-affected areas may be determined using HEC-RAS and ArcGIS, and mapping of those areas is simple, thanks to the use of Machine Learning. Plotting graphs between different return dates and inundated areas allows for assessing swamped areas at any return period. Suppose a graph is displayed at 0, 10, 20, 30, 40, 50, 60, 70, 80, 90, and 100 return periods with inundated areas matching these return periods. In that case, the inundated area at any return period in between can be examined using graphical methods. It may be stated that the GIS technique employing the HEC-RAS model proved to be a reasonable and appropriate method for obtaining a precise flood inundation map. This aids in detecting and rescuing endangered and vulnerable persons in that location. It was envisaged that scientists from developing nations would contribute to a better understanding of these various methodologies by delivering or utilising restricted information.

The findings revealed that uncertainties are inherent in all stages of the FIM process. These uncertainties arise from design flow estimation methods, model structure and configuration, spatial variability of precipitation and infiltration, and the selection of probability distributions for flood frequency analysis. It was demonstrated that high-resolution datasets, such as LiDAR, substantially improve the accuracy of flood modeling. However, challenges persist, such as limited data availability, stage–discharge rating curve errors, and assumptions of climatic variability and watershed alteration.

In addition, the hybrid of machine learning and modeling techniques assisted in enhancing flood susceptibility assessment by marrying diverse data and the intricate nonlinear dependence between environmental variables. Despite these efforts, challenges related to model interpretability and user accessibility persist. Therefore, it is recommended that future research should direct efforts into designing transparent, flexible, and user-friendly frameworks as decision support tools for practical flood management applications.

In conclusion, flood-risk management under a changing climate would call for an interdisciplinary approach, mixing the hydrologic-hydraulic and data-driven modeling methods and explicitly accounting for uncertainties by applying probabilistic methods and sensitivity analyses. Through this new approach, we can produce more reliable and actionable flood inundation maps. These tools support infrastructure and emergency preparedness planning and help devise adaptive strategies to mitigate the impacts wrought by future floods, both socially and environmentally.

## RECOMMENDATIONS

Following this study's conclusion, several recommendations have appeared to enhance the reliability and applicability of FIM and flood risk management. These recommendations seek to assist researchers, practitioners, and policymakers apply better modeling, data integration, and decision-making frameworks.

The first recommendation is prioritizing high-resolution topographic data, such as LiDAR-derived DEMs, for flood modeling applications. Coarser-scale terrain data was shown to have led to severe errors in hydraulic simulations, resulting in inaccurate delineations of inundation extents. Thus, accurate terrain mapping should be the highest priority in flood prediction and management.

Second, a systematic account of uncertainty should be present at all stages of the FIM. Amidst uncertainties coming from design flow estimation, model selection, and parameterization, probabilistic approaches, sensitivity analyses, and confidence intervals should be encouraged. This allows for communicating potential error margins to stakeholders and may further improve transparency in flood risk assessments.

Third, integrated hydrologic-hydraulic modeling approaches are also recommended. Coupled 1D-2D models can fully depict flood dynamics, especially in complex floodplains. Furthermore, incorporating structural features such as levees, bridges, and breaklines is necessary for an enhanced flow-pathway representation that avoids over- or underestimating flood extents.

Fourth, an expansion should be made in applying machine learning and hybrid modeling techniques. These methods boosted prediction ability by integrating heterogeneous datasets and modeling nonlinear relationships between environmental variables. However, it is important to emphasize and focus on building interpretable models accessible to decision-makers and emergency planners. This further extends to developing simple platforms and visualization tools that connect high-level analytics to practicalities.

Data sharing and collaborative platforms can be fostered to beat data scarcity. These platforms stand across agencies to share data and foster international cooperation in sharing hydrological, meteorological, and remote-sensing data, which are essential for proper modeling and validation.

Lastly, adaptive management strategies must be developed to deal with climate change and its impact on flood risk. Scenario-based modeling can be used to study the future patterns of floods to design resilient infrastructure and land-use planning policies. This future-oriented approach shall help communities prepare for increased flood hazards and provide solutions for the socioeconomic and environmental impacts.

These recommendations will thus foster future flood inundation mapping endeavors, gaining higher precision, reliability, and relevance for better flood risk management and disaster preparedness in a changing climate.

## ACKNOWLEDGMENTS

**Author Contributions:** Conceptualization, M.A.Ç. A.E., R.M., S.S., D.K.V., K.P., and M.E.A.; methodology, M.A.Ç. A.E., R.M., S.S., D.K.V., K.P. and M.E.A.; formal analysis, M.A.Ç., A.E., R.M., S.S., D.K.V., K.P.and M.E.A.; investigation, M.A.Ç., A.E., R.M., S.S., D.K.V., K.P. and M.E.A.; resources, review & editing, M.A.Ç., A.E., R.M., S.S., D.K.V., K.P. and M.E.A.; visualization, M.A.Ç., A.E., R.M., S.S., D.K.V., K.P. and M.E.A.; supervision, M.A.Ç., A.E., R.M., S.S., D.K.V., K.P. and M.E.A. All authors have read and agreed to the published version of the manuscript.

**Funding:** The Scientific and Technological Research Council of Turkey (TÜBİTAK)

## REFERENCES

Aalto, R., Dunne, T., & Guyot, J. L. (2006). Geomorphic Controls on Andean Denudation Rates. *The Journal of Geology, 114*(1), 85–99. DOI: 10.1086/498101

Agatonovic-Kustrin, S., & Beresford, R. (2000). Basic concepts of artificial neural network (ANN) modeling and its application in pharmaceutical research. *Journal of Pharmaceutical and Biomedical Analysis, 22*(5), 717–727. DOI: 10.1016/S0731-7085(99)00272-1

Akiner, M. E., & Lačnjevac, Č. (2022). Managing Saltwater Intrusion and Agricultural Practices along the Boğaçay River, Turkey: Effects from Excavation and Land Source Pollution. *Journal of Coastal Research, 38*(3), 567–577. DOI: 10.2112/JCOASTRES-D-21-00101.1

Alderman, K., Turner, L. R., & Tong, S. (2012). Floods and human health: A systematic review. *Environment International, 47*, 37–47. DOI: 10.1016/j.envint.2012.06.003

Bagja, I., Pangi, P., Mispaki, S., & Rahayu, S. (2025). Utilization of Sentinel-1 imagery for inundation flood mapping in northern part of Java Island. *IOP Conference Series. Earth and Environmental Science, 1438*(1), 012045. DOI: 10.1088/1755-1315/1438/1/012045

Beven, K. (2012). Rainfall-Runoff Modelling. DOI: 10.1002/9781119951001

Brasington, J., & Richards, K. (1998). Interactions between model predictions, parameters and DTM scales for TopModel. Computers &amp. *Computers & Geosciences, 24*(4), 299–314. DOI: 10.1016/S0098-3004(97)00081-2

Breiman, L. (2001). Random Forests. *Machine Learning*, *45*(1), 5–32. DOI: 10.1023/A:1010933404324

Breiman, L., Friedman, J. H., Olshen, R. A., & Stone, C. J. (2017). Regression Trees. Classification And Regression Trees, 216–265. DOI: 10.1201/9781315139470-8

Chapi, K., Singh, V. P., Shirzadi, A., Shahabi, H., Bui, D. T., Pham, B. T., & Khosravi, K. (2017). A novel hybrid artificial intelligence approach for flood susceptibility assessment. Environmental modelling &amp. *Environmental Modelling & Software*, *95*, 229–245. DOI: 10.1016/j.envsoft.2017.06.012

Chen, S.-J., & Hwang, C.-L. (1992). Fuzzy Multiple Attribute Decision Making Methods. Fuzzy Multiple Attribute Decision Making, 289–486. DOI: 10.1007/978-3-642-46768-4_5

Cortes, C., & Vapnik, V. (1995). Support-Vector Networks. *Machine Learning*, *20*(3), 273–297. DOI: 10.1023/A:1022627411411

Costa, J. E. (1987). Hydraulics and basin morphometry of the largest flash floods in the conterminous United States. *Journal of Hydrology (Amsterdam)*, *93*(3–4), 313–338. DOI: 10.1016/0022-1694(87)90102-8

Cutter, S. L., Ash, K. D., & Emrich, C. T. (2014). The geographies of community disaster resilience. *Global Environmental Change*, *29*, 65–77. DOI: 10.1016/j.gloenvcha.2014.08.005

Cvetković, V. M., Renner, R., Aleksova, B., & Lukić, T. (2024). Geospatial and temporal patterns of natural and man-made (Technological) disasters (1900–2024): Insights from different socioeconomic and demographic perspectives. *Applied Sciences (Basel, Switzerland)*, *14*(18), 8129. DOI: 10.3390/app14188129

Darko, A., Chan, A. P. C., Ameyaw, E. E., Owusu, E. K., Pärn, E., & Edwards, D. J. (2018). Review of application of analytic hierarchy process (AHP) in construction. *International Journal of Construction Management*, *19*(5), 436–452. DOI: 10.1080/15623599.2018.1452098

De Moel, H., Jongman, B., Kreibich, H., Merz, B., Penning-Rowsell, E., & Ward, P. J. (2015). Flood risk assessments at different spatial scales. *Mitigation and Adaptation Strategies for Global Change*, *20*(6), 865–890. DOI: 10.1007/s11027-015-9654-z

Dhotre, M. T., Niceno, B., Smith, B. L., & Simiano, M. (2009). Large-eddy simulation (LES) of the large scale bubble plume. *Chemical Engineering Science*, *64*(11), 2692–2704. DOI: 10.1016/j.ces.2009.02.040

Di Baldassarre, G., Castellarin, A., Montanari, A., & Brath, A. (2009). Probability-weighted hazard maps for comparing different flood risk management strategies: A case study. *Natural Hazards, 50*(3), 479–496. DOI: 10.1007/s11069-009-9355-6

Di Baldassarre, G., & Montanari, A. (2009). Uncertainty in river discharge observations: A quantitative analysis. *Hydrology and Earth System Sciences, 13*(6), 913–921. DOI: 10.5194/hess-13-913-2009

Dilley, M., Chen, R. S., Deichmann, U., Lerner-Lam, A. L., & Arnold, M. (2005). *Natural Disaster Hotspots*. The World Bank., DOI: 10.1596/0-8213-5930-4

Domeneghetti, A., Vorogushyn, S., Castellarin, A., Merz, B., & Brath, A. (2013). Probabilistic flood hazard mapping: effects of uncertain boundary conditions. Hydrology and Earth System Sciences, 17(8), 3127–3140. EEA Report "EEA Signals 2010": stories about people and the environment. (2010). International Journal of Sustainability in Higher Education, 11(3). https://doi.org/DOI: 10.5194/hess-17-3127-2013

Dullo, T., Darkwah, G., Gangrade, S., Morales-Hernández, M., Sharif, M., Kalyanapu, A., Kao, S.-C., Ghafoor, S., & Ashfaq, M. (2021). Assessing climate-change-induced flood risk in the Conasauga River watershed: An application of ensemble hydrodynamic inundation modeling. *Natural Hazards and Earth System Sciences, 21*(6), 1739–1757. DOI: 10.5194/nhess-21-1739-2021

Elith, J., Leathwick, J. R., & Hastie, T. (2008). A working guide to boosted regression trees. Journal of Animal Ecology, 77(4), 802–813. Portico. DOI: 10.1111/j.1365-2656.2008.01390.x

EM-DAT (Emergency Events Database). (2009) Monitoring hazards & disasters worldwide since 1988. Link: https://www.emdat.be/ last access: 045.09.2025

Endreny, T., & Higgins, M. (2008). Adding Radar Rainfall and Calibration to the TR-20 Watershed Model to Improve Dam Removal Flood Analysis. *Journal of Water Resources Planning and Management, 134*(3), 314–317. DOI: 10.1061/(ASCE)0733-9496(2008)134:3(314)

England, J. F.Jr. (2011). Flood Frequency and Design Flood Estimation Procedures in the United States: Progress and Challenges. *Australasian Journal of Water Resources, 15*(1), 33–46. DOI: 10.1080/13241583.2011.11465388

England, J. F., Cohn, T. A., Faber, B. A., Stedinger, J. R., Thomas, W. O., Veilleux, A. G., Kiang, J. E., & Mason, R. R. (2018). *Guidelines for determining flood flow frequency — Bulletin 17C*. Techniques and Methods., DOI: 10.3133/tm4B5

Fitri, A., Nurul Hadie, M. S., Agustina, A., Pratiwi, D., Susarman, Pramita, G., & Salah, H. R. A. (2021). Analyses of flood peak discharge in Cimadur River basin, Banten Province, Indonesia. E3S Web of Conferences, 331, 08006. DOI: 10.1051/e3sconf/202133108006

Follum, M. L., Tavakoly, A. A., Niemann, J. D., & Snow, A. D. (2016). AutoRAPID: A Model for Prompt Streamflow Estimation and Flood Inundation Mapping over Regional to Continental Extents. JAWRA Journal of the American Water Resources Association, 53(2), 280–299. Portico. DOI: 10.1111/1752-1688.12476

Foster, M., Fell, R., & Spannagle, M. (2000). The statistics of embankment dam failures and accidents. *Canadian Geotechnical Journal*, *37*(5), 1000–1024. DOI: 10.1139/t00-030

Freeze, R. A., & Cherry, J. A. (1979). Groundwater prentice-hall. Englewood Cliffs, NJ, 176, 161-177.

Friedman, J. H. (1991). Multivariate Adaptive Regression Splines. *Annals of Statistics*, *19*(1). Advance online publication. DOI: 10.1214/aos/1176347963

Fu, X., He, X., & Ding, L. (2023). Stochastic Flood Simulation Method Combining Flood Intensity and Morphological Indicators. *Sustainability (Basel)*, *15*(18), 14032. DOI: 10.3390/su151814032

Gaál, L., Szolgay, J., Kohnová, S., Hlavčová, K., Parajka, J., Viglione, A., Merz, R., & Blöschl, G. (2015). Dependence between flood peaks and volumes: A case study on climate and hydrological controls. *Hydrological Sciences Journal*, *60*(6), 968–984. DOI: 10.1080/02626667.2014.951361

Hall, J. W., Tarantola, S., Bates, P. D., & Horritt, M. S. (2005). Distributed Sensitivity Analysis of Flood Inundation Model Calibration. *Journal of Hydraulic Engineering (New York, N.Y.)*, *131*(2), 117–126. DOI: 10.1061/(ASCE)0733-9429(2005)131:2(117)

Hamidifar, H., Nones, M., & Rowinski, P. M. (2024). Flood modeling and fluvial dynamics: A scoping review on the role of sediment transport. *Earth-Science Reviews*, *253*, 104775. DOI: 10.1016/j.earscirev.2024.104775

Hancock, G. R. (2005). The use of digital elevation models in the identification and characterization of catchments over different grid scales. *Hydrological Processes*, *19*(9), 1727–1749. DOI: 10.1002/hyp.5632

Holtz, R. D. (1981). Discussion of "Particle Contacts in Discrete Materials.". *Journal of the Geotechnical Engineering Division*, *107*(10), 1437–1440. DOI: 10.1061/AJGEB6.0001199

Horritt, M. S., & Bates, P. D. (2002). Evaluation of 1D and 2D numerical models for predicting river flood inundation. *Journal of Hydrology (Amsterdam)*, *268*(1–4), 87–99. DOI: 10.1016/S0022-1694(02)00121-X

IACWD (1982). Water resources data. Annual Report. DOI: 10.3133/891

Ingle Smith, D. (1999). Floods: physical processes and human impacts by K. Smith and R. Ward, John Wiley, Chichester, 1998. No. of pages: 382. Earth Surface Processes and Landforms, 24(13), 1261–1261. Portico. https://doi.org/DOI: 10.1002/(sici)1096-9837(199912)24:13<1261:aid-esp22>3.0.co;2-#

Inman, V., & Lyons, M. (2020). Automated inundation mapping over large areas using Landsat data and Google Earth Engine. *Remote Sensing (Basel)*, *12*(8), 1348. DOI: 10.3390/rs12081348

Intergovernmental Panel on Climate Change (IPCC). (2007). https://www.ipcc.ch/report/ar4/syr/ last access: 05.09.2025

Intergovernmental Panel on Climate Change (IPCC). (2023). Climate Change 2022 – Impacts, Adaptation and Vulnerability. DOI: 10.1017/9781009325844

Iverson, R. M. (1997). The physics of debris flows. Reviews of Geophysics, 35(3), 245–296. Portico. DOI: 10.1029/97RG00426

Iverson, R. M. (2000). Landslide triggering by rain infiltration. Water Resources Research, 36(7), 1897–1910. Portico. DOI: 10.1029/2000WR900090

Jonkman, S. N. (2005). Global Perspectives on Loss of Human Life Caused by Floods. *Natural Hazards*, *34*(2), 151–175. DOI: 10.1007/s11069-004-8891-3

Jonkman, S. N., & Kelman, I. (2005). An Analysis of the Causes and Circumstances of Flood Disaster Deaths. *Disasters*, *29*(1), 75–97. DOI: 10.1111/j.0361-3666.2005.00275.x

Jonkman, S. N., & Penning-Rowsell, E. (2008). Human instability in flood flows 1. *Journal of the American Water Resources Association*, *44*(5), 1208–1218. DOI: 10.1111/j.1752-1688.2008.00217.x

Julien, P. Y. (2010). Erosion and Sedimentation. DOI: 10.1017/CBO9780511806049

Junk, W. J. (1999). The flood pulse concept of large rivers: Learning from the tropics. *River Systems*, *11*(3), 261–280. DOI: 10.1127/lr/11/1999/261

Kesavan, H., Choudhary, S., Khan, S., Samhita, R., & Sharma, A. (2024). AI-based urban resilience planning: Opportunities and challenges. *Journal of Environment and Earth Science*, *6*(2), 200–214. DOI: 10.30564/jees.v6i2.6681

Khosravi, K., Pham, B. T., Chapi, K., Shirzadi, A., Shahabi, H., Revhaug, I., Prakash, I., & Tien Bui, D. (2018). A comparative assessment of decision trees algorithms for flash flood susceptibility modeling at Haraz watershed, northern Iran. *The Science of the Total Environment*, *627*, 744–755. DOI: 10.1016/j.scitotenv.2018.01.266

Knighton, A. D. (1998). Fluvial Forms and Processes: A New Perspective, Arnold, London,383 p., ill., tabl, pl., 15,5 x 23,5 cm. ISBN 0 340 66313 8. *Géographie Physique et Quaternaire*, *53*(3), 413. DOI: 10.7202/004839ar

Kotlarski, S., Keuler, K., Christensen, O. B., Colette, A., Déqué, M., Gobiet, A., Goergen, K., Jacob, D., Lüthi, D., van Meijgaard, E., Nikulin, G., Schär, C., Teichmann, C., Vautard, R., Warrach-Sagi, K., & Wulfmeyer, V. (2014). Regional climate modeling on European scales: A joint standard evaluation of the EURO-CORDEX RCM ensemble. *Geoscientific Model Development*, *7*(4), 1297–1333. DOI: 10.5194/gmd-7-1297-2014

Kozlowski, T. T. (1997). Responses of woody plants to flooding and salinity. *Tree Physiology*, *17*(7), 490–490. DOI: 10.1093/treephys/17.7.490

Kundzewicz, Z. W., Kanae, S., Seneviratne, S. I., Handmer, J., Nicholls, N., Peduzzi, P., Mechler, R., Bouwer, L. M., Arnell, N., Mach, K., Muir-Wood, R., Brakenridge, G. R., Kron, W., Benito, G., Honda, Y., Takahashi, K., & Sherstyukov, B. (2013). Flood risk and climate change: Global and regional perspectives. *Hydrological Sciences Journal*, *59*(1), 1–28. DOI: 10.1080/02626667.2013.857411

Kvočka, D., Falconer, R. A., & Bray, M. (2016). Flood hazard assessment for extreme flood events. *Natural Hazards*, *84*(3), 1569–1599. DOI: 10.1007/s11069-016-2501-z

Lamb, R., Keef, C., Tawn, J., Laeger, S., Meadowcroft, I., Surendran, S., Dunning, P., & Batstone, C. (2010). A new method to assess the risk of local and widespread flooding on rivers and coasts. Journal of Flood Risk Management, 3(4), 323–336. Portico. DOI: 10.1111/j.1753-318X.2010.01081.x

Lee, J., & Choi, H. (2018). Comparison of Flood Vulnerability Assessments to Climate Change by Construction Frameworks for a Composite Indicator. *Sustainability (Basel)*, *10*(3), 768. DOI: 10.3390/su10030768

Locatelli, B., Rojas, V., & Salinas, Z. (2008). Impacts of payments for environmental services on local development in northern Costa Rica: A fuzzy multi-criteria analysis. *Forest Policy and Economics*, *10*(5), 275–285. DOI: 10.1016/j.forpol.2007.11.007

Marks, K., & Bates, P. (2000). Integration of high-resolution topographic data with floodplain flow models. *Hydrological Processes*, *14*(11–12), 2109–2122. DOI: 10.1002/1099-1085(20000815/30)14:11/12<2109::AID-HYP58>3.0.CO;2-1

Martínez-Gomariz, E., Gómez, M., Russo, B., & Djordjević, S. (2016). Stability criteria for flooded vehicles: a state-of-the-art review. Journal of Flood Risk Management, 11(S2). Portico. DOI: 10.1111/jfr3.12262

Massam, B. H. (1988). Multi-Criteria Decision Making (MCDM) techniques in planning. *Progress in Planning*, *30*, 1–84. DOI: 10.1016/0305-9006(88)90012-8

Merz, B., Hall, J., Disse, M., & Schumann, A. (2010). Fluvial flood risk management in a changing world. *Natural Hazards and Earth System Sciences*, *10*(3), 509–527. DOI: 10.5194/nhess-10-509-2010

Merz, R., & Blöschl, G. (2003). A process typology of regional floods. Water Resources Research, 39(12). Portico. DOI: 10.1029/2002WR001952

Miljkovic, B., Zizovic, M., Petojevic, A., & Damljanovic, N. (2017). New weighted sum model. *Filomat*, *31*(10), 2991–2998. DOI: 10.2298/FIL1710991M

Millar, R. G., & Quick, M. C. (1998). Stable Width and Depth of Gravel-Bed Rivers with Cohesive Banks. *Journal of Hydraulic Engineering (New York, N.Y.)*, *124*(10), 1005–1013. DOI: 10.1061/(ASCE)0733-9429(1998)124:10(1005)

Mir, Y. H., Mir, S., Ganie, M. A., Bhat, J. A., Shah, A. M., Mushtaq, M., & Irshad, I. (2025). Overview of land use and land cover change and its impacts on natural resources. In Ecologically Mediated Development: Promoting Biodiversity Conservation and Food Security (pp. 101-130). Singapore: Springer Nature Singapore. DOI: 10.1007/978-981-96-2413-3_5

Montgomery, D. R. (2007). Soil erosion and agricultural sustainability. *Proceedings of the National Academy of Sciences of the United States of America*, *104*(33), 13268–13272. DOI: 10.1073/pnas.0611508104

Murase, H., Nagashima, H., Yonezaki, S., Matsukura, R., & Kitakado, T. (2009). Application of a generalized additive model (GAM) to reveal relationships between environmental factors and distributions of pelagic fish and krill: A case study in Sendai Bay, Japan. *ICES Journal of Marine Science*, *66*(6), 1417–1424. DOI: 10.1093/icesjms/fsp105

Nagarajan, K., Narwade, R., Pathak, H., Panhalkar, S., Kulkarni, V. S., & Hingmire, A. P. (2022). Review paper for floodplain mapping with applications of HEC-HMS, HEC-RAS, and ArcGIS softwares a remote sensing and GIS approach. [IRJET]. *International Research Journal of Engineering and Technology*, *9*(6), 812–826.

Neal, J., Schumann, G., Fewtrell, T., Budimir, M., Bates, P., & Mason, D. (2011). Evaluating a new LISFLOOD-FP formulation with data from the summer 2007 floods in Tewkesbury, UK. Journal of Flood Risk Management, 4(2), 88–95. Portico. DOI: 10.1111/j.1753-318X.2011.01093.x

Nobre, A. D., Cuartas, L. A., Hodnett, M., Rennó, C. D., Rodrigues, G., Silveira, A., Waterloo, M., & Saleska, S. (2011). Height Above the Nearest Drainage – a hydrologically relevant new terrain model. *Journal of Hydrology (Amsterdam), 404*(1–2), 13–29. DOI: 10.1016/j.jhydrol.2011.03.051

Ouarda, T. B. M. J., Girard, C., Cavadias, G. S., & Bobée, B. (2001). Regional flood frequency estimation with canonical correlation analysis. *Journal of Hydrology (Amsterdam), 254*(1–4), 157–173. DOI: 10.1016/S0022-1694(01)00488-7

Ozturk, D., & Batuk, F. (2011). Technique for order preference by similarity to ideal solution (TOPSIS) for spatial decision problems. In Proceedings ISPRS (Vol. 1, No. 4).

Pappenberger, F., Beven, K., Horritt, M., & Blazkova, S. (2005). Uncertainty in the calibration of effective roughness parameters in HEC-RAS using inundation and downstream level observations. *Journal of Hydrology (Amsterdam), 302*(1–4), 46–69. DOI: 10.1016/j.jhydrol.2004.06.036

Paul, M. J., & Meyer, J. L. (2001). Streams in the Urban Landscape. *Annual Review of Ecology and Systematics, 32*(1), 333–365. DOI: 10.1146/annurev.ecolsys.32.081501.114040

Penning-Rowsell, E. C., & Green, C. (2000). New Insights into the Appraisal of Flood-Alleviation Benefits: (1) Flood Damage and Flood Loss Information. Water and Environment Journal, 14(5), 347–353. Portico. DOI: 10.1111/j.1747-6593.2000.tb00272.x

Permana, S., Risandi, J., Tito, C., Ramdhan, M., Sufyan, A., Solihuddin, T., Purbani, D., Pryambodo, D. G., Hidayat, R., Rifai, H., & Lukman, K. (2022). Predicting coastal inundation triggered by the oceanic forcing across Jakarta Bay. *IOP Conference Series. Earth and Environmental Science, 1109*(1), 012005. DOI: 10.1088/1755-1315/1109/1/012005

Phillips, S. J., Anderson, R. P., & Schapire, R. E. (2006). Maximum entropy modeling of species geographic distributions. *Ecological Modelling, 190*(3–4), 231–259. DOI: 10.1016/j.ecolmodel.2005.03.026

Pielke, R. A.Jr, & Downton, M. W. (2000). Precipitation and Damaging Floods: Trends in the United States, 1932–97. *Journal of Climate*, *13*(20), 3625–3637. DOI: 10.1175/1520-0442(2000)013<3625:PADFTI>2.0.CO;2

Plate, E. J. (2002). Flood risk and flood management. *Journal of Hydrology (Amsterdam)*, *267*(1–2), 2–11. DOI: 10.1016/S0022-1694(02)00135-X

Priestley, C. H. B., & Taylor, R. J. (1972). On the Assessment of Surface Heat Flux and Evaporation Using Large-Scale Parameters. *Monthly Weather Review*, *100*(2), 81–92. DOI: 10.1175/1520-0493(1972)100<0081:OTAOSH>2.3.CO;2

Rahmati, O., Zeinivand, H., & Besharat, M. (2015). Flood hazard zoning in Yasooj region, Iran, using GIS and multi-criteria decision analysis. *Geomatics, Natural Hazards & Risk*, *7*(3), 1000–1017. DOI: 10.1080/19475705.2015.1045043

Rawls, W. J., Brakensiek, D. L., & Saxtonn, K. E.W. J. RawlsD. L. BrakensiekK. E. Saxtonn. (1982). Estimation of Soil Water Properties. *Transactions of the ASAE. American Society of Agricultural Engineers*, *25*(5), 1316–1320. DOI: 10.13031/2013.33720

Reggiani, P., & Weerts, A. H. (2008). A Bayesian approach to decision-making under uncertainty: An application to real-time forecasting in the river Rhine. *Journal of Hydrology (Amsterdam)*, *356*(1–2), 56–69. DOI: 10.1016/j.jhydrol.2008.03.027

Rong, Y., Zhang, T., Zheng, Y., Hu, C., Peng, L., & Feng, P. (2020). Three-dimensional urban flood inundation simulation based on digital aerial photogrammetry. *Journal of Hydrology (Amsterdam)*, *584*, 124308. DOI: 10.1016/j.jhydrol.2019.124308

Saaty, T. L. (2008). Decision making with the analytic hierarchy process. *International Journal of Services Sciences*, *1*(1), 83. DOI: 10.1504/IJSSCI.2008.017590

Sauer, V. B., Thomas, W. O.Jr, Stricker, V. A., & Wilson, K. V. (1983). *Flood characteristics of urban watersheds in the United States (No. 2207)*. USGPO.

Schumann, G., Di Baldassarre, G., & Bates, P. D. (2009). The Utility of Spaceborne Radar to Render Flood Inundation Maps Based on Multialgorithm Ensembles. *IEEE Transactions on Geoscience and Remote Sensing*, *47*(8), 2801–2807. DOI: 10.1109/TGRS.2009.2017937

Scott, K. M., Macias, J. L., Naranjo, J. A., Rodriguez, S., & McGeehin, J. P. (2001). *Catastrophic debris flows transformed from landslides in volcanic terrains : mobility, hazard assessment and mitigation strategies*. Professional Paper., DOI: 10.3133/pp1630

Senthil Kumar, A. R., Goyal, M. K., Ojha, C. S. P., Singh, R. D., & Swamee, P. K. (2013). Application of artificial neural network, fuzzy logic and decision tree algorithms for modelling of streamflow at Kasol in India. *Water Science and Technology*, *68*(12), 2521–2526. DOI: 10.2166/wst.2013.491

Shaw, H. F. (1994). M. J. Wilson, Ed. Clay Mineralogy: Spectroscopic and Chemical Determinative Methods. Hall, London. 1994. 367 pp. ISBN 0-412-533804. Mineralogical Magazine, 58(393), 701–701. DOI: 10.1180/minmag.1994.058.393.30

Shustikova, I., Domeneghetti, A., Neal, J. C., Bates, P., & Castellarin, A. (2019). Comparing 2D capabilities of HEC-RAS and LISFLOOD-FP on complex topography. *Hydrological Sciences Journal*, *64*(14), 1769–1782. DOI: 10.1080/02626667.2019.1671982

Sigafoos, R. S. (1964). *Botanical evidence of floods and floodplain deposition*. Professional Paper., DOI: 10.3133/pp485A

Singh, V. P. (1997). Effect of spatial and temporal variability in rainfall and watershed characteristics on stream flow hydrograph. *Hydrological Processes*, *11*(12), 1649–1669. DOI: 10.1002/(SICI)1099-1085(19971015)11:12<1649::AID-HYP495>3.0.CO;2-1

Smith, D. P., & Bedient, P. B. (1981). Preliminary model of an urban floodplain under changing land use. *Journal of Hydrology (Amsterdam)*, *51*(1–4), 179–185. DOI: 10.1016/0022-1694(81)90127-X

Sørensen, R., Zinko, U., & Seibert, J. (2006). On the calculation of the topographic wetness index: Evaluation of different methods based on field observations. *Hydrology and Earth System Sciences*, *10*(1), 101–112. DOI: 10.5194/hess-10-101-2006

Spencer, K. (2024). As floodwaters rise, toxic contaminants released from old landfills pose more of a hazard to nature and to us. DOI: 10.64628/AB.vajrrfwsd

Stedinger, J. R., & Griffis, V. W. (2008). Flood Frequency Analysis in the United States: Time to Update. *Journal of Hydrologic Engineering*, *13*(4), 199–204. DOI: 10.1061/(ASCE)1084-0699(2008)13:4(199)

Stuckey, M. H., & Reed, L. A. (2000). *Techniques for estimating magnitude and frequency of peak flows for Pennsylvania streams (No. 4189)*. US Department of the Interior, US Geological Survey.

Tehrany, M. S., Lee, M. J., Pradhan, B., Jebur, M. N., & Lee, S. (2014). Flood susceptibility mapping using integrated bivariate and multivariate statistical models. *Environmental Earth Sciences*, *72*(10), 4001–4015. DOI: 10.1007/s12665-014-3289-3

Teng, J., Jakeman, A. J., Vaze, J., Croke, B. F. W., Dutta, D., & Kim, S. (2017). Flood inundation modelling: A review of methods, recent advances and uncertainty analysis. Environmental Modelling &amp. *Environmental Modelling & Software, 90*, 201–216. DOI: 10.1016/j.envsoft.2017.01.006

Tien Bui, D., Hoang, N.-D., Martínez-Álvarez, F., Ngo, P.-T. T., Hoa, P. V., Pham, T. D., Samui, P., & Costache, R. (2020). A novel deep learning neural network approach for predicting flash flood susceptibility: A case study at a high frequency tropical storm area. *The Science of the Total Environment, 701*, 134413. DOI: 10.1016/j.scitotenv.2019.134413

Tockner, K., & Stanford, J. A. (2002). Riverine flood plains: Present state and future trends. *Environmental Conservation, 29*(3), 308–330. DOI: 10.1017/S037689290200022X

UNDRR. (2021) The human cost of disasters: an overview of the last 20 years (2000-2019) Link: https://www.undrr.org/publication/human-cost-disasters-overview-last-20-years-2000-2019?utm_source=chatgpt.com Last access: 05.09.2025

Walling, D. E., & Fang, D. (2003). Recent trends in the suspended sediment loads of the world's rivers. *Global and Planetary Change, 39*(1–2), 111–126. DOI: 10.1016/S0921-8181(03)00020-1

Ward, P. J., Blauhut, V., Bloemendaal, N., Daniell, J. E., de Ruiter, M. C., Duncan, M. J., Emberson, R., Jenkins, S. F., Kirschbaum, D., Kunz, M., Mohr, S., Muis, S., Riddell, G. A., Schäfer, A., Stanley, T., Veldkamp, T. I. E., & Winsemius, H. C. (2020). Review article: Natural hazard risk assessments at the global scale. *Natural Hazards and Earth System Sciences, 20*(4), 1069–1096. DOI: 10.5194/nhess-20-1069-2020

Werner, A. D., & Simmons, C. T. (2009). Impact of Sea-Level Rise on Sea Water Intrusion in Coastal Aquifers. Groundwater, 47(2), 197–204. Portico. DOI: 10.1111/j.1745-6584.2008.00535.x

Werner, M. G. F. (2001). Impact of grid size in GIS based flood extent mapping using a 1D flow model. *Physics and Chemistry of the Earth. Part B: Hydrology, Oceans and Atmosphere, 26*(7–8), 517–522. DOI: 10.1016/S1464-1909(01)00043-0

Werner, M. G. F. (2004). A comparison of flood extent modelling approaches through constraining uncertainties on gauge data. *Hydrology and Earth System Sciences, 8*(6), 1141–1152. DOI: 10.5194/hess-8-1141-2004

Wing, O. E. J., Bates, P. D., Smith, A. M., Sampson, C. C., Johnson, K. A., Fargione, J., & Morefield, P. (2018). Estimates of present and future flood risk in the conterminous United States. *Environmental Research Letters*, *13*(3), 034023. DOI: 10.1088/1748-9326/aaac65

Winter, B., Schneeberger, K., Förster, K., & Vorogushyn, S. (2020). Event generation for probabilistic flood risk modelling: Multi-site peak flow dependence model vs. weather-generator-based approach. *Natural Hazards and Earth System Sciences*, *20*(6), 1689–1703. DOI: 10.5194/nhess-20-1689-2020

Yilmaz, K. K., Gupta, H. V., & Wagener, T. (2008). A process-based diagnostic approach to model evaluation: Application to the NWS distributed hydrologic model. Water Resources Research, 44(9). Portico. DOI: 10.1029/2007WR006716

Zhao, J., Jin, J., Guo, Q., Liu, L., Chen, Y., & Pan, M. (2014). Dynamic risk assessment model for flood disaster on a projection pursuit cluster and its application. *Stochastic Environmental Research and Risk Assessment*, *28*(8), 2175–2183. DOI: 10.1007/s00477-014-0881-8

Zheng, B., & Agresti, A. (2000). Summarizing the predictive power of a generalized linear model. *Statistics in Medicine*, *19*(13), 1771–1781. DOI: 10.1002/1097-0258(20000715)19:13<1771::AID-SIM485>3.0.CO;2-P

Zhou, Y., Ma, Z., & Wang, L. (2002). Chaotic dynamics of the flood series in the Huaihe River Basin for the last 500 years. *Journal of Hydrology (Amsterdam)*, *258*(1–4), 100–110. DOI: 10.1016/S0022-1694(01)00561-3

# Chapter 2
# Shrinkage of the Lake Dayet Er-Roumi, Morocco: A Decadal Remote Sensing Analysis

**Adil Moumane**
https://orcid.org/0000-0003-0296-2679
*Ibn Tofail University, Morocco*

## ABSTRACT

*Freshwater lakes in semi-arid regions are critically vulnerable to climate change and anthropogenic pressures. Lake Dayet Er-Roumi, a permanent natural lake in Morocco recognized as a Site of Biological and Ecological Interest (SIBE), exemplifies such threats. This study quantifies the spatiotemporal dynamics of its water surface area from 2015 to 2025. Using Google Earth Engine, we analyzed Landsat 8/9 TOA imagery. The NDWI and MNDWI were applied to derive water masks. Seasonal water areas were computed by averaging cloud-filtered (5%) images per year. The lake underwent a 50% surface area reduction over the decade. Critical declines occurred in 2016–2017, 2019–2020, and most severely in 2023–2024 (-0.137 km$^2$), with recoveries too brief to reverse the trend.*

## I. INTRODUCTION

Freshwater lakes are among the most vulnerable ecosystems to impacts climate change and increasing anthropogenic pressures (Bouazzati et al. 2025 ; Meinam et al. 2025 ; Moumane et al. 2024 ; Mukherjee et al. 2023 ; Zhang et al. 2024). In semi-

DOI: 10.4018/979-8-3373-6608-1.ch002

arid and Mediterranean environments, they act as critical reservoirs of biodiversity, provide water regulation services, and sustain local livelihoods. However, they are undergoing rapid transformations due to rising temperatures, and decreasing precepitation, and intensified human exploitation of water resources. Globally, numerous studies have reported shrinking lake areas, declining water quality, and reduced ecological resilience, highlighting an urgent need for monitoring and adaptive management strategies (Bucak et al. 2017; Moumane et al. 2025; Sahbani et al. 2022).

In Morocco, freshwater ecosystems are particularly scarce and fragile (Chrif El Idrissi et al. 2025; El Orfi et al. 2025). While significant number of researches have focused on large reservoirs and wetlands, comparatively little attention has been given to small ones, which face equally pressing hydrological and ecological threats. This knowledge gap is especially concerning given their dual role in preserving biodiversity and development. Recent studies show that Moroccan lakes and ponds are under threat from reduced precipitation, increasing evapotranspiration, groundwater overuse, and pollution (Addou et al. 2025; Grillas et al. 2021).

Lake Dayet Er-Roumi (33.744379°N, 6.192217°W) provides a critical case study for understanding wetland dynamics in semi-arid North Africa. Situated in Morocco's Khémisset Province within the Rabat-Salé-Kénitra region, this ecosystem represents the only permanent natural continental lake of low altitude, the lake supports unique biodiversity and provides essential ecosystem services.

Previous research has documented severe environmental degradation, including physicochemical pollution exceeding multiple regulatory thresholds (Ougrad et al. 2024), declining ichthyofauna diversity (Baladia et al. 2025), and intensifying anthropogenic pressures from agricultural expansion, groundwater extraction, and domestic wastewater inputs (El Ghizi et al. 2021). These compounding stressors have accelerated the lake's ecological deterioration, threatening its functional integrity and conservation status.

Our study addresses quantifying the spatiotemporal dynamics of Lake Dayet Er-Roumi's water surface area between 2015 and 2025 using multi-temporal remote sensing analysis. This research not only documents the alarming hydrological trajectory of a nationally significant ecosystem but also contributes to the broader understanding of freshwater vulnerability in semi-arid North African contexts, where climate change and human activities increasingly threaten wetland sustainability.

## II. STUDY AREA: LAKE DAYET ER-ROUMI

Lake Dayet Er-Roumi is a unique natural feature in Morocco (Figure 1 and 2), recognized as a Site of Biological and Ecological Interest (SIBE*)* due to its ecological importance and biodiversity (El Ghizi et al. 2023; Sadki et al. 2022), making it an exceptional hydrological system in an otherwise semi-arid to Mediterranean landscape. The lake lies at an elevation of 497 m on the alluvial coastal plain, positioned between the Atlantic coastal capital Rabat to the west and the foothills of the Middle Atlas to the east. Administratively, its basin extends across three rural communes: Ait Ouribel, Ait Ouahi, and Ait Houderrane.

Morphologically, Dayet Er-Roumi is an elongated water body aligned WSW–ENE, with an approximate length of 2 km and a variable width ranging from 400 to 700 m, covering a surface area of around 90 ha. Its maximum depth reaches 13.5 m in the central basin (Elabidi et al. 2023). The lake is primarily fed by a combination of groundwater inflows and three tributaries: Oued Derbane to the southeast, Oued Assighaou to the northwest, and a seasonal stream to the northeast that flows only during winter months. This combination of permanent and intermittent hydrological inputs highlights the sensitivity of the lake to both climatic fluctuations and anthropogenic pressures on groundwater resources.

Climatically, the region experiences a typical Mediterranean climate, characterized by mild, wet winters and hot, dry summer. Rainfall follows a Mediterranean regime, concentrated in the winter months, with significant interannual variability.

Ecologically, Lake Dayet Er-Roumi hosts a rich and diverse biota (Sadki et al. 2024). Its riparian and aquatic vegetation includes more than 35 species of vascular plants such as *Phragmites australis*, *Typha angustifolia*, *Juncus spp.*, *Tamarix sp.*, and *Potamogeton pectinatus*. The phytoplankton community shows strong seasonal succession: dinoflagellates (*Peridinium willei*, *Ceratium hirundinella*) dominate in late winter, whereas cyanobacteria (*Microcystis aeruginosa*, *Oscillatoria sp.*) flourish in summer months. The ichthyofauna is equally diverse, with both native and introduced species, including *Micropterus salmoides*, *Lepomis gibbosus*, *Esox lucius*, *Cyprinus carpio*, *Scardinius erythrophthalmus*, *Coptodon zillii*, *Gambusia affinis*, and *Anguilla anguilla* (Berthon, 1984; Mouslih, 1987; Moreau et al., 1988; Yaakoubi, 1994; Azeroual, 2003). This ecological diversity underpins the lake's role as both a habitat for wildlife and a resource for human activities.

From a socio-economic perspective, Dayet Er-Roumi is a central location within the regional lake tourism circuit of Khémisset. Its scenic landscape, recreational potential, and ecological value have long attracted visitors for activities such as birdwatching, fishing, and leisure tourism. On a broader scale, the lake contributes to local livelihoods, strengthens eco-tourism opportunities, and enhances the cultural and natural heritage of the region. Its ecological status at the national and international level makes it a critical site for balancing conservation priorities with socio-economic development.

*Figure 1. A. Location of the Rabat-Salé-Kénitra Region in Morocco B. Location of Dayat Erroumi within the Rabat-Salé-Kénitra Region Dayat Erroumi is situated in the Khémisset Province of the Rabat-Salé-Kénitra region, approximately 80 kilometers southeast of Rabat.* **C.** *Delimitation of the Lake during its Last Major Extension in 2017*

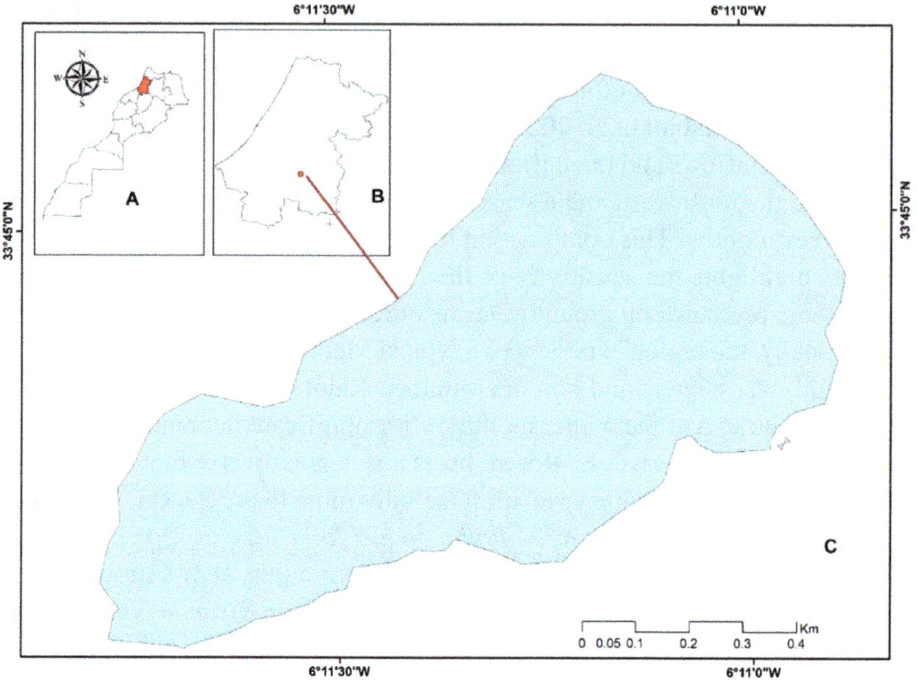

*Figure 2. Photographic Documentation of Lake Dayat Erroumi's Hydrological Decline (Khemisset Province, Morocco - October 2025)*

## III. METHODOLOGY

This section describes, step by step, the exact processing logic implemented in the Google Earth Engine (GEE).

### 3.1. Study Area and Temporal Bounds

We defined the study area as a polygon representing the spatial limits of the region under investigation and centred the analysis on this area to facilitate visualization and processing. The temporal framework was established by specifying a start year and an end year, thereby setting the time span of the study. The analysis was conducted on an annual basis, with the code iterating year by year within these defined bounds to generate a representative value for each selected seasonal window. This approach ensured temporal consistency and allowed for the assessment of interannual variations across the study period.

## 3.2. User-Defined Seasonal Window

We defined a specific seasonal time window to focus the analysis on periods of particular hydrological relevance. The selected months, representing the desired season (for example, autumn or winter), were identified numerically from 1 to 12. For each year, we constructed an explicit start and end date corresponding to these months, ensuring that the approach correctly handled wrap-around seasons that extend across two calendar years (such as December to February). This method provided a consistent temporal framework for interannual comparison of surface water dynamics.

## 3.3. Input Imagery: Landsat 8 and Landsat 9

We used imagery from Landsat 8 and Landsat 9, both derived from the Tier-1 Top-of-Atmosphere (TOA) reflectance collections. The two datasets were merged to ensure continuity and compatibility across the full observation period. To improve data reliability, we applied a cloud cover filter that retained only images with less than 5% cloud cover. Each image was subsequently clipped to the area of interest (AOI) to confine the analysis spatially to the study site and minimize unnecessary computation.

## 3.4. Water index and per-pixel water area calculation

To delineate surface water, two spectral indices were applied: The Normalized Difference Water Index (NDWI) (McFEETERS 1996) and the Modified Normalized Difference Water Index (MNDWI) (Xu 2006).

The NDWI was computed as:

$$\text{NDWI} = \frac{(\text{Green} - \text{NIR})}{(\text{Green} + \text{NIR})} \tag{1}$$

where, for Landsat 8/9 imagery, the Green band corresponds to B3 and the Near-Infrared (NIR) band to B5.

Similarly, the MNDWI was calculated as:

$$\text{MNDWI} = \frac{(\text{Green} - \text{SWIR})}{(\text{Green} + \text{SWIR})} \tag{2}$$

where the Green band corresponds to B3 and the Shortwave Infrared (SWIR1) band to B6.

A binary water mask was generated by applying a threshold of NDWI ≥ 0.0 and MNDWI ≥ 0.0 (Guo et al. 2017; Qi et al. 2022). This mask was multiplied by the per-pixel surface area (using the pixelArea() function) to derive a raster layer representing per-pixel water area .

### 3.5. Water Surface Area Quantification

We processed each satellite image within the designated seasonal timeframe by summing all water-classified pixels across our area of interest using a sum reduction algorithm at 30-meter spatial resolution to calculate total water coverage in square meters per image.

### 3.6. Seasonal and Annual Summary Metrics

For each calendar year, we compiled all qualified images meeting seasonal date criteria and cloud coverage filters, then computed image counts and average seasonal water area using the formula: Average Water Area ($km^2$) = Total Water Area ($m^2$) / (1,000,000 × Image Count). When no suitable imagery existed for specific year-season combinations, we recorded null values to maintain statistical integrity. We selected this per-image averaging approach to prevent bias from unequal image availability and minimize the impact of anomalous observations.

### 3.7. Data Output and Visualization

We generated a structured dataset containing annual entries with calendar year, seasonal boundaries, image counts, and mean water surface area measurements, which we exported in CSV format to facilitate statistical analysis and archival. We also created automated temporal visualizations to enable preliminary trend assessment of inter-annual water surface area variations.

### 3.8. Accuracy Assessment

The accuracy assessment of water classification from Landsat 8 and Landsat 9 satellite imagery employed a validation framework following established remote sensing protocols. A stratified random sampling approach was implemented to ensure representative coverage across the study area, accounting for variations in land cover heterogeneity and water body distribution. Field data collection was conducted during October 2025, during which 197 reference points were systematically acquired through ground surveys. The reference dataset comprised 95 non-water points encompassing various land cover types including, vegetation, bare soil, and agricultural fields, along with 102 water points.

## IV. RESULTS

### 4.1. A Decade of Accelerating Decline

Based on NDWI data (Figure 4 and 5), the comprehensive statistical analysis of Lake Dayet Er-Roumi's water area dynamics from 2015 to 2025, the data reveal an alarming hydrological regression characterized by consistent cross-seasonal decline and substantial interannual variability. The lake experienced a dramatic 49.7% reduction in overall water area over the eleven-year period, equivalent to an average annual loss of 0.041 $km^2$. All four seasons exhibited statistically significant decreasing trends ($p < 0.01$ for Autumn, Winter, and Summer; $p < 0.05$ for Spring), with Summer showing the most rapid decline at -0.0431 $km^2$/year, followed closely by Autumn at -0.0410 $km^2$/year. The trend strength varied seasonally, with Winter demonstrating the strongest linear relationship ($R^2 = 0.9328$) while Spring showed more moderate correlation ($R^2 = 0.5849$), indicating differing sensitivity to environmental drivers across seasons.

*Figure 4. Seasonal Water Area Dynamics of Lake Dayet Er-Roumi Using NDWI (2015-2025)*

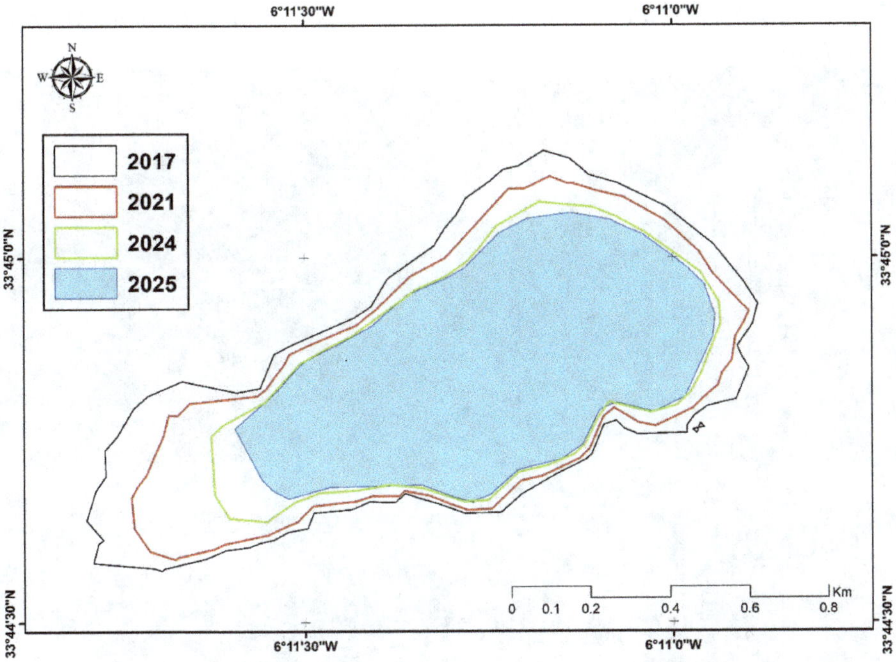


The interannual pattern reveals critical stress periods, particularly the severe reductions observed between 2016-2017 (-0.116 km$^2$), 2019-2020 (-0.130 km$^2$), and most notably 2023-2024 (-0.137 km$^2$), representing the largest single-year decline in the dataset. Although brief recovery periods occurred in 2017-2018, 2018-2019, and 2020-2021, these positive anomalies were insufficient to counter the dominant declining trajectory. Descriptive statistics highlight the lake's diminished state, with current water areas (2024-2025) operating at approximately half of their 2015 values across all seasons. The data collectively indicate that Lake Dayet Er- Roumi has undergone a fundamental hydrological regime shift, transitioning from a relatively stable system to one characterized by rapid, persistent decline that threatens its ecological functionality and long-term sustainability. The data extracted from Sentinel-2A corroborates this declining trend, revealing a consistent reduction in water surface area that aligns with the Landsat-derived observations (Figure 6).

*Figure 5. Temporal Variation of Dayet Er-Roumi Water Extent Derived from Landsat 8–9 (2017–2025) The map shows a clear progressive shrinkage of Dayet Er-Roumi from 2017 to 2025.The black contour (2017) outlines the largest extent, followed successively by red (2021)*

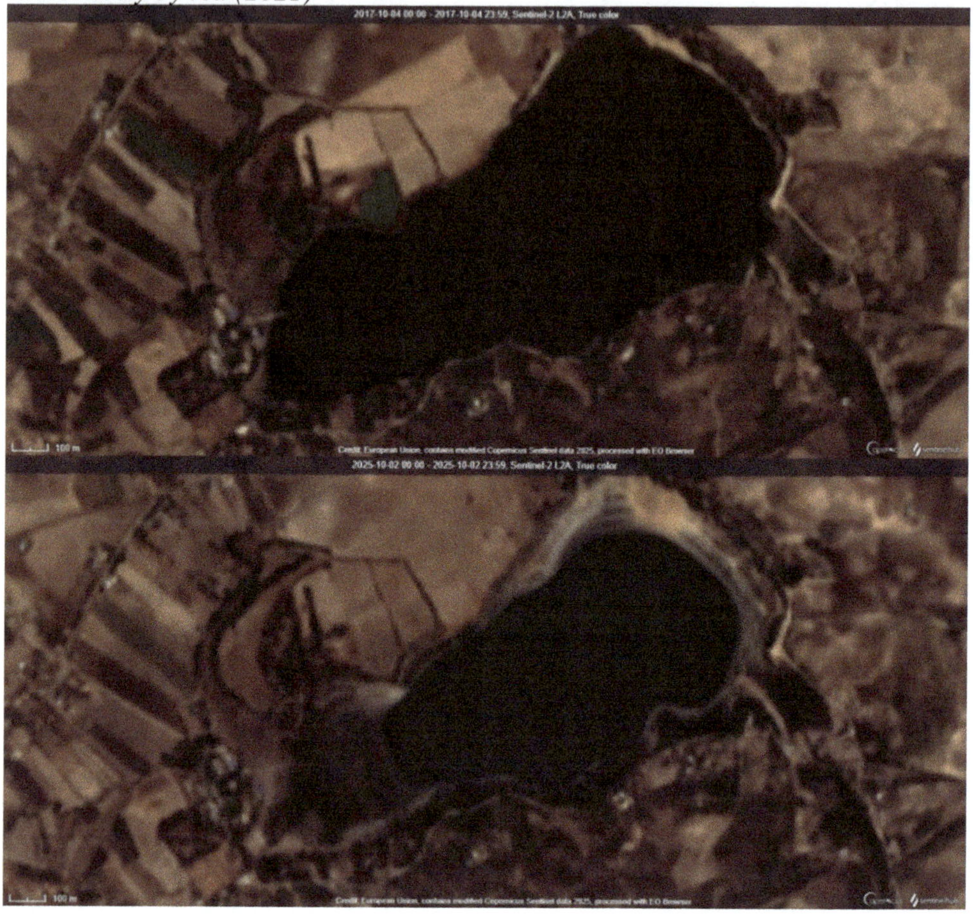

*Figure 6. Visual Evidence of Hydrological Decline in Lake Dayat Erroumi: A Sentinel-2A Imagery Comparison (October 2017 vs. October 2025).*

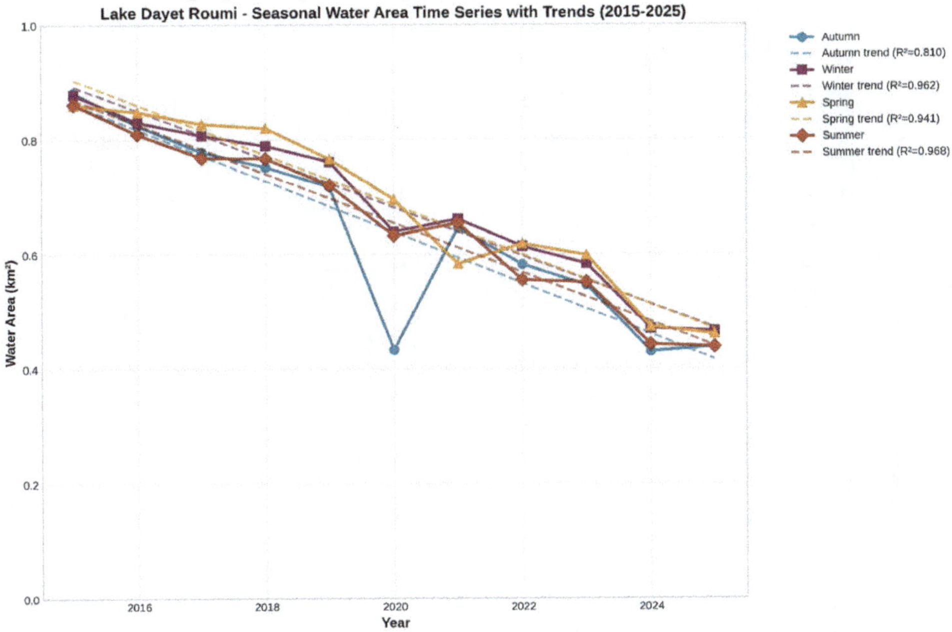


The data extracted from the Modified Normalized Difference Water Index (MNDWI) further confirms this pattern, demonstrating a parallel decline in water extent that is consistent with both the NDWI and Sentinel-2A observations (Figure 7).

*Figure 7. Seasonal Water Area Dynamics of Lake Dayet Er-Roumi Using MNDWI*

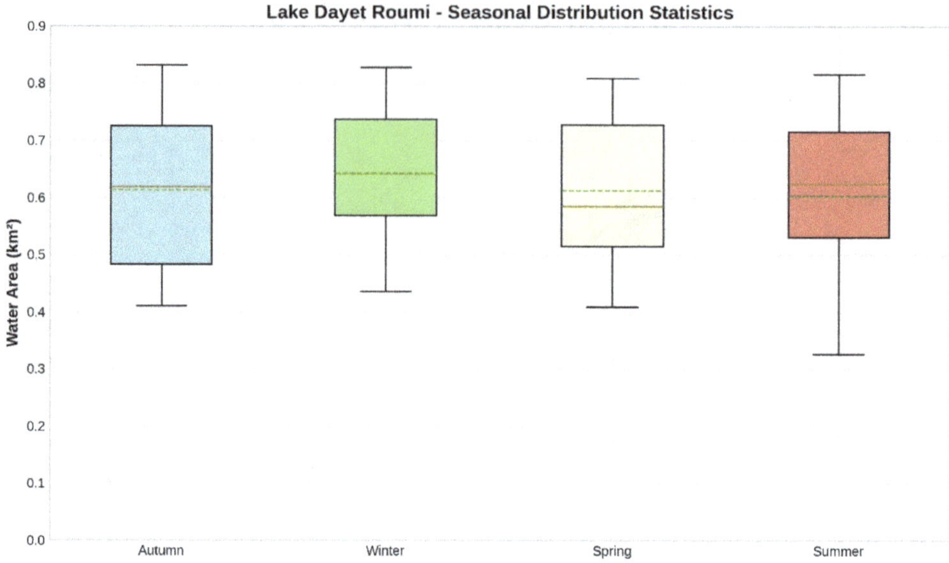


## 4.2. Descriptive Statistical Analysis

### 4.2.1. NDWI-Based Water Area Statistics

The NDWI-derived analysis reveals consistent seasonal patterns in Lake Dayet Roumi's water surface area dynamics (Table 1 and Figure 8). Winter exhibits the highest mean water area at 0.643 km$^2$, followed by Autumn at 0.614 km$^2$, Spring at 0.613 km$^2$, and Summer at 0.603 km$^2$. This seasonal hierarchy reflects the region's Mediterranean climate regime, with winter precipitation contributing to higher water levels that gradually decline through the dry summer months.

The variability analysis shows substantial interannual fluctuations across all seasons, with standard deviation values ranging from 0.131 km$^2$ (Winter) to 0.154 km$^2$ (Summer). Autumn demonstrates the highest dispersion with an interquartile range of 0.243 km$^2$, indicating considerable year-to-year variability in post-summer water retention. The minimum recorded values reveal critically low levels across all seasons, particularly in Summer 2025 (0.327 km$^2$) and Spring 2025 (0.409 km$^2$), representing approximately 40% and 50% of their respective maximum recorded values.

*Table 1. Box Plot Metrics by Season (NDWI)*

| Season | Mean ($km^2$) | Median ($km^2$) | Std Dev ($km^2$) | Min ($km^2$) | Max ($km^2$) | Q1 (25%) | Q3 (75%) | IQR ($km^2$) |
|---|---|---|---|---|---|---|---|---|
| Autumn | 0.614 | 0.619 | 0.153 | 0.411 | 0.832 | 0.483 | 0.726 | 0.243 |
| Winter | 0.643 | 0.642 | 0.131 | 0.436 | 0.828 | 0.569 | 0.737 | 0.168 |
| Spring | 0.613 | 0.585 | 0.140 | 0.409 | 0.809 | 0.515 | 0.728 | 0.213 |
| Summer | 0.603 | 0.625 | 0.154 | 0.327 | 0.817 | 0.531 | 0.716 | 0.185 |

*Figure 8. Box-plot visualization showing median, quartiles, and range of water surface area across four seasons according to NDWI data.*

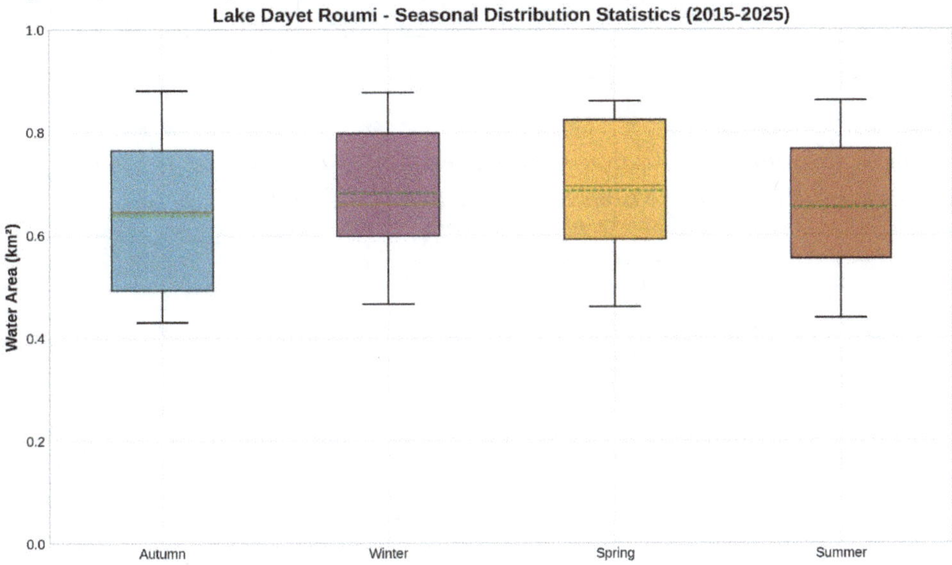


## 4.2.2. MNDWI-Based Water Area Statistics

The MNDWI analysis yields systematically higher water area estimates while maintaining similar seasonal patterns (Table 2 and Figure 9). Spring shows the highest mean water area at 0.686 $km^2$, followed by Winter at 0.681 $km^2$, Autumn at 0.639 $km^2$, and Summer at 0.654 $km^2$. This methodological difference results in an average 4.1% increase in detected water area compared to NDWI measurements.

MNDWI exhibits broader distribution characteristics across all seasons, with interquartile ranges exceeding NDWI values by 7-12%. Autumn shows the most substantial variability with an IQR of 0.273 $km^2$, while maintaining the highest coefficient of variation at 25.7%. The minimum values detected by MNDWI are consistently higher, particularly in Summer where it records 0.438 km² compared to NDWI's 0.327 km², representing a 34% increase in minimum water area detection during critical dry periods.

*Table 2. Box Plot Metrics by Season (MNDWI)*

| Season | Mean (km²) | Median (km²) | Std Dev (km²) | Min (km²) | Max (km²) | Q1 (25%) | Q3 (75%) | IQR (km²) |
|---|---|---|---|---|---|---|---|---|
| **Autumn** | 0.639 | 0.645 | 0.164 | 0.430 | 0.881 | 0.492 | 0.765 | 0.273 |
| **Winter** | 0.681 | 0.661 | 0.142 | 0.465 | 0.878 | 0.598 | 0.798 | 0.200 |
| **Spring** | 0.686 | 0.695 | 0.148 | 0.460 | 0.860 | 0.591 | 0.823 | 0.233 |
| **Summer** | 0.654 | 0.654 | 0.145 | 0.438 | 0.861 | 0.553 | 0.768 | 0.215 |

*Figure 9. Box-plot visualization showing median, quartiles, and range of water surface area across four seasons according to MNDWI data.*

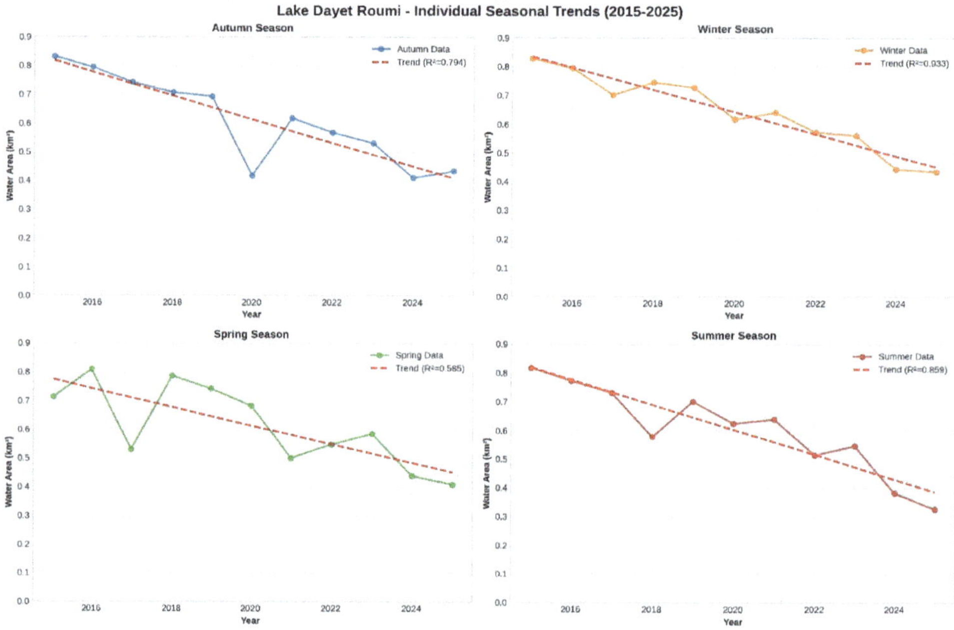

### 4.2.3. Trend Analysis and Seasonal Patterns

The trend analysis reveals statistically significant declining trajectories across all four seasons (Figure 10), with p-values well below the 0.05 significance threshold (Autumn: p=0.0002, Winter: p=0.0000, Spring: p=0.0061, Summer: p=0.0000). The Summer season demonstrates the most rapid rate of decline at -0.0431 $km^2$/year, followed closely by Autumn at -0.0410 $km^2$/year, suggesting heightened vulnerability during warmer periods potentially due to increased evaporation rates. The Winter season, while showing the strongest linear relationship ($R^2 = 0.9328$), still experiences substantial annual loss (-0.0382 $km^2$/year). The varying R-squared values across seasons (Spring: 0.5849, Autumn: 0.7944, Summer: 0.8590) indicate differential sensitivity to environmental drivers, with Spring showing the most complex, non-linear behavior potentially influenced by variable precipitation patterns and snowmelt timing.

*Figure 10. Seasonal decline of the lake*

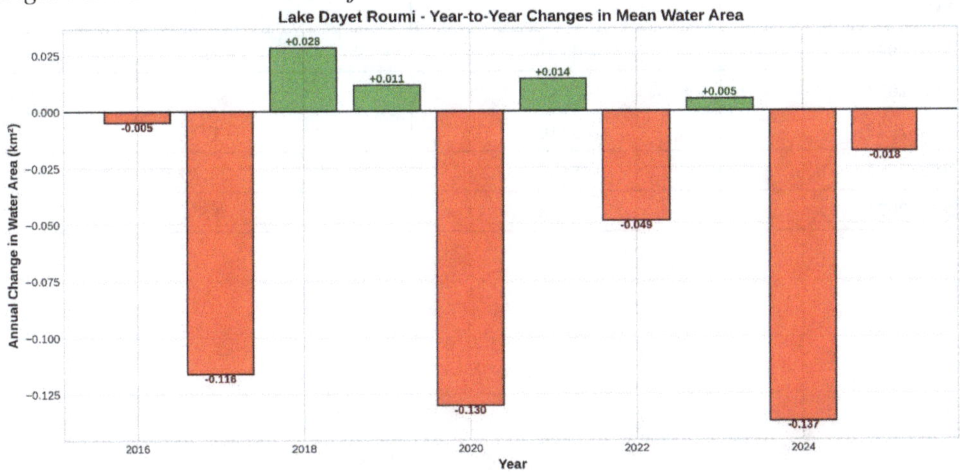


## 4.3. Interannual Variability and Critical Decline Periods

### 4.3.1. NDWI-Based Annual Changes

The NDWI analysis (Table 3 and Figure 9) identifies three critical decline periods that underscore the lake's vulnerability. The most severe reduction occurred during 2023-2024, with a dramatic decrease of 0.137 $km^2$ (-24.7%), representing the largest single-year percentage decline in the study period. This was closely followed by the

2019-2020 period, which saw a reduction of 0.130 km$^2$ (-18.2%), and 2016-2017 with a decline of 0.116 km$^2$ (-14.7%). These three episodes collectively account for 65% of the total absolute decline observed over the decade.

The data reveal a pattern of abrupt declines followed by modest recoveries. The period 2020-2021 showed a slight recovery of +0.014 km$^2$ (+2.4%), while 2017-2018 exhibited a more substantial increase of +0.028 km$^2$ (+4.2%). However, these recovery phases were insufficient to counteract the overall negative trajectory, with the most recent period (2024-2025) continuing the decline at -0.018 km$^2$ (-4.3%).

*Table 3. Year-to-Year Changes in Mean Water Area (NDWI)*

| Period | Change (km$^2$) | Percentage Change |
|---|---|---|
| **2015-2016** | -0.005 | -0.6% |
| **2016-2017** | -0.116 | -14.7% |
| **2017-2018** | +0.028 | +4.2% |
| **2018-2019** | +0.011 | +1.6% |
| **2019-2020** | -0.130 | -18.2% |
| **2020-2021** | +0.014 | +2.4% |
| **2021-2022** | -0.049 | -8.1% |
| **2022-2023** | +0.005 | +1.0% |
| **2023-2024** | -0.137 | -24.7% |
| **2024-2025** | -0.018 | -4.3% |

*Figure 11. Bar chart showing annual water balance changes, highlighting critical decline periods (2016-2017: -0.116 km², 2019-2020: -0.130 km², 2023-2024: -0.137 km²) and brief recovery phases.*

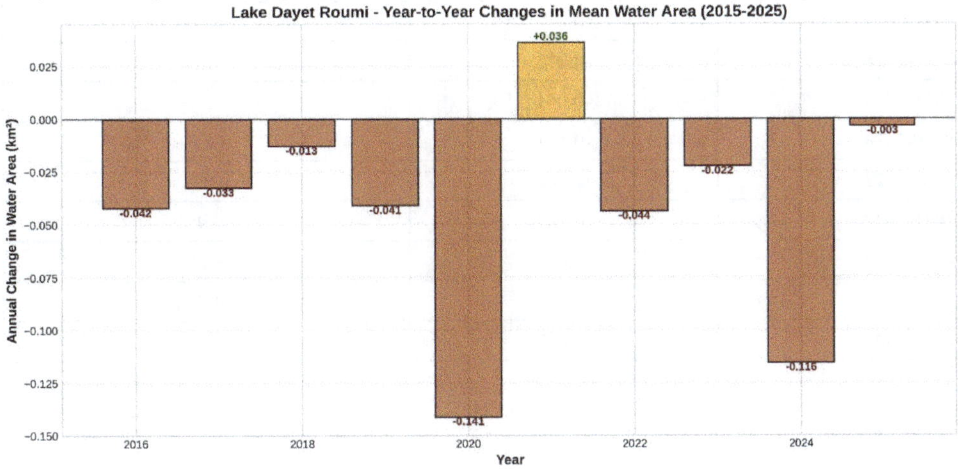


## 4.3.2. MNDWI-Based Annual Changes

The MNDWI analysis corroborates the critical decline periods (Table 4 and Figure 12), while revealing some methodological differences in magnitude and timing. The most severe reduction was recorded during 2019-2020 at 0.141 km² (-19.1%), followed by 2023-2024 with a decrease of 0.116 km² (-20.3%). Notably, MNDWI detects more consistent year-to-year declines, with only one substantial recovery period in 2020-2021 (+0.036 km², +6.0%).

Both methodologies identify the 2019-2020 and 2023-2024 periods as the most critical, though with varying severity rankings. The MNDWI data show a more gradual progression of declines throughout the study period, with only two instances of positive change compared to three in the NDWI record. This pattern suggests MNDWI may be less sensitive to short-term fluctuations while capturing the same fundamental declining trend.

*Table 4. Year-to-Year Changes in Mean Water Area (MNDWI)*

| Period | Change (km²) | Percentage Change |
|---|---|---|
| **2015-2016** | -0.042 | -4.9% |
| **2016-2017** | -0.033 | -4.0% |
| **2017-2018** | -0.013 | -1.6% |
| **2018-2019** | -0.041 | -5.2% |
| **2019-2020** | -0.141 | -19.1% |
| **2020-2021** | +0.036 | +6.0% |
| **2021-2022** | -0.044 | -6.9% |
| **2022-2023** | -0.022 | -3.8% |
| **2023-2024** | -0.116 | -20.3% |
| **2024-2025** | -0.003 | -0.8% |

*Figure 12. Annual water balance changes in Lake Dayet Roumi using MNDWI (2015–2025).*

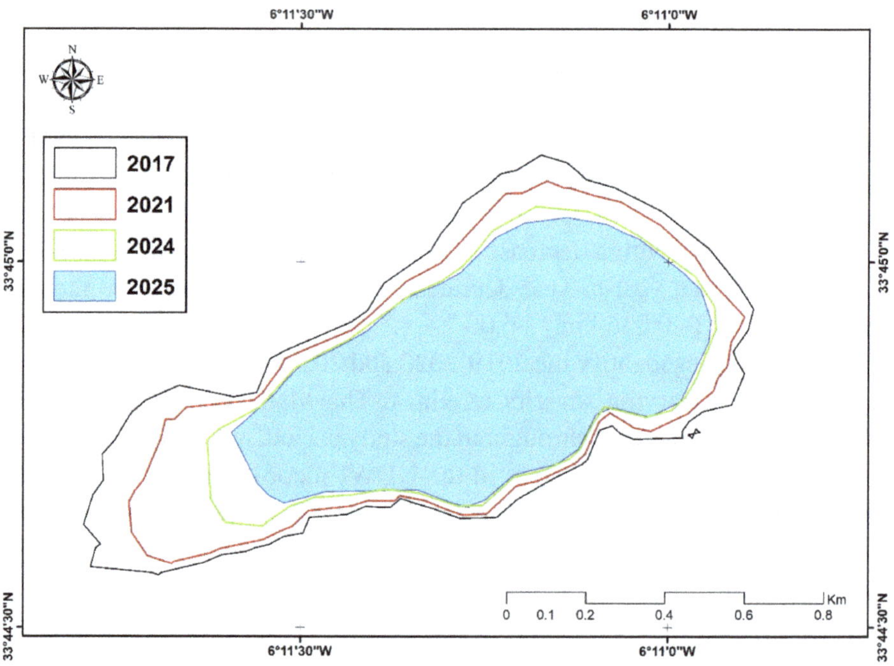

### 4.4. Decadal Change Assessment (2015-2025)

#### 4.4.1. NDWI-Based Decadal Changes

The NDWI analysis reveals catastrophic declines in Lake Dayet Roumi's water surface area over the study period (Figure 13). Summer emerges as the most severely affected season, suffering a 60.0% reduction equivalent to 0.490 $km^2$ of lost water area. Autumn and Winter show nearly identical decline rates of 47.8% and 47.3% respectively, while Spring demonstrates a slightly lower but still substantial reduction of 42.7%. Collectively, these seasonal declines result in an overall reduction of 49.7%, representing a loss of nearly half the lake's 2015 water surface area.

The absolute losses range from 0.305 $km^2$ in Spring to 0.490 $km^2$ in Summer, indicating that while all seasons experienced dramatic declines, Summer conditions have deteriorated most severely. The lake's total water area has been reduced from 0.798 $km^2$ to 0.401 $km^2$, highlighting the critical state of this ecosystem.

#### 4.4.2. MNDWI-Based Decadal Changes

The MNDWI analysis confirms the severe degradation trend while providing marginally different quantification. Autumn shows the highest percentage decline at 50.3%, followed by Summer at 49.1%, Winter at 47.0%, and Spring at 46.5%. The overall reduction of 48.2% aligns closely with NDWI findings, demonstrating methodological consistency in capturing the fundamental declining trend.

Notably, MNDWI detects higher absolute losses across all seasons, with total decline reaching 0.420 $km^2$ compared to NDWI's 0.397 $km^2$. This pattern is consistent with MNDWI's systematic tendency to estimate larger water areas, yet both methods converge on the same alarming conclusion: Lake Dayet Roumi has lost approximately half its water surface area within a single decade, representing an unprecedented ecological crisis.

*Figure 13. Temporal Variation of Dayet Er-Roumi Water Extent Derived from Landsat 8–9 (2017–2025) The map shows a clear progressive shrinkage of Dayet Er-Roumi from 2017 to 2025.The black contour (2017) outlines the largest extent, followed successively by red (2021), green (2024), and the blue-filled area (2025) representing the smallest extent. The most significant reduction occurs in the westeren and southwestern part of the lake, where the 2025 outline has retreated markedly compared to previous years.*

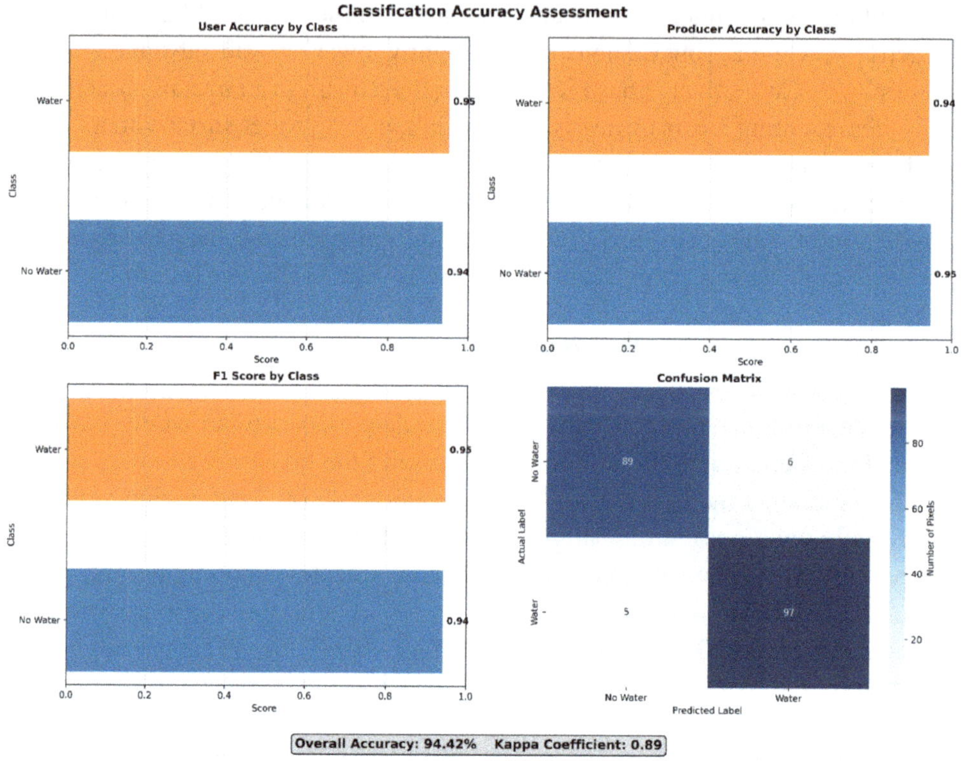


## 4.5. Hydrological Implications and Ecosystem Consequences

The 49.7% overall reduction in water area represents a catastrophic loss of aquatic habitat with severe implications for biodiversity, water quality, and ecosystem services. The parallel declines across all seasons suggest systemic rather than seasonal drivers, potentially including: reduced groundwater recharge, changing precipitation patterns, increased evaporation due to rising temperatures, and possible anthropogenic water extraction. The Summer season's particular vulnerability (-0.0431 $km^2$/year) raises concerns about the lake's ability to maintain minimum ecological levels during drought periods. The high variability in Spring measurements may reflect the lake's dependence on unpredictable seasonal rainfall, making it increasingly vulnerable to climate variability.

## 4.6. Classification Performance

The classification algorithm achieved a high level of accuracy for the October 2025 dataset (Figure 14). Key validation metrics include an Overall Accuracy of 94.42% and a Kappa Coefficient of 0.89. Class-specific metrics confirm consistent performance:

- **Water:** User's Accuracy (Precision) = 0.94, Producer's Accuracy (Recall) = 0.95, F1-Score = 0.95
- **No Water:** User's Accuracy (Precision) = 0.95, Producer's Accuracy (Recall) = 0.94, F1-Score = 0.94

These results, derived from 197 samples, confirm the model's reliability in distinguishing water bodies for this temporal snapshot.

*Figure 14. Accuracy Assessment of Water Classification.*

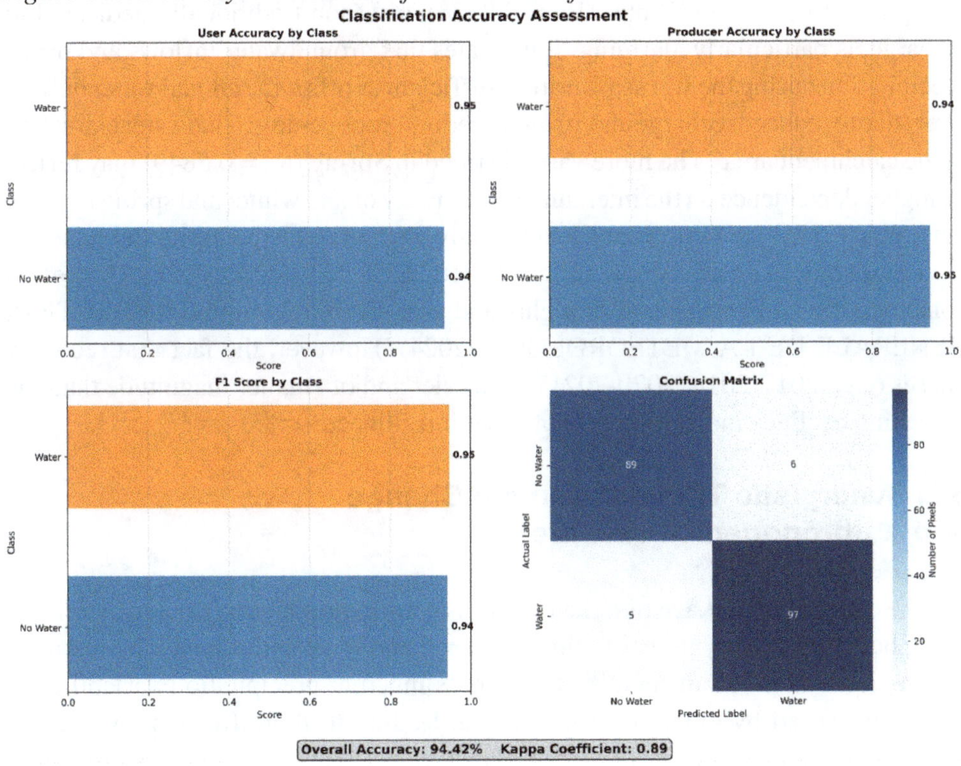

# V. DISCUSSION

## 5.1. A Trajectory of Accelerated Decline: Linking Observations to Drivers

The most salient finding of this research is the consistent and statistically significant decline in water surface area across all seasons, culminating in a 49.7% overall reduction. This aligns with global observations of freshwater lakes in semi-arid and Mediterranean climates facing intensified water stress (Baba et al. 2025; Moumane et al. 2024). The rate of decline observed at Dayet Er-Roumi, particularly the record single-year loss in 2023-2024, suggests an acceleration that exceeds natural climatic variability and points to a regime shift.

The seasonal trends offer critical insights into the primary drivers. The most rapid decline occurred during Summer (-0.0431 $km^2$/year), a period characterized by high temperatures and minimal precipitation. This strongly implicates increased evapotranspiration as a major contributing factor, a phenomenon exacerbated by regional warming trends (DIRECTION GENERALE DE LA METEOROLOGIE 2024). The parallel, though slightly slower, decline in Autumn suggests a carry-over effect, where the summer deficit is not fully replenished by early autumn rains. The strong linear decline in Winter ($R^2 = 0.9328$), a season traditionally dedicated to recharge, is particularly alarming. It indicates that groundwater inflows and direct precipitation during the wet season are insufficient to offset the annual water deficit, likely due to reduced recharge and/or increased anthropogenic groundwater extraction in the catchment area . The more variable trend in Spring ($R^2 = 0.5849$) may reflect its higher dependence on the interannual variability of late winter and spring rainfall, making it a sensitive indicator of increasingly unpredictable precipitation patterns.

The identified critical stress periods (2016-2017, 2019-2020, 2023-2024) often coincide with known regional droughts and periods of low rainfall (DIRECTION GENERALE DE LA METEOROLOGIE 2024). However, the fact that recovery phases (e.g., 2017-2018, 2020-2021) were brief and of smaller magnitude than the preceding declines indicates a loss of system resilience.

## 5.2. Synergistic Threats: Climate Change and Anthropogenic Pressures

The dramatic shrinkage of Lake Dayet Er-Roumi cannot be attributed to a single cause but rather to a synergy of climate change and local anthropogenic pressures. The regional Mediterranean climate, with its inherent precipitation variability, is being intensified by global climate change, leading to more frequent and severe droughts (Elmotawakkil et al. 2025; Kenawy et al. 2025). This reduces direct surface runoff and groundwater recharge, critically impacting a lake known to be fed by both tributaries and aquifers.

Simultaneously, anthropogenic pressures compound these climatic effects. The lake's catchment is situated in a region of active agriculture, which likely relies on intensive groundwater irrigation. Over-abstraction of groundwater would directly intercept the hydrological pathways that sustain the lake, effectively decoupling it from its natural recharge sources. This aligns with previous studies on the lake that have documented physicochemical pollution from agricultural sources (El Ghizi et al. 2021), confirming significant human activity within its watershed. Thus, the declining water area is likely a direct consequence of a growing imbalance between water extraction (both surface and subsurface) and climatic water input.

## 5.3. Ecological and Socio-Economic Implications

The ecological consequences of a 50% reduction in aquatic habitat are profound. The loss of surface area directly translates to a loss of littoral zones, which are critical for the riparian vegetation documented in the study area. This habitat destruction threatens the entire trophic web, from invertebrates to the diverse ichthyofauna, including both native and introduced species. A smaller, shallower lake is also more susceptible to water quality deterioration, including increased salinity, eutrophication, and the dominance of toxic cyanobacterial blooms, which were already noted to flourish in summer. This creates a negative feedback loop, where ecological degradation further diminishes the lake's value and functionality.

From a socio-economic perspective, the decline undermines the lake's recognized value as a Site of Biological and Ecological Interest (SIBE) and its role in the regional tourism circuit of Khémisset. Activities such as birdwatching, fishing, and leisure tourism depend on a healthy and aesthetically pleasing water body. The degradation of Lake Dayet Er-Roumi risks collapsing this eco-tourism sector, impacting local livelihoods and diminishing the cultural and natural heritage of the region. The lake's transition from a permanent to a potentially ephemeral system would represent a significant loss of a unique and nationally important freshwater ecosystem.

## 5.4. Methodological Considerations and Study Limitations

The methodology employed, utilizing Google Earth Engine and a standardized NDWI threshold, provided a robust and reproducible means of tracking surface water change. However, certain limitations must be acknowledged. The use of a simple NDWI threshold of 0.0, while transparent, may misclassify turbid water or mixed water-vegetation pixels. Future work could benefit from a multi-index approach or locally tuned thresholds to improve accuracy. Most significantly, this study quantifies the symptom surface water loss but not the precise proportions of the underlying drivers. While we can infer the roles of climate and groundwater extraction, a conclusive assessment requires integrated hydrological modeling that combines climate data, land use maps, and groundwater level measurements with the remote sensing data presented here.

## VI. CONCLUSION AND MANAGEMENT IMPLICATIONS

In conclusion, the spatiotemporal analysis of Lake Dayet Er-Roumi from 2015 to 2025 reveals a system in a state of acute crisis. The consistent, cross-seasonal decline and the accelerating rate of water loss signal that the lake may be approaching an ecological tipping point.

This research underscores the critical vulnerability of low-altitude natural lakes in North Africa, which have historically received less attention than mountain wetlands or large reservoirs. The case of Dayet Er-Roumi serves as a stark warning for similar ecosystems across the Maghreb. Effective conservation and management must move from passive observation to active intervention. This should include:

- **Implementing Integrated Water Resource Management:** Establishing a water budget for the catchment to regulate groundwater abstraction, particularly for agriculture, is paramount.
- **Enhanced Monitoring:** Complementing remote sensing with in-situ monitoring of groundwater levels, lake depth, and water quality to better understand the hydrological connections.
- **Developing Climate Adaptation Strategies:** Exploring measures such as managed aquifer recharge and the promotion of water-efficient agricultural practices in the watershed.
- **Urgent Conservation Action:** Re-evaluating the lake's protected status and implementing a robust action plan to prevent its irreversible loss.

Without such targeted and immediate efforts, the continued degradation of Lake Dayet Er-Roumi appears inevitable, with irreversible consequences for regional biodiversity, water security, and socio-economic stability.

## REFERENCES


Addou, R., Bhiry, N., & Achiban, H. (2025). Dynamics of Wetlands in Ifrane National Park, Morocco: An Approach Using Satellite Imagery and Spectral Indices. *Water (Basel)*, *17*(13), 1869. DOI: 10.3390/w17131869

Baba, W. M., Chehbouni, A., Ouassanouan, Y., Gascoin, S., Paganini, M., Ottavianelli, G., & Szantoi, Z. (2025). Monitoring water crisis from space across a Mediterranean region. *Scientific Reports*, *15*(1), 23262. DOI: 10.1038/s41598-025-06240-1 PMID: 40603462

Baladia, Y., Ben-Haddad, M., Laadel, N., & Oualid, J. A. (2025). Catastrophic fish mass mortality events in Moroccan freshwater ecosystems: Alarming trends and impacts on biodiversity. *Environmental Monitoring and Assessment*, *197*(3), 290. DOI: 10.1007/s10661-025-13711-5 PMID: 39946026

Bouazzati, M. E., Batchi, M., Fatah, F., Moumane, A., & Motib, I. (2025). Assessing the Spatio-Temporal Evolution of Lake Aguelmam Azigza Using Remote Sensing and GIS: Middle Atlas – Morocco. In *Remote Sensing and GIS Techniques in Hydrology* (pp. 233–258). IGI Global Scientific Publishing. DOI: 10.4018/979-8-3693-9651-3.ch009

Bucak, T., Trolle, D., Andersen, H. E., Thodsen, H., Erdoğan, Ş., Levi, E. E., Filiz, N., Jeppesen, E., & Beklioğlu, M. (2017). Future water availability in the largest freshwater Mediterranean lake is at great risk as evidenced from simulations with the SWAT model. *The Science of the Total Environment*, *581–582*, 413–425. DOI: 10.1016/j.scitotenv.2016.12.149 PMID: 28069301

Chrif El Idrissi, M., Saber, E.-R., & Al Mashoudi, A. (2025). Hydro-climatic stress and ecosystem vulnerability in a highland lake: The case of Dayet Aoua (1935–2022), Middle Atlas, Morocco. *Modeling Earth Systems and Environment*, *11*(5), 312. DOI: 10.1007/s40808-025-02496-8

El Ghizi, S., El Aadel, N., Sadik, M., El Bouch, M., & Hasnaoui, M. (2021). The physicochemical characteristics and the pollution level of Dayet Er-Roumi Lake in Morocco. *E3S Web of Conferences*, *234*, 00036. DOI: 10.1051/e3sconf/202123400036

El Ghizi, S., Sadik, M., El Bouch, M., & Hasnaoui, M. (2023). Study of the Ichthyological Populations of Lake Dayet Er-Roumi (Morocco): Use of Indices to Analyze the Ichthyofaunistic Biodiversity. *Egyptian Journal of Aquatic Biology and Fisheries*, *27*(2), 841–851. DOI: 10.21608/ejabf.2023.298312

El Orfi, T., El Ghachi, M., Lebaut, S., & Haidu, I. (2025). Monitoring the water surface dynamics of four Mediterranean mountain lakes in the Middle Atlas (Morocco) using Landsat imagery in a global change context. *Journal of Hydrology. Regional Studies*, *60*, 102516. DOI: 10.1016/j.ejrh.2025.102516

Elabidi, K., Remmal, T., Amraoui, F., Jounaid, H., Amar, M., & Bensalmia, A. (2023). *Master spécialisé : Gestion et Valorisation des Géoressources Projet de Fin d'Etudes L'apport de la Bathymetrie Pour L'etude Hydrogeologique de Dayet Roumi.*

Elmotawakkil, A., Moumane, A., Sadiki, A., Zahi, A., Al Karkouri, J., Batchi, M., Kumar Bhagat, S., & Enneya, N. (2025). Forecasting short-term rainfall patterns in arid and semi-arid regions using machine learning and deep learning models: A case study from Morocco. *Theoretical and Applied Climatology*, *156*(8), 436. DOI: 10.1007/s00704-025-05677-8

Generale de la Meteorologie, D. (2024). Etat Du Climat Maroc., https://www.marocmeteo.ma/fr/climat-2024, Retrieved September 12, 2024, from.

Grillas, P., Rhazi, L., Lefebvre, G., El Madihi, M., & Poulin, B. (2021). Foreseen impact of climate change on temporary ponds located along a latitudinal gradient in Morocco. *Inland Waters*, *11*(4), 492–507. DOI: 10.1080/20442041.2021.1962688

Guo, Q., Pu, R., Li, J., & Cheng, J. (2017). A weighted normalized difference water index for water extraction using Landsat imagery. *International Journal of Remote Sensing*, *38*(19), 5430–5445. DOI: 10.1080/01431161.2017.1341667

Kenawy, A. E., Al-Awadhi, T., Abdullah, M., Ostermann, F. O., & Abulibdeh, A. (2025). A Multidecadal Assessment of Drought Intensification in the Middle East and North Africa: The Role of Global Warming and Rainfall Deficit. *Earth Systems and Environment*. Advance online publication. DOI: 10.1007/s41748-025-00576-4

McFeeters, S. K. (1996). The use of the Normalized Difference Water Index (NDWI) in the delineation of open water features. *International Journal of Remote Sensing*, *17*(7), 1425–1432. DOI: 10.1080/01431169608948714

Meinam, M., Lahiri, B., Ghosh, A., Singh, S. K., Bharati, H., Upadhyay, A. D., Pal, P., Sanjenbam, B., & Ngasotter, S. (2025). Aquatic Ecosystem, Fish Diversity and Conservation Strategies of the Fragile Eastern Himalayas: A Comprehensive Review with a Focus on Loktak Lake. *Wetlands*, *45*(7), 96. DOI: 10.1007/s13157-025-01978-w

Moumane, A., Al Karkouri, J., & Mouhcine, B. (2024). Detailed JavaScript API for Google Earth Engine to Enhance Monitoring of Surface Water Dynamics in the Drought-Affected Mediterranean Region. In Batchi, M., & Moumane, A. (Eds.), (pp. 105–132). Advances in Geospatial Technologies. IGI Global., DOI: 10.4018/979-8-3693-9651-3.ch004

Moumane, A., Bahouq, T., Karmaoui, A., Laghfiri, D., Yassine, M., Karkouri, J. A., Batchi, M., Faouzi, M., Boulakhbar, M., & Youssef, A. A. (2025). Lake Iriqui's Remarkable Revival: Field Observations and a Google Earth Engine Analysis of Its Recovery After over Half a Century of Desiccation. *Land (Basel)*, *14*(1), 104. DOI: 10.3390/land14010104

Mukherjee, S., Rizvi, S. S., Biswas, G., Paswan, A. K., Vaiphei, S. P., Warsi, T., & Mitran, T. (2023). Aquatic Eco-systems Under Influence of Climate Change and Anthropogenic Activities. In *Hydrogeochemistry of Aquatic Ecosystems* (pp. 307–331). John Wiley & Sons, Ltd., DOI: 10.1002/9781119870562.ch14

Ougrad, I., Elassassi, Z., Mrabet, A., Mssillou, I., Lim, A., Shahat, A. A., Rezouki, S., & Moubchir, T. (2024). Health Risk Assessment of Trace Elements in Surface Water from Dayat Roumi Lake, Morocco. *Water (Basel)*, *16*(22), 3231. DOI: 10.3390/w16223231

Qi, Y., Dou, H., & Wang, Z. (2022). An Adaptive Threshold Selected Method from Remote Sensing Image based on Water Index. *Journal of Physics: Conference Series*, *2228*(1), 012001. DOI: 10.1088/1742-6596/2228/1/012001

Sadki, M., Chahlaoui, A., Belkhiri, A., Maliki, A., Saidi, A., & Khaffou, M. (2024). Ornithological characterisation of the natural habitats's Lake Dayet Er Roumi (Khemisset, Morocco). *IOP Conference Series. Earth and Environmental Science*, *1398*(1), 012003. DOI: 10.1088/1755-1315/1398/1/012003

Sadki, M., Chahlaoui, A., Maliki, A., Khaffou, M., & Lahrouz, S. (2022). Mapping the natural habitats of a continental wetland: Dayet Er Roumi lake (Khémisset, Morocco). *IOP Conference Series. Earth and Environmental Science*, *1090*(1), 012030. DOI: 10.1088/1755-1315/1090/1/012030

Sahbani, S., Béjaoui, B., Benabdallah, S., Toujani, R., Fathalli, A., Zaaboub, N., Aouissi, J., Kassouk, Z., Hamdi, N., Mbarek, N. B., Missaoui, H., Basti, L., Blake, R., & Norouzi, H. (2022). Systematic review of a RAMSAR wetland and UNESCO biosphere reserve in a climate change hotspot (Ichkeul Lake, Tunisia). *Journal of Sea Research*, *190*, 102288. DOI: 10.1016/j.seares.2022.102288

Xu, H. (2006). Modification of normalised difference water index (NDWI) to enhance open water features in remotely sensed imagery. *International Journal of Remote Sensing*, *27*(14), 3025–3033. DOI: 10.1080/01431160600589179

Zhang, Y., Shen, J., He, L., Feng, J., Chi, L., & Wang, X. (2024). Challenge to Lake Ecosystems: Changes in Thermal Structure Triggered by Climate Change. *Water (Basel)*, *16*(6), 888. DOI: 10.3390/w16060888

# Chapter 3
# Flood Risk Assement Using AHP, Frequency Ratio, Logistic Regression, and Random Forest in Naraipur Municipality

**Rubi Chaulagain**
https://orcid.org/0009-0000-1901-6641
*Westernal Regional Campus, Nepal*

**Rekha Paudel**
*Westernal Regional Campus, Nepal*

**Fatima Ezzahra El Ghazali**
https://orcid.org/0000-0002-1980-3028
*Geosciences, Georesources, and Environment, Semlalia Laboratory, Department of Geology, Faculty of Sciences Semlalia, Cadi Ayyad University, Morocco*

**Adil Moumane**
https://orcid.org/0000-0003-0296-2679
*Ibn Tofail University, Morocco*

## ABSTRACT

*Floods, a natural hazard characterized by the overflow of water from its normal riverbed onto dry land, are a recurrent problem. The study area we choose is Narainapur Municipality. This study aims to develop a detailed flood risk assessment*

DOI: 10.4018/979-8-3373-6608-1.ch003

*using the Analytic Hierarchy Process, Frequency Ratio, Logistic Regression,and Random Forest models. A flood inventory map was created, with 102 flooded and 95 non-flooded points Flood conditioning factors include slope, NDVI, soil type, distance from streams and roads, rainfall, TWI, aspect, and curvature. The Logistic Regression model (AUC = 0.944) achieved the highest AUC value, indicating superior predictive accuracy compared to other models. The Random Forest model (AUC = 0.936) also performed well, followed by the Frequency Ratio model (AUC = 0.855) and the AHP model (AUC = 0.822). These results highlight the effectiveness of machine learning-based models in our study area.The findings of this study provide valuable insights for flood risk management, offering a scientific basis for better planning and decision-making in vulnerable regions.*

## I. INTRODUCTION

Natural disasters triggered by heavy rainfall influence on environment which affects day to day human life inviting various socio-economic consequnces. (Djalante, 2018). Climate change acts as driving factor for flood conditioning factors such as land use patterns, geology, terrain slope, rainfall intensity, elevation, and land cover (Stefanidis and Stathis, 2013; Gacu et al., 2022). This trend points to a future where seasonal floods may become more frequent, larger in scale, and harder to predict. It highlights the urgent need for strong flood risk management strategies, with hazard assessment at the heart of efforts to protect not just ecosystems, but also the communities and economies that depend on them (Liu et al., 2016; Danumah et al., 2016).To effectively identify flood prone areas, physically based models like HEC_RAS and MIKE11 become expensive, suggesting more appropriate statistical methods and Multi-criteria decision analysis (MCDA) techniques like FR, LR and AHP respectively. However, newer machine learning approaches like RF, Artificial Neural Networks (ANN), and Support Vector Machines (SVM), stand to be the present time best alternatives for analysis of complex spatial relationships. 80% of population at considerable risk shows, how vulnerable is Nepal from geographically and climate invited challenges (multiple hazards, including floods, landslides, and earthquake s(Khatakho et al., 2021).Floods are one of the most occurring disasters causing several impacts on human life and livelihoods. The causing source for flood risk susceptibility can be tough topography like mountainous and steep slopes, unpredictable monsoon seasons and variations in rainfall pattern. These conditions create an environment ripe for catastrophic flooding, resulting in tragic annual losses of life and property (Khanal et al., 2007). Nepal should have clear strategies over flood risk causes, effect and control measures by thorough reviews of existing research. Given with the factors contribution for structured decision- making tools like AHP becomes more authentic and proven multi-criteria framework for evaluating hazard factors and informing risk assessment decisions (Khatakho et al., 2021). The challenging topography, high relief and unpredicted monsoon of Nepal make the environment vulnerable to serious flood disasters and this has contributed to large loss of human life and property every year (Khanal et al., 2007). Nepal has consistently experienced flood problems, aggravated by its land formations, steep terrain, and high rainfall in monsoons since floods have in the past led to high rates of victims and economic damages, with about 7,599 deaths and economic losses of about 10.6 billion USD reported between 1954 and 2018, yet such influences as misuse of land, unplanned settlements, and deforestation have increased the reliability of the population and therefore effective flood control systems and preventive actions are imperative to assess flood vulnerability that cont Equally, Community based flood damage assessment approach of the lower West Rapti River basin in Nepal under the influence of climate change examines the impact of climate change on flooding in the West Rapti River (WRR) basin which is a major agricultural area using community-based surveys and hydrological modelling it estimates the current and future flood damages, especially in reference to the extensive flood experienced in 2007 (Perera et al., 2015). Although it was discovered that the model of flood risk assessment in Wuhan City by means of RF includes annual precipitation as the main influential variable, then vegetation coverage, the elevation, and the river network density, with the areas at the risk of floods being mainly located in Jianghan, Jiangan, and Wuchang Districts and being highly accurate when compared to the inundation risk map of 2020 (Wu and Jiang, 2024). The geospatial analysis of the Relative FR of the region incorporating the study area in the eastern Hindu Kush, in Khyber Pakhtunkhwa, Pakistan (34.33 71.0ully 72.0-) confirmed 15% of the area as highly susceptible to floods and the high-risk zones are concentrated in the low-elevation, high-drainage-density floodplains around rivers with an AUC accuracy of more than 82 percent (Ullah and Zhang, 2020). The research was carried out in Jamaica, a Caribbean Island, and it employed a hybrid method, which combined LR, PCA, and FD methods, to better predict the flood hazards in which PCA was used to simplify the data, LR was used to determine the important geospatial variables, and FD was used to determine the frequency of occurrence of the floods, and finally, the flood hazard mapping was improved (Nandi et al., 2016). The principal goal of the research is to prepare an elaborate flood risk assessment based on the AHP and the RF algorithm. To justify this, we see the validity of the chosen parameters to make sure that they are really the indicators of the flood risk conditions. The creation of accurate thematic maps is done using satellite images in variables like slope, curvature, soil type, rainfall, distance to road and stream, TWI, NDVI, and aspect. With these datasets, the study area is classified into three types of flood risks namely high, medium, and low. All the models are then tested using another independent set of data to determine their accuracy and reliability. Hence, the outcomes are compared to determine which model is best method fitted for the study.

## II. METHODOLOGY

### 2.1. Study Area

*Figure 1. Study Area*

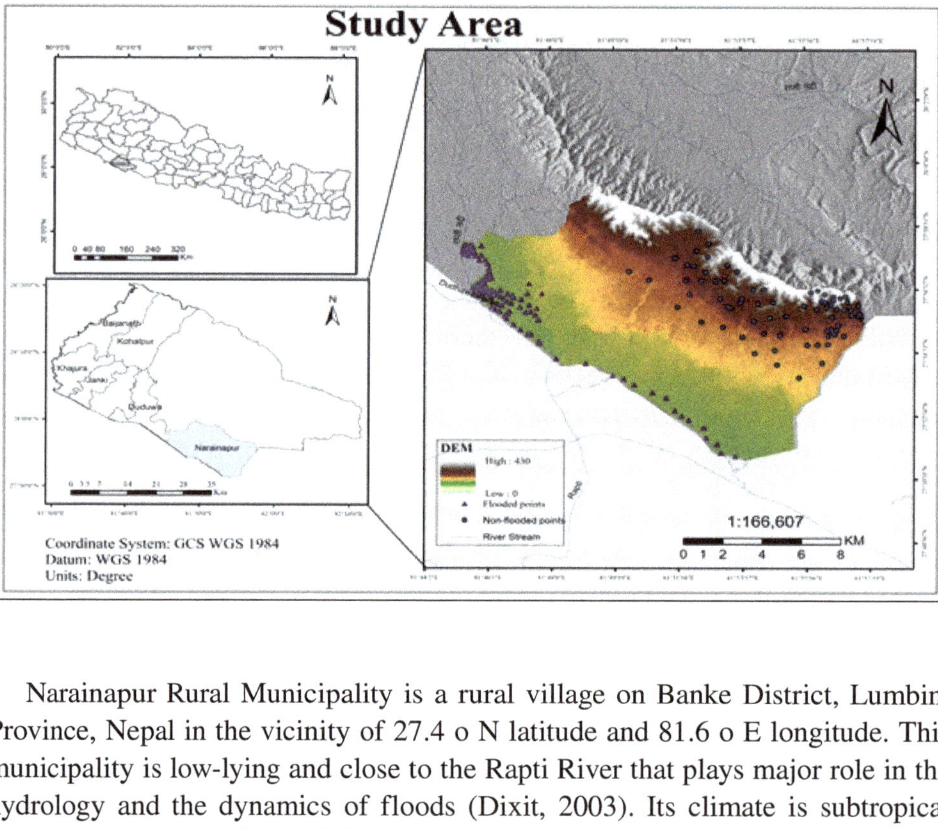


Narainapur Rural Municipality is a rural village on Banke District, Lumbini Province, Nepal in the vicinity of 27.4 o N latitude and 81.6 o E longitude. This municipality is low-lying and close to the Rapti River that plays major role in the hydrology and the dynamics of floods (Dixit, 2003). Its climate is subtropical monsoon, and most of the rainfall is received between June and September, which contributes to the high risks of floods during the monsoon season (Shrestha et al., 2020). According to BIPAD (Building Information Platform Against Disaster) portal, an average of 1,500 mm is considered to be the annual rainfall in Narainapur, which is just enough to cause flood conditions and especially in low-lying regions bordering the river (BIPAD Portal, 2023; Paudel and Andersen, 2013).

## 2.2. Data Collection

Table 1 shows the data we collected for our Region of interest.

*Table 1. Data Source*

| Category | Type | Source | Resolution/Scale |
|---|---|---|---|
| DEM | Raster | United States Survey(USGS) | 30m |
| Landsat Images | Raster | United States Survey(USGS) | 30m |
| Administrative boundary | Vector | National Geoportal | 1: 1000000 |
| Rainfall | Raster | CHIRPS | 5500m |
| Soil | Vector | National Soil Science Research Center NARC | 1:10000 |

## 2.3 Preperation of Flood Inventory Map and Training Dataset

A flood inventory map and a training dataset was created extracted 102 flooded points and 95 non-flooded points by use of Google Earth Pro (Komi et al., 2017; Sanyal and Lu, 2004). These data were combined into one dataset in order to give a detailed picture of flood-prone and non flood areas. The dataset was subsequently divided into two parts, 70 percent of which was used in training so that the model could learn the patterns, and 30 percent of which was used in testing in order to determine how well the model predicts flood (Zhang et al., 2019).Such systematic treatment guarantees a balanced and credible dataset used in the risk assessment of floods is as below in figure 2.

*Figure 2. Methodology Chart*

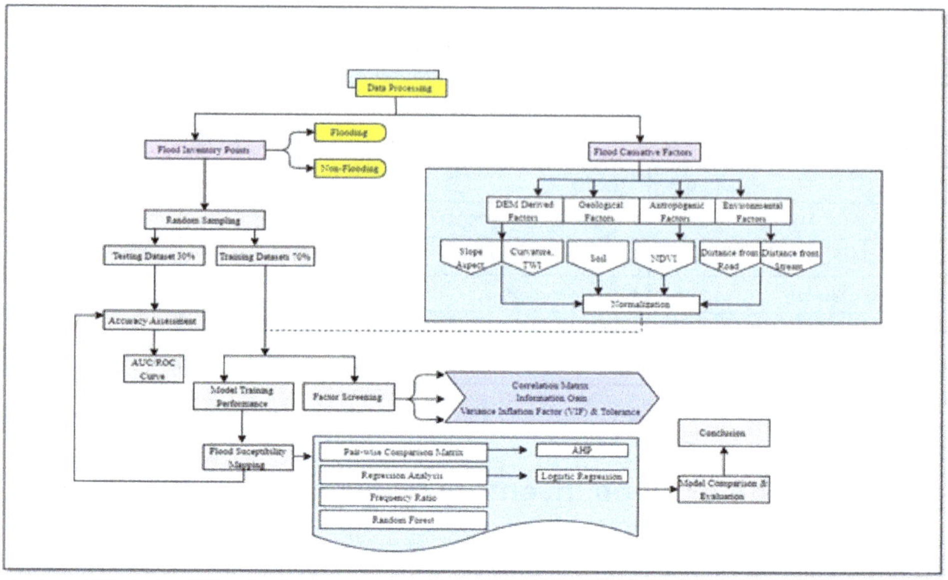


## 2.4 Screening Factor

Screening factors consist of the choice of the most pertinent, non-redundant, and independent variables (Rahmati et al., 2019; Tehrany et al., 2014). It is done to make sure that only meaningful factors are utilized as they are evaluated on their relevance (through Information Gain), redundancy (through Kendall-Pearson Correlation), and multicollinearity (through Variance Inflation Factor) (Choubin et al., 2019).

### 2.4.1. Correlation Matrix

Correlation table A correlation table is a table displayed with correlation coefficients among two or more variables, which measures the direction and strength of the correlation. The values are between -1 and +1 with 1 being the perfect positive correlation, -1 being the perfect negative correlation, and 0 denoting no correlation (Dormann et al., 2013).

### 2.4.2. Variance Inflation Factor (VIF) and Tolerance

They are the measure of multicollinearity in regression models to determine the degree to which a variable is increasing the variance of another variable (O'Brien,

2007). The value of VIF (usually above 5 or 10) is a measure of high multicollinearity, i.e. the variable is very redundant and can be dropped. On the other hand, low values of VIF (< 2) indicate that there is weak correlation between predictors, which is desirable in regression models. Another measure of multicollinearity is the tolerance which is the inverse of VIF (Tolerance = 1/VIF). A tolerance value of less than 0.2 or 0.1 is a very high level of collinearity, indicating that the variable does not provide much unique information and may be re-evaluated as to whether it should be included in the model or not (Kutner et al., 2005).

### 2.4.3. Information Gain

The Information Gain (IG) is a metric applied in decision trees and feature selection to estimate the ability of a single feature to reduce uncertainty in predicting a target variable (Quinlan, 1993; Han et al., 2011). It is computed based on entropic functions where a feature with high information gain is an effective way of splitting the data resulting in a better classification. The greater the IG the more practical the feature is to the performance of the models (Kotsiantis, 2013).

## 2.5. Flood Conditioning Factor

### 2.5.1 Slope

Slope is one of the most significant topographical attributes in a hydrological research, and the slope is a crucial element in the regulation of surface water flow (Sharifi et al., 2019; Pourghasemi et al., 2012). One of such effective factors in floods is land surface slope: the more the slope is low, the more intense is the flooding, and vice versa (Choubin et al., 2019).

### 2.5.2. Normalized Difference Vegetation Index(NDVI)

Another variable which is a good index when evaluating the vegetation cover and its effects in flooding in a study area is the NDVI (Rouse et al., 1974; Xu, 2006). The values of the NDVI normally vary between -1 and +1. Satellite images of LANDSAT 8 OLI were used to prepare an NDVI map and compute the value of NDVI using the following equation (Tucker, 1979).

$$NDVI = \frac{\mathrm{NIR} - \mathrm{RED}}{\mathrm{NIR} + \mathrm{RED}} \quad (1)$$

### 2.5.3. Soil

Two primary factors like soil type and soil texture determine the water holding capacity and surface infiltration properties of an area (Hillel, 2003; Rawls et al., 2003). In the present study, 3 soil classes were used in the study area, which included Fluvial Non-Calcareous, Fluvial Calcareous, and Sandstone/Greywacke/Arkose. The map of soils type was acquired at NARC (National Agricultural Research Council, 2020).

### 2.5.4 Distance from Stream

The proximity to the river is among the most factorial in flood hazard map (Pourghasemi et al., 2012). The elevation and slope rise as the distance is greater. In addition, a stream is usually the lowest point in that specific area. This means that regions distant to the river are not prone to the occurrence of floods. We have in this paper categorized the distance with river under five classes, namely, very high (0 - 882 m), high (882 -1,764 m), moderate (1,764- 2,647 m), low (2,647- 3,529 m) and very low (3,529- 4,412 m). The closer the river the more vulnerable the floods will be and vice versa (Chow et al., 2010).

### 2.5.5 Rainfall

The most striking factor is rainfall because the coastal area districts have sufficient rainfall due to northeast monsoon and southwest monsoon, although the northeast monsoon season (October to December) is regarded as wetter than the southwest one (Shrestha et al., 2020). To obtain the precipitation of the rainfall distribution map, the rainfall data of all rain gauge stations were computed using the Inverse Distance Weighted interpolation method in ArcGIS 10.8.1 of Climate Hazards Group InfraRed Precipitation with Station data (Funk et al., 2015).

### 2.5.6 Topography Wetness Index (TWI)

TWI describes how a topography influences the runoff creation and the quantity of the accumulated flow (Beven and Kirkby, 1979; Srensen et al., 2006). The equation that calculates the TWI may be given as:

$$\mathrm{TWI} = \mathrm{In}\frac{\mathrm{As}}{\tan \beta} \quad (2)$$

Where, As is the area and $\beta$ is the local slope gradient in degree. There is greater potential of Vulnerability to flooding in higher TWI regions. On the contrary, the high TWI areas possess diminished potential of Vulnerability (Gessler et al., 1995).

### 2.5.7 Distance from Road

The distance to the roads is another significant variable used in the measurement of flood suseptibility. When the waters go on the river banks and lowlands during floods, the flood water finds its way into the roads and streams destroying the property of people, roads and houses (Chow et al., 2010).

### 2.5.8 Aspect

Direction that a slope is facing, known as aspect, is another crucial aspect of flood risk assessment because it influences the sun exposure, plant cover and runoff patterns (Moore et al., 1991; Zhang et al., 2008). The down-slope slope is also prone to more moisture and this will augment flood risks during intense rainfalls.

### 2.5.9 Curvature

In flood risk assessment, topographical parameter is a crucial parameter because it defines the processes of water flow and water accumulation (Zevenbergen and Thorne, 1987; Moore et al., 1991). The concave formations are also prone to accumulating water, which enhances flood exposures whilst the convex features allow water to run off, which minimizes flood hazards.

## 2.6. Flood Susceptibility Models

### 2.6.1 RF

RF is a supervised learning algorithm. Its constructed forest is an assembly of decision trees, typically trained by the bagging technique (Breiman, 2001; Liaw and Wiener, 2002). The overall principle of the bagging technique is that having a combination of learning models augments the sum total outcome. The large utility of RF is that it applies to classification and regression problems that constitute most of the existing machine learning systems (Rodriguez-Galiano et al., 2012). The hyperparameters of RF are almost equivalent to that of a decision tree or a bagging classifier (Prasad et al., 2006).

### 2.6.2 AHP

AHP is a decision making methodology involving multi-criteria and it was proposed by Saaty (Saaty, 1980; Saaty, 2008). The AHP is a decision support tool that is applied to address complex decision problems. It involves the use of multi-level hormonized hierarchy of objectives, criteria, sub-criteria and alternatives. It identifies the weights and ranks of various parameters of the flood vulnerable areas (FVZ). The model uses different weights to compare the ten individual factors in question and on the weight of relative importance of the factors, these nine parameters are graded between 1 and 9 in absolute number scale (Malczewski, 2006).

#### *2.6.2.1. Pairwise Comparison Ratio*

Pairwise Comparison Ratio is a method used in multi-criteria decision-making to compare factors relative to each other based on their importance (Saaty, 1980; Ishizaka & Labib, 2011) as shown in table 2.

*Table 2. Pair-wise comparison scale weights on the bases of AHP scale*

| Scale | Judge | Explaination |
|---|---|---|
| 1 | Equally Important | Two factors contribute equally to the objective |
| 3 | Moderate Important | Experience and judgment slightly favour one over the other |
| 5 | Important | Experience and judgment strongly important favour one over the other |
| 7 | Very strongly Important | Experience and judgment strongly important favour one over the other |
| 9 | Extremely Important | The evidence favoring one over the other is of the highest possible validity |
| 2,4,6,8 | Intermediate preference between adjacent scales | When compromised is needed |

This scale is widely used in AHP to assign relative weights to criteria based on expert judgment (Malczewski, 2006; Moumane et al. 2024, 2025).

#### *2.6.2.2. Consistency Ratio(CR)*

To rectify the constructed pairwise matrix and its given weightage, evaluation through the CR) is performed, where the acceptable CR must be below 0.1 (Saaty, 1980; Ishizaka & Labib, 2011). The CR is calculated as:

$$\mathrm{CR} = \frac{CI}{RI} \tag{3}$$

Where CR represents the consistency ratio, CI means consistency index, RI indicates the random index, $\lambda max$ represents the principal eigenvalue of the comparison matrix, and n is the number of components or factors in the matrix. When compromise is needed, 12 is involved in AHP. Based on the nine variables in this study, the derived RI is 1.49 and the obtained CR is 0.063 (Malczewski, 2006).

*Table 3. Random Index (RI) value*

| Number of Criteria | 2 | 3 | 4 | 5 | 6 | 7 | 8 | 9 | 10 | 11 |
|---|---|---|---|---|---|---|---|---|---|---|
| RI | 0 | 0.58 | 0.9 | 1.12 | 1.24 | 1.32 | 1.41 | 1.45 | 1.49 | 1.5 |

### 2.6.3. LR

LR is one of the most widely used techniques for analyzing a set of well-known data due to its convenience in application, especially in the field of water resources and environmental engineering (Hosmer et al., 2013; Menard, 2002). Regression analysis is widely applied for prediction using data interpolation and extrapolation, as well as for estimation of missing data (Peng et al., 2002). In this specific study within the file content, LR was used among several regression techniques to predict inundated grid cells in non-parametric models. The LR model is mathematically expressed by logarithmically converting Equation (5), which represents the Odds Ratio (OR) for a specific event. This resulting logarithmic expression is called a logit function (Agresti, 2013), and is presented as Equation (8) in the study.

$$OR = \frac{p}{1-p} \quad (4)$$

$$\mathrm{fx} = \log(\mathrm{OR}) = \log\frac{p}{1-p} \quad (5)$$

$$log\frac{p}{1-p} = b1 + b2x \quad (6)$$

$$p = \frac{1}{e^{b1+b2x}} \quad (7)$$

In LR, $p$ explain about the probability of the target event. $b2$x represents the constant term of the LR, and $b1$ represents the parameter for an independent variable (Hosmer et al., 2013; Agresti, 2013). Equation (4) represents the relationship of the logit function to a continuous variable. The logistic function transforms the continuous variable into values between 0 and 1, serving as the inverse of the logit function (Menard, 2002; Peng et al., 2002)

## 2.7 Relative Frequency Model

One of the important techniques in hydrological studies is flood hazard assessment. The PRB is a region that is flood-prone, and in this study, FR model is applied to map it (Tehrany et al., 2014; Rahmati et al., 2019). FR is a bivariate statistical analysis tool, which presupposes the spatial distribution (probability) of the dependent variable and triggers and causes of floods (Choubin et al., 2019). The probability of each independent factor of flood triggering, when put against flood occurrence, was established by the relationship between the two variables to give out the bivariate probability. The larger the bivariate probability (more than 1), the stronger is the correlation between flood incidence and flood triggering factors; the smaller the probability, the weaker the correlation.

Once normalization is carried out, the RF model further assumes that the causative factors are equivalent. In order to address this shortcoming as well as in order to capture the interrelationships between the causative factors of floods, a rate predictor (PR) or a weight was obtained by rating all the flood causative factors using the training dataset. Lastly, the flood susceptibility index was derived using these formula (Pourghasemi et al., 2012).

$$FSI = \sum \mathrm{PR} \times \mathrm{RF} \tag{8}$$

## 2.8 Accuracy Assessment

The Receiver Operating Characteristic (ROC) curve and the Area Under the Curve (AUC) are widely used performance evaluation metrics for classification models (Fawcett, 2006; Han et al., 2011). The ROC curve examine the True Positive Rate (TPR) against the False Positive Rate (FPR) at different thresholds, providing information about the model's ability to distinguish between classes. The AUC value quantifies the overall ability of the model with values closer to 1 indicating better performance (Bradley, 1997).

- TPR:

$$TPR = \frac{TP}{FN + TP} \tag{9}$$

- FPR:

$$FPR = \frac{FP}{TN + FP} \tag{10}$$

- TNR:

$$TNR = \frac{TN}{TN + FP} \tag{11}$$

# III RESULT

## 3.1. Correlation and Variance Inflation

Correlation values relationship between different parameters are shown and described in figure 3.

*Figure 3. Correlation*

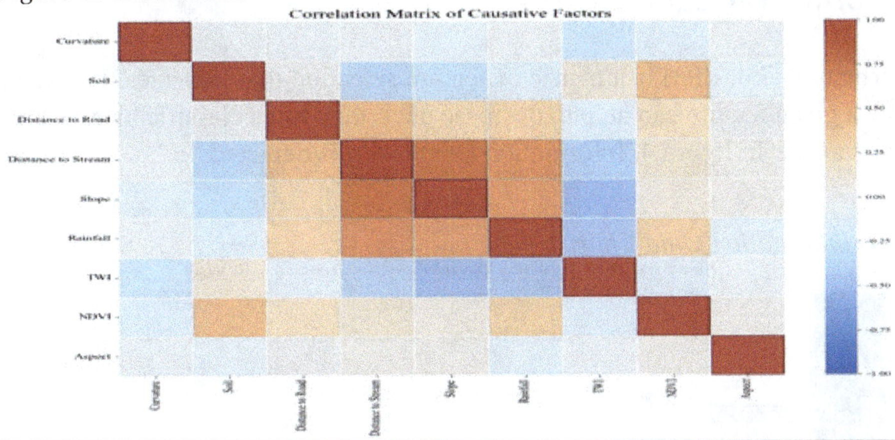


## 3.2.Variance Inflation Factor (VIF) and Tolerance

Table 4 presents VIF and Tolerance values for different variables, which are used to check for multicollinearity in regression analysis.

*Table 4. VIF and Tolerance Value*

| Variable | VIF | Tolerance |
|---|---|---|
| Constant | 22772.31 | 4.39-05 |
| Curvature | 1.220483 | 0.819348 |
| Soil | 1.6796 | 0.59538 |
| Distance to Road | 1.277972 | 0.78249 |
| Distance to Stream | 3.256382 | 0.307089 |
| Slope | 2.886561 | 0.346433 |
| Rainfall | 2.122063 | 0.47124 |

continued on following page

*Table 4. Continued*

| Variable | VIF | Tolerance |
|---|---|---|
| TWI | 1.692257 | 0.590927 |
| NDVI | 1.649253 | 0.606335 |
| Aspect | 1.092923 | 0.914977 |

## 3.3. Information Gain

Information Gain (IG) is a metric of feature selection that indicates the extent to which a given feature can help to decrease uncertainty in a classification or prediction problem. In figure 4, Information Gain Chart is depicted.

*Figure 4. Information Gain Chart*

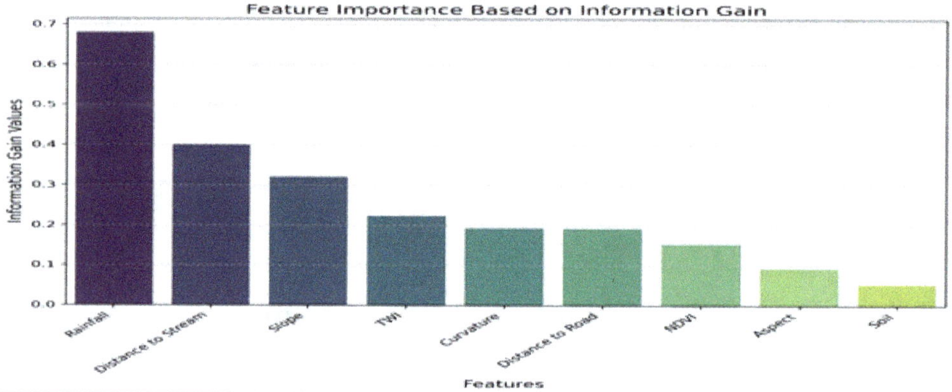


## 3.4. Flood Conditioning Factor Map

The Flood Conditioning Factor map is presented in Figure 5.

*Figure 5. Parameters Map*

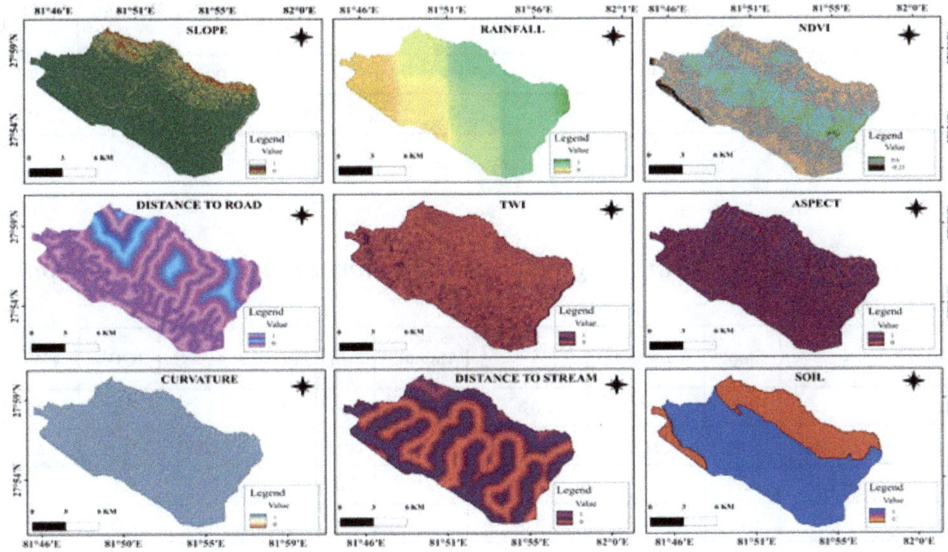


## 3.5.Flood Susceptibility Map

### 3.5.1. AHP

The results are presented in Table 6 and Table 7 as shown below.

*Table 5. Pair-wise comparison of 9 9 decision matrix*

| Factors | Slope | Curvature | Distance Stream | NDVI | Soil | Road | Aspect | Rainfall | TWI |
|---|---|---|---|---|---|---|---|---|---|
| **Slope** | 1.00 | 7.00 | 3.00 | 4.00 | 5.00 | 9.00 | 8.00 | 2.00 | 6.00 |
| **Curvature** | 0.14 | 1.00 | 0.20 | 0.25 | 0.33 | 3.00 | 2.00 | 0.17 | 0.50 |
| **Distance to Stream** | 0.33 | 5.00 | 1.00 | 2.00 | 3.00 | 7.00 | 6.00 | 0.50 | 4.00 |
| **NDVI** | 0.25 | 4.00 | 0.50 | 1.00 | 2.00 | 6.00 | 5.00 | 0.33 | 3.00 |
| **Soil** | 0.20 | 3.00 | 0.33 | 0.50 | 1.00 | 5.00 | 4.00 | 0.25 | 2.00 |
| **Road** | 0.11 | 0.33 | 0.14 | 0.17 | 0.20 | 1.00 | 0.33 | 0.13 | 0.25 |
| **Aspect** | 0.13 | 0.50 | 0.17 | 0.20 | 0.25 | 3.00 | 1.00 | 0.14 | 0.33 |
| **Rainfall** | 0.50 | 6.00 | 2.00 | 3.00 | 4.00 | 8.00 | 7.00 | 1.00 | 5.00 |

continued on following page

*Table 5. Continued*

| Factors | Slope | Curvature | Distance Stream | NDVI | Soil | Road | Aspect | Rainfall | TWI |
|---|---|---|---|---|---|---|---|---|---|
| TWI | 0.17 | 2.00 | 0.25 | 0.33 | 0.50 | 4.00 | 3.00 | 0.20 | 1.00 |
| ColumnTotal | 2.83 | 28.83 | 7.59 | 11.45 | 16.28 | 46.00 | 36.33 | 4.72 | 22.08 |

*Table 6. Normalized and the weight values in the standardized pair-wise comparison matrix*

| Factors | Slope | Curvature | Distance to Stream | Rainfall | Soil | Road | Aspect | NDVI | TWI |
|---|---|---|---|---|---|---|---|---|---|
| Slope | 0.35 | 0.24 | 0.4 | 0.35 | 0.31 | 0.2 | 0.22 | 0.42 | 0.27 |
| Curvature | 0.05 | 0.03 | 0.03 | 0.02 | 0.02 | 0.07 | 0.06 | 0.04 | 0.02 |
| Distance to Stream | 0.12 | 0.17 | 0.13 | 0.17 | 0.18 | 0.15 | 0.17 | 0.11 | 0.18 |
| Rainfall | 0.09 | 0.14 | 0.07 | 0.09 | 0.12 | 0.13 | 0.14 | 0.07 | 0.14 |
| Soil | 0.07 | 0.1 | 0.04 | 0.04 | 0.06 | 0.11 | 0.11 | 0.05 | 0.09 |
| Road | 0.04 | 0.01 | 0.02 | 0.01 | 0.01 | 0.02 | 0.01 | 0.03 | 0.01 |
| Aspect | 0.04 | 0.02 | 0.02 | 0.02 | 0.02 | 0.07 | 0.03 | 0.03 | 0.02 |
| NDVI | 0.18 | 0.21 | 0.26 | 0.26 | 0.25 | 0.17 | 0.19 | 0.21 | 0.23 |
| TWI | 0.06 | 0.07 | 0.03 | 0.03 | 0.03 | 0.09 | 0.08 | 0.04 | 0.05 |

### 3.5.2. FR

The FR model was applied to analyze the relationship between flood occurrences and conditioning factors. Firstly, the FR values for all factors were calculated using training data and those parameter maps and evaluated to ensure their relevance in flood susceptibility analysis. Based on the computed FR values, weightages were assigned to each factor, as shown in the table 8. Once the weightages were determined, they were applied to the factor maps to develop the final flood susceptibility map. The resultant maps were then classified into five categories: Very Low, Low, Moderate, High, and Very High as shown in figure 6, providing a clear and systematic representation of flood-prone areas for further analysis and decision-making.

*Table 7. FR*

| Data Layer | Class | Class Pixel | Flood Pixels | FR | RF | RF[%] | RF | MaxMin | PR |
|---|---|---|---|---|---|---|---|---|---|
| Rainfall | 0 - 0.17 | 24251 | 86 | 0.0035 | 0.8961 | 90 | 89 | | |
| | 0.17-0.36 | 54116 | 5 | 0.0001 | 0.0233 | 2 | 2 | | |
| | 0.36-0.53 | 44483 | 7 | 0.0002 | 0.0398 | 4 | 3 | 0.00366 | 243.19 |
| | 0.53 - 0.71 | 61454 | 0 | 0.0000 | 0.0041 | 0 | 0 | | |
| | 0.71-1 | 6890 | 0 | 0.0001 | 0.0367 | 4 | 3 | | |
| Total | | 191194 | 98 | 0.0040 | | | | | |
| NDVI | 0-0.05 | 38462 | 4 | 0.0001 | 0.0052 | 1 | 196 | | |
| | 0.05-0.16 | 55885 | 11 | 0.0002 | 0.0098 | 1 | 0 | | |
| | 0.16-0.27 | 92726 | 43 | 0.0005 | 0.0230 | 2 | 2 | 0.00251 | 193.55 |
| | 0.27-0.36 | 3272 | 31 | 0.0095 | 0.4707 | 47 | 47 | | |
| | 0.36-0.66 | 910 | 9 | 0.0099 | 0.4913 | 49 | 49 | | |
| Total | | 191255 | 98 | 0.0201 | | | | | |
| TWI | 0-0.05 | 38462 | 4 | 0.0001 | 0.0052 | 1 | 196 | 0.0075 | 40.096 |
| | 0-0.24 | 46415 | 3 | 0.0001 | 0.0249 | 2 | 2 | | |
| | 0.24-0.33 | 78085 | 44 | 0.0006 | 0.2174 | 22 | 21 | | |
| | 0.33-0.44 | 40838 | 33 | 0.0008 | 0.3118 | 31 | 31 | | |
| | 0.44-0.6 | 17697 | 15 | 0.0008 | 0.3271 | 33 | 32 | | |
| | 0.6-1 | 6497 | 2 | 0.0003 | 0.1188 | 12 | 11 | | |
| Total | | 189532 | 97 | 0.0026 | | | | | |
| Distance from Stream | 0-0.12 | 9552 | 0 | 0.0001 | 0.0611 | 6 | 6 | | |
| | 0.12-0.25 | 21244 | 0 | 0.0000 | 0.0275 | 3 | 2 | | |
| | 0.25-0.41 | 37023 | 1 | 0.0000 | 0.0158 | 2 | 1 | | |
| | 0.41-0.62 | 58720 | 20 | 0.0003 | 0.1989 | 20 | 19 | | |
| | 0.62-1 | 64540 | 77 | 0.0012 | 0.6967 | 70 | 69 | 0.01074 | 63.399 |
| Total | | 191079 | 98 | 0.0017 | | | | | |
| Soil | fluvial NonCalvareous | 12685 | 5 | 0.0004 | 0.0096 | 1 | 0 | | |
| | fluvial Calvareous | 1751 | 70 | 0.0400 | 0.9688 | 97 | 96 | 0.0091 | 104.68 |
| | Sandstone | 32507 | 29 | 0.0009 | 0.0216 | 2 | 2 | | |
| Total | | 14436 | 75 | 0.0413 | | | | | |
| Distance from Road | 0-0.1 | 7708 | 1 | 0.0001 | 0.0832 | 8 | 8 | | |
| | 0.1-0.25 | 12119 | 1 | 0.0001 | 0.0529 | 5 | 5 | | |
| | 0.25-0.43 | 30672 | 1 | 0.0000 | 0.0209 | 2 | 2 | 0.0095 | 47.845 |
| | 0.43-0.66 | 40638 | 23 | 0.0006 | 0.3628 | 36 | 36 | | |
| | 0.66-1 | 100124 | 75 | 0.0007 | 0.4802 | 48 | 48 | | |
| Total | | 191261 | 101 | 0.0016 | | | | | |

continued on following page

*Table 7. Continued*

| Data Layer | Class | Class Pixel | Flood Pixels | FR | RF | RF[%] | RF | MaxMin | PR |
|---|---|---|---|---|---|---|---|---|---|
| Slope | 0- 0.05 | 719 | 1 | 0.0014 | 0.0691 | 7 | 6 | 0.00026 | 24.74 |
| | 0.05 - 0.12 | 4933 | 1 | 0.0002 | 0.0101 | 1 | 1 | | |
| | 0.12-0.22 | 12407 | 1 | 0.0001 | 0.0040 | 0 | 0 | | |
| | 0.22 -0.54 | 32130 | 5 | 0.0002 | 0.0077 | 1 | 0 | | |
| | 0.54-1 | 139261 | 92 | 0.0007 | 0.0328 | 3 | 3 | | |
| Total | | 189450 | 100 | 0.0201 | | | | | |
| Aspect | 0-0.23 | 32632 | 12 | 0.0004 | 0.1360 | 14 | 13 | 0.0262 | 7.368 |
| | 0.23-0.41 | 46331 | 17 | 0.0004 | 0.1357 | 14 | 13 | | |
| | 0.41-0.57 | 48903 | 25 | 0.0005 | 0.1891 | 19 | 18 | | |
| | 0.57-0.74 | 36923 | 21 | 0.0006 | 0.2104 | 21 | 21 | | |
| | 0.74-1 | 24743 | 22 | 0.0009 | 0.3289 | 33 | 32 | | |
| Total | | 189532 | 97 | 0.0027 | | | | | |
| Curvature | 0-0.25 | 110 | 1 | 0.0091 | 0.3257 | 33 | 32 | 0.00771 | 54.226 |
| | 0.25-0.41 | 371 | 1 | 0.0027 | 0.0966 | 10 | 9 | | |
| | 0.41-0.54 | 190410 | 98 | 0.0005 | 0.0184 | 2 | 1 | | |
| | 0.54-0.68 | 293 | 1 | 0.0034 | 0.1223 | 12 | 12 | | |
| | 0.68-1 | 82 | 1 | 0.0122 | 0.4370 | 44 | 43 | | |
| Total | | 191266 | 102 | 0.0279 | | | | | |

## 3.5.3. LR

The LR model was applied to analyze the relationship between flood occurrences and various conditioning factors. Initially, training data and factor maps were used to determine the coefficients for each factor, The coefficients were examine along with their CI and other statistical parameters which is shown in table 9. Once the coefficients were established, they were applied to the factor maps to generate the final flood susceptibility map. The classified maps were divided into five categories: Very Low, Low, Moderate, High, and Very High, providing a clear spatial representation of flood-prone areas. The corresponding tables are presented below for further reference and analysis.

*Table 8. LR*

| Factors | Coefficient | Standard Error | Z-value | P-value | Lower 95% CI | Upper 95% CI | Lower 97% CI | Upper 97% CI |
|---|---|---|---|---|---|---|---|---|
| **Constant** | -0.07 | 1.83 | -0.04 | 0.97 | -3.7 | 3.56 | -4.09 | 3.95 |
| **Curvature** | 2.62 | 3.67 | 0.71 | 0.48 | -4.64 | 9.87 | -5.43 | 10.66 |
| **Soil** | 0.1 | 0.04 | 2.83 | 0.01 | 0.03 | 0.17 | 0.02 | 0.18 |
| **Distance to Road** | -0.9 | 0.13 | -6.88 | 0 | -1.16 | -0.64 | -1.19 | -0.61 |
| **Distance to Stream** | -0.31 | 0.13 | -2.37 | 0.02 | -0.57 | -0.05 | -0.6 | -0.02 |
| **Slope** | 0.8 | 0.4 | 1.98 | 0.05 | 0 | 1.59 | -0.09 | 1.68 |
| **Rainfall** | -1.12 | 0.05 | -22.42 | 0 | -1.22 | -1.02 | -1.23 | -1.01 |
| **TWI** | 0.08 | 0.16 | 0.51 | 0.61 | -0.23 | 0.4 | -0.27 | 0.43 |
| **NDVI** | -0.63 | 0.15 | -4.09 | 0 | -0.94 | -0.33 | -0.97 | -0.29 |
| **Aspect** | -0.13 | 0.06 | -2.17 | 0.03 | -0.24 | -0.01 | -0.25 | 0 |

### 3.5.4. RF

To understand how different factors influence flood risk, scientists used the RF model. This approach works by combining many individual decision trees, all built using the training data and flood-influencing factor maps. The model not only predicts flood likelihood but also reveals which factors matter most by ranking their importance. After thoroughly testing the model, these importance rankings were applied to the factor maps across the study area. This process generated a detailed flood susceptibility map. Finally, this map was divided into five distinct risk zones – ranging from the safest spots to the highest-danger areas – clearly visualizing flood-prone locations, as illustrated in Figure 6.

*Figure 6. Final Susceptibility Map*

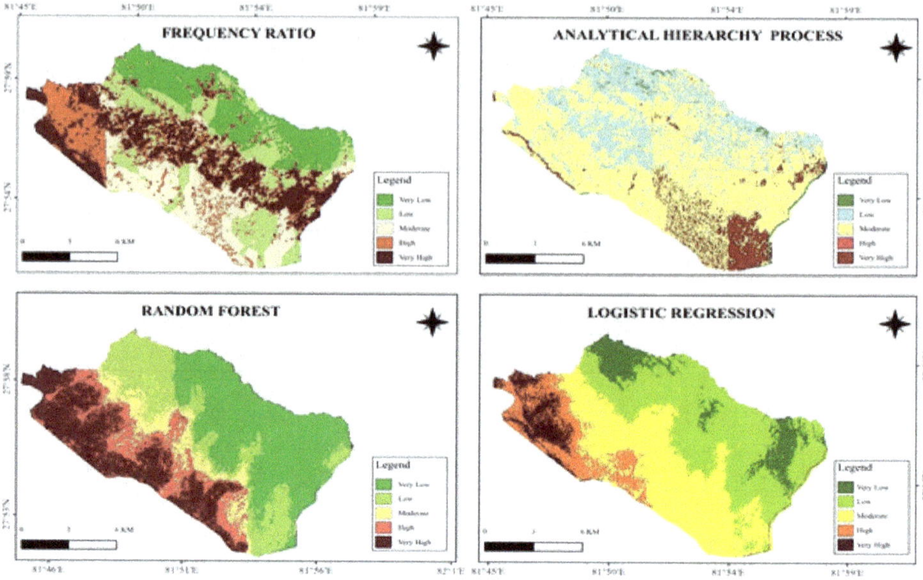


## 3.5.5. AUC/ROC Curve

To determine which models did the best job telling flooded and non-flooded areas apart, we looked at their ROC curves and calculated the AUC scores. The standout performer was the LR model, achieving an impressive AUC of 0.944. This high score means it made the most accurate predictions overall. The RF model was also very strong, with an AUC of 0.936. The FR (AUC = 0.855) and AHP (AUC = 0.822) models followed behind. These results clearly show that the machine learning models—LR and RF—were the most effective at predicting flood risk in our specific study area. Because LR had the highest AUC score, it proved to be the best fit for our location. Its excellent performance suggests it reliably understands how different landscape features contribute to flooding, making it the most dependable choice for creating flood susceptibility maps. Its ability to accurately distinguish between flood and non-flood zones, confirmed by testing on 30% of the data, also makes it particularly valuable for assessing flood risk here area.

*Figure 8. AUC/ROC Curve*

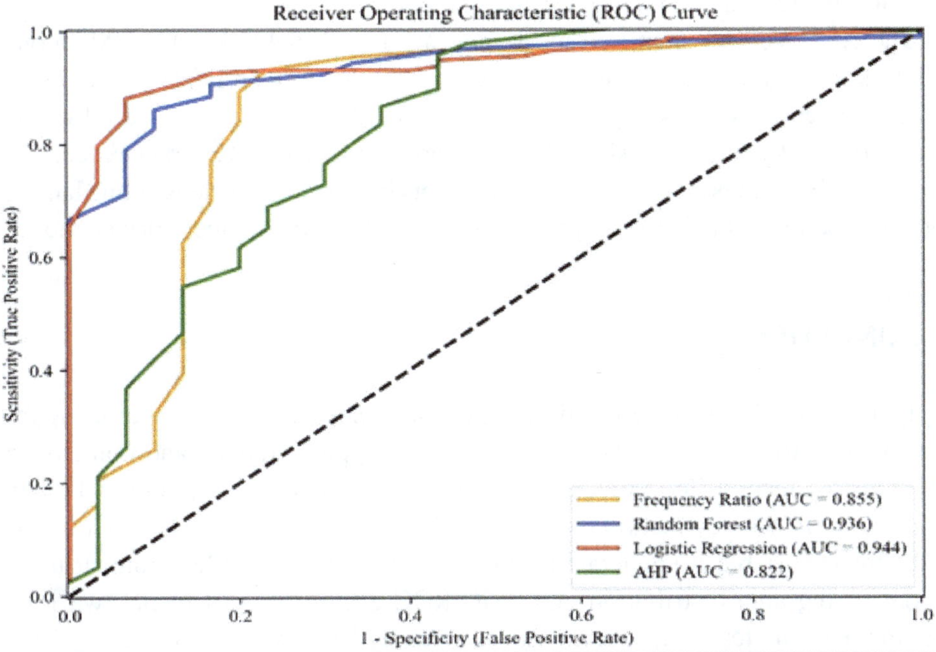


# IV DISCUSSION

## 4.1 Comparison of Flood Risk Assessment Results

We put our LR, RF, FR, and AHP flood risk maps side by side with other research to see how they stacked up against each approach. (Bui et al., 2022; Farhadi & Najafzadeh, 2021). The LR model gained the highest AUC of 0.94 and classified 4,214 km$^2$ as "Moderate" risk, while RF followed closely with an AUC of 0.93 and identified a significant "High" risk area of 11,059 km$^2$. These results align with Bui et al. (2022), who used machine learning models integrated with AHP to generate flood susceptibility maps for Quang Binh Province, Vietnam. Their study reported that deep learning models achieved an AUC of 0.984, outperforming other ensemble models, similar to how our LR and RF models demonstrated high predictive accuracy.

Our RF model also flagged large areas as 'High' risk, backing up what Farhadi & Najafzadeh (2021) found in Iran. They used RF with satellite data to pinpoint flood zones accurately, noting how well it handles different factors like elevation and land use. Like their study, ours shows RF shines when capturing how these variables interact to paint a true picture of flood risk.

The AHP method identified most areas as “Low” (6,423 $km^2$) or “Moderate” risk (5,964 $km^2$) but had the lowest AUC of 0.82 due to its reliance on subjective weighting schemes. This aligns with Kazakis et al. (2015), who used GIS-based multi-criteria decision analysis for flood hazard assessment in Central Greece. They analyzed AHP brings in useful real-world factors like population density and land use for flood mapping, but it doesn't predict as well as the more numbers-driven models.

## V CONCLUSION

In this study, We tackled flood susceptibility mapping in Narainapur using four models: LR, RF, FR, and AHP. we started by mapping flooded and non-flooded locations to create a flood inventory, then split this data into training and testing sets then analyzed key factors like topography, land use, distance to water bodies, and climate variables to generate flood susceptibility maps. When comparing the models using area distribution and AUC/ROC accuracy assessments, we found that the machine learning approaches—especially LR—were the most helpful for forecasting flood-prone areas. LR emerged as the most appropriate model for our research area. These results offer valuable insights for flood risk management, providing a clear understanding of how flood-triggering factors correlate with actual flooding events. Ultimately, this work creates a scientific foundation for smarter planning and safer decision-making in vulnerable communities.

## REFERENCES

Agresti, A. (2013). *Categorical Data Analysis* (3rd ed.). Wiley.

Beven, K., & Kirkby, M. J. (1979). A physically based, variable contributing area model of basin hydrology. *Hydrological Sciences Bulletin*, *24*(1), 43–69. DOI: 10.1080/02626667909491834

Bradley, A. P. (1997). The use of the area under the ROC curve in the evaluation of machine learning algorithms. *Pattern Recognition*, *30*(7), 1145–1159. DOI: 10.1016/S0031-3203(96)00142-2

Breiman, L. (2001). Random Forests. *Machine Learning*, *45*(1), 5–32. DOI: 10.1023/A:1010933404324

Bui, D. T., Nguyen, T. T., Hoang, N. D., & Pradhan, B. (2022). Flood susceptibility mapping using deep learning, machine learning, and AHP models in Quang Binh Province, Vietnam. *Environmental Monitoring and Assessment*, *194*(5), 349. DOI: 10.1007/s10661-022-10012-0 PMID: 35394223

Choubin, B., Moradi, E., Golshan, M., Adamowski, J., Sajedi-Hosseini, F., & Mosavi, A. (2019). An ensemble prediction of flood susceptibility using multivariate discriminant analysis, classification and regression trees, and support vector machines. *The Science of the Total Environment*, *651*, 2087–2096. DOI: 10.1016/j.scitotenv.2018.10.064 PMID: 30321730

Chow, V. T., Maidment, D. R., & Mays, L. W. (2010). *Applied Hydrology* (2nd ed.). McGraw-Hill Education.

Danumah, J. H., Odai, S. N., Saley, B. M., Szarzynski, J., Thiel, M., Kwaku, A., Kouame, F. K., & Akpa, L. Y. (2016). Flood risk assessment and mapping in Abidjan district using multi-criteria analysis model and geoinformation techniques, (Côte d'Ivoire). *Geoenvironmental Disasters*, *3*(1), 10. DOI: 10.1186/s40677-016-0044-y

Dixit, A. (2003). Floods and vulnerability in Nepal: Need for a paradigm shift. *Natural Hazards*, *28*(1), 155–179. DOI: 10.1023/A:1021134218121

Djalante, R. (2018). Review article: A systematic literature review of research trends and authorships on natural hazards, disasters, risk reduction and climate change in Indonesia, 3137–3153. *Natural Hazards and Earth System Sciences*, *18*(11), 3137–3153. DOI: 10.5194/nhess-18-3137-2018

Dormann, C. F., Elith, J., Bacher, S., Buchmann, C., Carl, G., Carré, G., Marquéz, J. R. G., Gruber, B., Lafourcade, B., Leitão, P. J., Münkemüller, T., McClean, C., Osborne, P. E., Reineking, B., Schröder, B., Skidmore, A. K., Zurell, D., & Lautenbach, S. (2013). Collinearity: A review of methods to deal with it and a simulation study evaluating their performance. *Ecography, 36*(1), 27–46. DOI: 10.1111/j.1600-0587.2012.07348.x

Farhadi, A., & Najafzadeh, M. (2021). Flood susceptibility mapping using Random Forest and remote sensing data in Iran. *Remote Sensing Applications: Society and Environment, 21*, 100475. DOI: 10.1016/j.rsase.2021.100475

Fawcett, T. (2006). An introduction to ROC analysis. *Pattern Recognition Letters, 27*(8), 861–874. DOI: 10.1016/j.patrec.2005.10.010

Funk, C., Peterson, P., Landsfeld, M., Pedreros, D., Verdin, J., Shukla, S., Husak, G., Rowland, J., Harrison, L., Hoell, A., & Michaelsen, J. (2015). The climate hazards infrared precipitation with stations—A new environmental record for monitoring extremes. *Scientific Data, 2*(1), 150066. DOI: 10.1038/sdata.2015.66 PMID: 26646728

Gessler, P. E., Moore, I. D., McKenzie, N. J., & Ryan, P. J. (1995). Soil-landscape modeling and spatial prediction of soil attributes. *International Journal of Geographical Information Systems, 9*(4), 421–432. DOI: 10.1080/02693799508902047

Han, J., Kamber, M., & Pei, J. (2011). *Data Mining: Concepts and Techniques* (3rd ed.). Morgan Kaufmann.

Hillel, D. (2003). *Introduction to Environmental Soil Physics*. Academic Press.

Hosmer, D. W., Lemeshow, S., & Sturdivant, R. X. (2013). *Applied Logistic Regression* (3rd ed.). Wiley. DOI: 10.1002/9781118548387

Ishizaka, A., & Labib, A. (2011). Review of the main developments in the Analytic Hierarchy Process. *Expert Systems with Applications, 38*(11), 14336–14345. DOI: 10.1016/j.eswa.2011.04.143

Kazakis, N., Giannakis, E., & Verveniotis, G. (2015). Assessment of flood hazard areas at a regional scale using an index-based approach and Analytical Hierarchy Process: Application in Rhodope–Evros region, Greece. *The Science of the Total Environment, 538*, 555–563. DOI: 10.1016/j.scitotenv.2015.08.055 PMID: 26318691

Khanal, N. R., Shrestha, M., & Ghimire, M. (2007). Preparing for floods: Mapping and assessing hazard in the Ratu watershed, Nepal. *ICIMOD Technical Paper, 7*, 1–34.

Khatakho, R., Kayastha, P., & Pathak, D. (2021). Multi-hazard risk assessment and mapping in Nepal using geospatial tools and AHP. *Geoscience Frontiers*, *12*(1), 101104. DOI: 10.1016/j.gsf.2020.08.006

Komi, K., Neal, J., Trigg, M. A., & Diekkrüger, B. (2017). Modelling of flood hazard extent in data sparse areas: A case study of the Oti River basin, West Africa. *Journal of Hydrology (Amsterdam)*, *555*, 330–345. DOI: 10.1016/j.jhydrol.2017.10.004

Kotsiantis, S. B. (2013). Decision trees: A recent overview. *Artificial Intelligence Review*, *39*(4), 261–283. DOI: 10.1007/s10462-011-9272-4

Kutner, M. H., Nachtsheim, C., Neter, J., & Li, W. (2005). *Applied Linear Statistical Models* (5th ed.). McGraw-Hill Education.

Liaw, A., & Wiener, M. (2002). Classification and regression by randomForest. *R News*, *2*(3), 18–22.

Liu, R., Chen, Y., Wu, J., Gao, L., Barrett, D., Xu, T., Li, L., Huang, C., & Yu, J. (2016). Assessing spatial likelihood of flooding hazard using naïve Bayes and GIS: A case study in Bowen Basin, Australia. *Stochastic Environmental Research and Risk Assessment*, *30*(6), 1575–1590. DOI: 10.1007/s00477-015-1198-y

Malczewski, J. (2006). GIS-based multicriteria decision analysis: A survey of the literature. *International Journal of Geographical Information Science*, *20*(7), 703–726. DOI: 10.1080/13658810600661508

Menard, S. (2002). *Applied Logistic Regression Analysis* (2nd ed.). Sage Publications. DOI: 10.4135/9781412983433

Moore, I. D., Grayson, R. B., & Ladson, A. R. (1991). Digital terrain modeling: A review of hydrological, geomorphological, and biological applications. *Hydrological Processes*, *5*(1), 3–30. DOI: 10.1002/hyp.3360050103

Moumane, A., Al Karkouri, J., & Batchi, M. (2025). Utilizing GIS, remote sensing, and AHP-multi-criteria decision analysis for optimal landfill site selection in Kenitra Province, Morocco: A step towards sustainable development goals. *Discover Environment*, *3*(1), 4. DOI: 10.1007/s44274-025-00183-0

Moumane, A., Enajar, A. A., Ghazali, F. E. E., Khouz, A., Karmaoui, A., Al Karkouri, J., & Batchi, M. (2024). *GIS, remote sensing, and analytical hierarchy process (AHP) approach for rainwater harvesting site selection in arid regions: Feija Plain case study, Zagora*. Applied Geomatics., DOI: 10.1007/s12518-024-00585-4

Nandi, A., Mandal, A., Wilson, M., & Smith, D. (2016). Flood hazard mapping in Jamaica using principal component analysis and logistic regression. *Environmental Earth Sciences, 75*(6), 465. DOI: 10.1007/s12665-016-5323-0

National Agricultural Research Council (NARC). (2020). *Soil Map Data for Nepal*. NARC.

O'Brien, R. M. (2007). A caution regarding rules of thumb for variance inflation factors. *Quality & Quantity, 41*(5), 673–690. DOI: 10.1007/s11135-006-9018-6

Paudel, Y., & Andersen, T. J. (2013). Monitoring flood and river dynamics in Nepal using remote sensing. *Remote Sensing of Environment, 136*, 198–210. DOI: 10.1016/j.rse.2013.05.010

Peng, C. Y. J., Lee, K. L., & Ingersoll, G. M. (2002). An introduction to logistic regression analysis and reporting. *The Journal of Educational Research, 96*(1), 3–14. DOI: 10.1080/00220670209598786

Perera, E. D. P., Hiroe, A., Shrestha, D., Fukami, K., Basnyat, D. B., Gautam, S., Hasegawa, A., Uenoyama, T., & Tanaka, S. (2015). Community-based flood damage assessment approach for lower West Rapti River basin in Nepal under the impact of climate change. *Natural Hazards, 75*(1), 669–699. DOI: 10.1007/s11069-014-1339-5

Portal, B. I. P. A. D. (2023). Building Information Platform Against Disaster. Government of Nepal. Retrieved from https://bipadportal.gov.np

Pourghasemi, H. R., Moradi, S., & Pradhan, B. (2012). Flood susceptibility mapping using GIS-based support vector machine model. *Journal of Hydrology (Amsterdam), 460–461*, 103–113. DOI: 10.1016/j.jhydrol.2012.06.033

Prasad, A. M., Iverson, L. R., & Liaw, A. (2006). Newer classification and regression tree techniques: Bagging and random forests for ecological prediction. *Ecosystems (New York, N.Y.), 9*(2), 181–199. DOI: 10.1007/s10021-005-0054-1

Quinlan, J. R. (1993). *C4.5: Programs for Machine Learning*. Morgan Kaufmann.

Rahmati, O., Pourghasemi, H. R., & Zeinivand, H. (2019). Flood susceptibility mapping using frequency ratio and weights-of-evidence models in the Golastan Province, Iran. *Geocarto International, 31*(1), 42–70. DOI: 10.1080/10106049.2015.1041559

Rawls, W. J., Pachepsky, Y. A., Ritchie, J. C., Sobecki, T. M., & Bloodworth, H. (2003). Effect of soil organic carbon on soil water retention. *Geoderma, 116*(1–2), 61–76. DOI: 10.1016/S0016-7061(03)00094-6

Rodriguez-Galiano, V. F., Ghimire, B., Rogan, J., Chica-Olmo, M., & Rigol-Sanchez, J. P. (2012). An assessment of the effectiveness of a random forest classifier for land-cover classification. *ISPRS Journal of Photogrammetry and Remote Sensing*, *67*, 93–104. DOI: 10.1016/j.isprsjprs.2011.11.002

Rouse, J. W., Haas, R. H., Schell, J. A., & Deering, D. W. (1974). Monitoring vegetation systems in the Great Plains with ERTS. *NASA Special Publication*, *351*, 309.

Saaty, T. L. (1980). *The Analytic Hierarchy Process*. McGraw-Hill.

Saaty, T. L. (2008). Decision making with the Analytic Hierarchy Process. *International Journal of Services Sciences*, *1*(1), 83–98. DOI: 10.1504/IJSSCI.2008.017590

Sanyal, J., & Lu, X. X. (2004). Application of remote sensing in flood management with special reference to monsoon Asia: A review. *Natural Hazards*, *33*(2), 283–301. DOI: 10.1023/B:NHAZ.0000037035.65105.95

Sharifi, E., Pradhan, B., & Pourghasemi, H. R. (2019). GIS-based flood susceptibility mapping using multi-criteria decision analysis. *Catena*, *173*, 501–512. DOI: 10.1016/j.catena.2018.10.030

Shrestha, M., Acharya, S. C., & Shrestha, P. K. (2020). Climate change impact on water resources in Nepal: A review. *Journal of Water and Climate Change*, *11*(2), 281–301. DOI: 10.2166/wcc.2019.080

Stefanidis, S., & Stathis, D. (2013). Assessment of flood hazard based on natural and anthropogenic factors using AHP. *Natural Hazards*, *68*, 569–585. DOI: 10.1007/s11069-013-0639-5

Tehrany, M. S., Pradhan, B., & Jebur, M. N. (2014). Flood susceptibility mapping using a novel ensemble weights-of-evidence and support vector machine models in GIS. *Journal of Hydrology (Amsterdam)*, *512*, 332–343. DOI: 10.1016/j.jhydrol.2014.03.008

Tucker, C. J. (1979). Red and photographic infrared linear combinations for monitoring vegetation. *Remote Sensing of Environment*, *8*(2), 127–150. DOI: 10.1016/0034-4257(79)90013-0

Ullah, S., & Zhang, J. (2020). GIS-based flood hazard mapping using relative frequency ratio method: A case study from eastern Hindu Kush, Pakistan. *Journal of Flood Risk Management*, *13*(2), e12587. DOI: 10.1111/jfr3.12587 PMID: 32210424

Wu, Y., & Jiang, Q. (2024). Flood risk assessment using Random Forest: A case study of Wuhan City, China. *Water (Basel)*, *16*(2), 321. DOI: 10.3390/w16020321

Xu, H. (2006). Modification of normalised difference water index (NDWI) to enhance open water features in remotely sensed imagery. *International Journal of Remote Sensing*, *27*(14), 3025–3033. DOI: 10.1080/01431160600589179

Zevenbergen, L. W., & Thorne, C. R. (1987). Quantitative analysis of land surface topography. *Earth Surface Processes and Landforms*, *12*(1), 47–56. DOI: 10.1002/esp.3290120107

Zhang, D., Xu, H., Wang, T., & He, Q. (2019). Flood susceptibility mapping using machine learning models based on remote sensing big data. *Remote Sensing*, *11*(19), 2283. DOI: 10.3390/rs11192283

Zhang, L., Dawes, W. R., & Walker, G. R. (2008). Response of mean annual evapotranspiration to vegetation changes at catchment scale. *Water Resources Research*, *37*(3), 701–708. DOI: 10.1029/2000WR900325

# Chapter 4
# Land Use and Environmental Dynamics of the Moroccan Coastal Strip 1984–2025:
## A Google Earth Engine Approach for Monitoring Vegetation, Water, and Urban Transformation – Case Study

**Tarik Bahouq**
https://orcid.org/0009-0000-6678-693X
*Ibn Tofail University, Morocco*

**Amina Moumane**
https://orcid.org/0009-0001-9188-6964
*Faculté des Sciences Humaines et Sociales, Departement de Geographie, Ibn Tofail University, Morocco*

**Nadia Touhami**
*Faculté des Sciences Humaines et Sociales, Departement de Geographie, Ibn Tofail University, Morocco*

## ABSTRACT

*This study examines land use, vegetation cover, and water body changes in the Moulay Bousselham area of Kenitra Province, focusing on the impact of agricultural activities on the Merja Zerga wetland. Using multi-source satellite imagery (Landsat*

DOI: 10.4018/979-8-3373-6608-1.ch004

*and Sentinel) processed in Google Earth Engine, vegetation dynamics were tracked from 2014 to 2024, and water surface changes from 1984 to 2024. Results reveal a steady annual increase of about 9.5 km² in vegetation cover, driven by intensive farming and modern irrigation that boosted productivity and mitigated desertification. Conversely, the water surface of Lake Merja Zerga declined by 0.0352 km² per year, threatening aquatic habitats, water quality, and local climate regulation. Land use analysis highlights urban expansion, a shift to intensive production systems, and forest degradation. These findings stress the need for sustainable land management and wetland conservation to balance development with ecosystem protection.*

## I. INTRODUCTION

Morocco, by virtue of its strategic geographic location at the crossroads of the Mediterranean Sea and the Atlantic Ocean (Sharma et al. 2025), and thanks to its diverse climatic and natural characteristics (Boyer 2025), possesses an important network of wetlands that represent a significant environmental and ecological asset. These wetlands include rivers, lakes, and dams, and are characterized by their rich biological diversity. They provide a natural habitat for numerous plant and animal species, including rare and endangered ones (Ennabili et al. 2025). Wetlands also play a vital role in maintaining ecological balance by regulating the water cycle, recharging groundwater aquifers, and mitigating floods. Moreover, they support local economic activities such as agriculture, livestock breeding, and traditional fishing.

At the same time, Morocco experiences intensive agricultural development, particularly along its coastal strip. The Gharb region, for instance, benefits from fertile soils, substantial water resources supplied by major rivers such as the Sebou and aquifers like the Gharb aquifer (Amina Moumane et al. 2025), and a mild climate with relatively abundant rainfall and moderate temperatures. These natural and climatic advantages have made the region one of Morocco's most productive agricultural zones, supporting diverse systems of production such as vegetable cultivation, fruit tree plantations, red berry farming, and livestock breeding (Sharma et al. 2025). Agriculture in this area has increasingly shifted toward intensive and market-oriented systems, contributing significantly to the national economy.

However, this agricultural intensification has generated mounting environmental challenges. The expansion of irrigated farming and the widespread use of chemical fertilizers and pesticides have exerted significant pressure on fragile wetland ecosystems (Asawra and Koli 2025). In the case of the Merja Zerga wetland in Moulay Bousselham, agricultural encroachment has led to groundwater depletion(Elmotawakkil and Enneya 2024; Moumane et al. 2021), pollution of water bodies(Rad et al. 2022), and the conversion of wetland margins into farmland (Bahouq et al.

2024). These processes threaten biodiversity, degrade water quality, and diminish the wetland's capacity to provide essential ecological services such as water purification, flood regulation, and habitat provision.

The growing tension between agricultural expansion and wetland conservation underscores a pressing environmental issue in Morocco (Arabi et al. 2024): how to reconcile food production needs with the preservation of vital ecosystems((Purba et al. 2024) Despite the ecological and socio-economic importance of Merja Zerga, systematic monitoring of land use change and its impacts remains limited (Thompson et al. 2009). This knowledge gap hinders the development of effective management strategies and policies aimed at ensuring the sustainable coexistence of agricultural development and wetland protection (He et al. 2025).

To address this challenge, the present study focuses on monitoring and analyzing land use changes in the Moulay Bousselham area, within Kenitra Province. Using multi-source satellite imagery such as Landsat and Sentinel, processed through the Google Earth Engine platform (Adil Moumane et al. 2025; Chafiq et al. 2024), the research tracks the evolution of agricultural activities and wetland transformations over recent decades. The goal is to assess the extent and nature of human-induced changes in Merja Zerga, provide insights into the dynamics of the area, and evaluate the implications for ecological sustainability and environmental management (Rehman et al. 2021).

## II. METHODOLOGY WORK FLAW

### 2.1 Study Area

Merja Zerga (also called the Moulay Bousselham Lagoon) is located in northwestern Morocco on the Atlantic coast. It lies about 70 km north of Kenitra and approximately 130 km north of Rabat (Alaoui et al. 2010), with the village of Moulay Bousselham as its closest settlement. The lagoon covers about 7,300 hectares (73 km$^2$), measuring around 9 km in length and 5 km in width. About 30% of this area is open water.

Merja Zerga was declared a Permanent Biological Reserve in 1978 by ministerial decree, and in 1980 it was designated a Ramsar site (Erraj Chahid et al. 2023), recognizing its global importance as a wetland. It hosts a rich ecological system with high biodiversity: more than 100 bird species regularly use the site, especially in winter, when it supports tens of thousands of waterfowl and wader birds such as ducks, flamingos, and other migratory species (Ayache et al. 2009). Among its fauna are threatened species like the Moroccan spadefoot toad (Pelobates varaldii) and the European eel (Anguilla anguilla). The vegetation is dominated by salt-tolerant and marsh plants, while aquatic species, mollusks, and fish are found particularly in the channels and areas connected to the sea. The lagoon is also influenced strongly by tidal movement and is connected to the ocean via a permanent channel. Deep ecological, hydrological, and socio-economic functions are under pressure from agricultural expansion (Ayache et al. 2009), overuse of water resources, pollution, and habitat alteration, underscoring the urgent need for sustainable management.

*Figure 1. Location of the study area.*

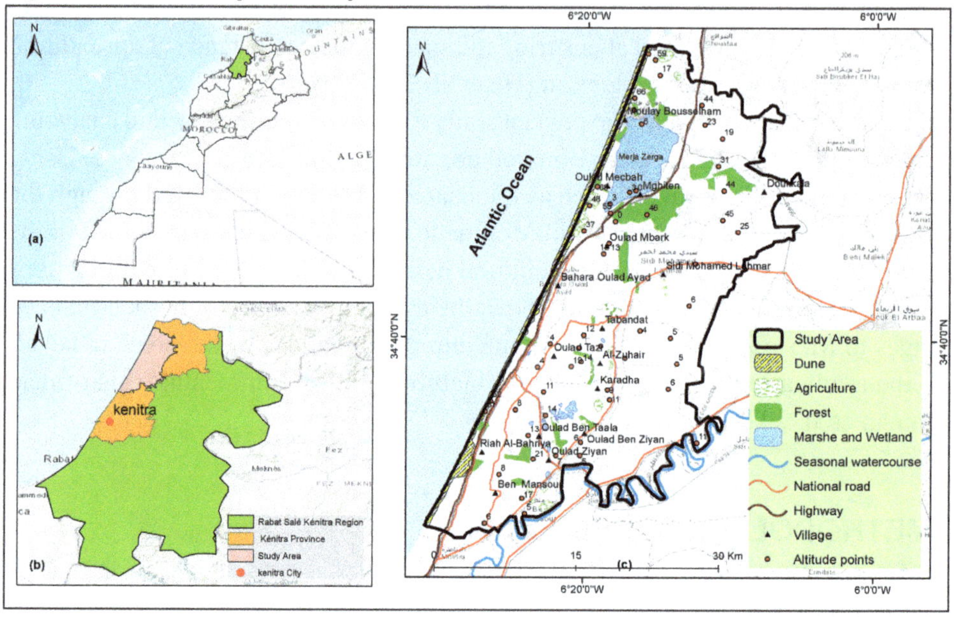


## 2.2 Data Collection and Preprocessing

The Google Earth Engine platform was utilized for efficient geospatial data processing. Satellite imagery, such as Landsat 8/9 or Sentinel-2, available on the platform, was employed. Preliminary processing included applying temporal and spatial filters to delineate the coastal study area, along with atmospheric correction and cloud removal to ensure data quality.

### Extraction of Spectral Indices and Classification

Specialized spectral indices were directly extracted on the platform, including:

- NDVI for vegetation cover
- NDWI for water bodies
- MNDWI for precise water discrimination

Classification models (Random) were trained using training samples collected through Google Earth Engine, leveraging the platform's immense computational power for processing large-scale data.

## Accuracy Validation and Output of Results

Accuracy assessment techniques were applied using independent sample data, and performance metrics such as Overall Accuracy and Confusion Matrix were calculated. Finally, the final maps and area statistics for each land use class were exported, with results available for download to Google Drive or direct visualization on the platform.

*Figure 2. Methodology Work Flaw*

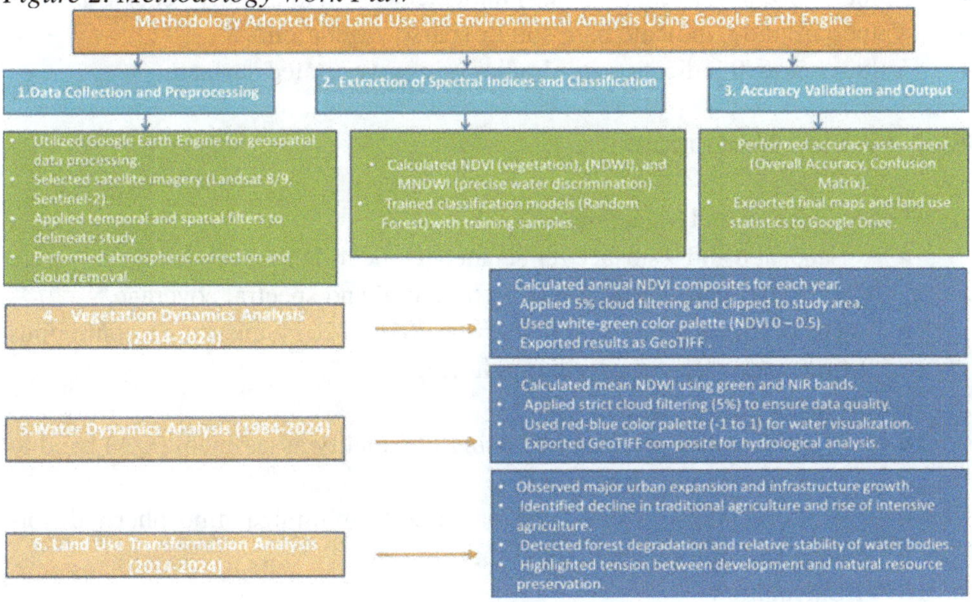


## 2.3 vegetation dynamics

We adopted a comprehensive analytical methodology to study vegetation dynamics in the study area by calculating the Normalized Difference Vegetation Index

(NDVI) over the period 2014–2024 using the Google Earth Engine (GEE) platform. The NDVI is a widely recognized spectral index in remote sensing that measures vegetation vigor and density by exploiting the difference in reflectance between the near-infrared (NIR) and the red (RED) spectral bands.

The NDVI is calculated according to the following standard formula:

$$NDVI = \frac{(NIR - Green\ )}{(NIR + RED)} \tag{1}$$

Where:

- NIR represents the reflectance in the near-infrared band (for Landsat 8/9: SR_B5)
- RED represents the reflectance in the red band (for Landsat 8/9: SR_B4)

This formula produces values ranging between -1 and +1:

- Values close to +1 (e.g., 0.6–0.9) indicate dense and healthy vegetation.
- Values around 0 indicate bare soil or sparse vegetation.
- Values below 0 indicate water bodies or non-vegetated surfaces.

## Step-by-Step Methodology

1. Data Collection and Selection
   - Landsat 8 and 9 surface reflectance (Collection 2 Level-2) imagery were selected to ensure high-quality temporal and spectral coverage.
   - The period of analysis was set from 2014 to 2024, ensuring a continuous time series for vegetation monitoring.
2. Preprocessing and Filtering
   - Cloud masking was applied using the quality assessment (QA) band to remove contaminated pixels.
   - A 5% cloud cover threshold was used to minimize atmospheric distortions and improve data reliability.
3. NDVI Calculation
   - For each image, the NDVI was computed using the above formula, with SR_B5 (NIR) and SR_B4 (RED) bands as inputs.
   - The resulting NDVI raster layers were validated for consistency across multiple time steps.
4. Annual Mean Composites
   - NDVI values were aggregated for each year to create annual mean composites, which represent average vegetation conditions over time.

- These composites minimized the impact of seasonal fluctuations and isolated long-term vegetation dynamics.

5. Geographical Clipping
   - All composites were spatially clipped to the Moulay Bousselham study area, focusing analysis on the Merja Zerga wetland and surrounding agricultural lands.
6. Visualization
   - NDVI maps were visualized using a white-to-green color palette, where white represented low NDVI values (0), and deep green represented higher NDVI values (0.5 or more).
   - This visual scheme highlighted spatial variations in vegetation vigor.
7. Data Export and Storage
   - Final annual NDVI composites were exported in GeoTIFF format, allowing integration with GIS software for further spatial analysis and environmental assessment.

## 2.4 Water Dynamics

We adopted a comprehensive analytical methodology to study water dynamics using the Google Earth Engine platform, through which we calculated and exported the mean Normalized Difference Water Index (NDWI) for the specified study area for the year 2018.

The methodology involved integrating multi-sensor data from Landsat 8 and Landsat 9 satellites, applying cloud filtering techniques at 5% to ensure data quality, and implementing the standardized NDWI calculation formula using the green (SR_B3) and near-infrared (SR_B5) spectral bands. The results were visualized using a color palette ranging from red to blue, reflecting NDWI values from -1 to 1, with the final composite exported as a GeoTIFF file for further hydrological analysis.

The technical approach featured several important methodological aspects in remote sensing-based water monitoring:

- Integration of multiple Landsat missions (8 and 9) to enhance temporal coverage
- Application of rigorous cloud filtering to minimize atmospheric interference
- Implementation of the standardized NDWI algorithm for water body detection
- Utilization of spatial clipping to focus on the specified study area
- Employment of automated export functionality for data preservation

This analytical approach provided valuable insights into water presence and distribution patterns, contributing to monitoring surface water dynamics, assessing water body extent, and tracking hydrological changes within the study area. The NDWI composite served as fundamental input for numerous hydrological applications, including water resource management, flood monitoring, drought assessment, and aquatic ecosystem studies.

The color-coded visualization palette enabled clear interpretation of results, with blue tones indicating water presence, green representing vegetation with high water content, yellow showing moist surfaces, and red indicating dry land areas. This methodological framework demonstrated the effective application of cloud-based remote sensing platforms for sustainable water resource monitoring and management.

## Mathematical Basis of NDWI

The NDWI was calculated using the standard McFEETERS (1996)

$$NDWI = \frac{(Green - NIR)}{(Green + NIR)} \quad (2)$$

- GREEN = reflectance in the green band (Landsat 8/9: SR_B3)
- NIR = reflectance in the near-infrared band (Landsat 8/9: SR_B5)
- The resulting NDWI values range from -1 to +1:
- Values close to +1 indicate open water bodies.
- Values around 0 indicate moist soil or transitional areas.
- Negative values indicate non-water features such as vegetation, bare soil, or built-up surfaces.

### *Step-by-Step Implementation in GEE*

1. Data Collection: Landsat 8 and Landsat 9 images (Collection 2, Level-2 SR) for the year 2018 were imported.
2. Cloud Filtering: A 5% cloud cover threshold was applied, and QA bands were used to mask residual cloud contamination.
3. NDWI Calculation: The index was computed per image using the SR_B3 (GREEN) and SR_B5 (NIR) bands.
4. Annual Composite: An annual mean NDWI composite was generated to represent water body distribution during 2018.
5. Spatial Clipping: The results were clipped to the Moulay Bousselham study area, focusing on the Merja Zerga wetland.
6. Visualization: A red-to-blue color ramp was applied, with blue highlighting water bodies and red indicating dry land.
7. Export: The composite was exported as a GeoTIFF file for integration into GIS platforms and hydrological analysis.

*Figure 3. Land Cover Dynamics (2013-2024)*

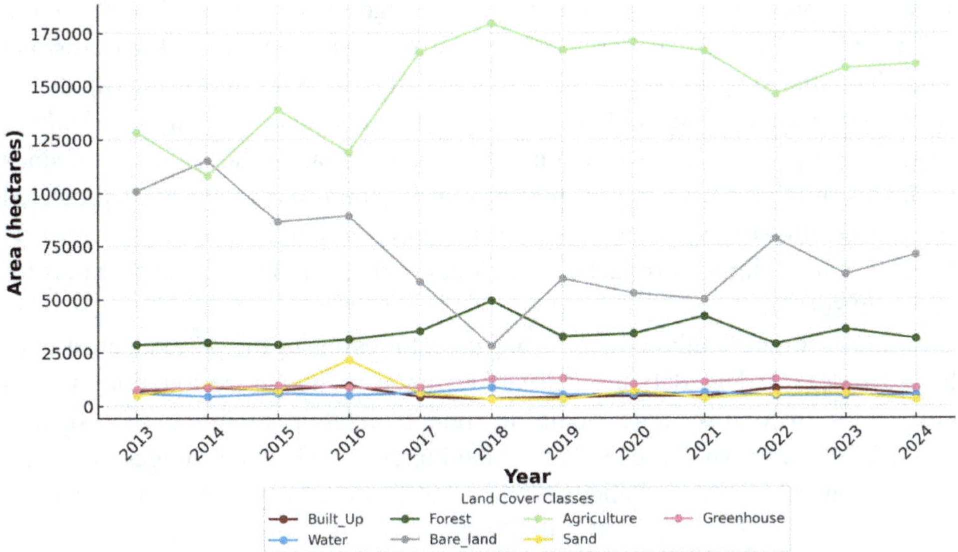


## 2.5 Data Collection of Validation Points and Accuracy Assessment Procedure

A total of 200 reference points were collected during 2024 from the study area using field surveys, high-resolution satellite imagery, and expert local knowledge. The sampling strategy was proportional, assigning more points to classes covering large extents (agriculture, forest, built-up) and fewer to minor classes (water, greenhouse).

Each reference point was cross-compared with the classified LULC map for 2024. The comparison produced a confusion matrix, from which key accuracy indices were calculated, including:

- Overall Accuracy
- Producer's and User's Accuracy (per class)

## 2.6 Land Use Transformation Analysis (2013-2024)

The period from 2013 to 2024 witnessed a fundamental transformation in land use patterns, characterized by significant shifts in urban and agricultural landscapes. Urban expansion emerged as the dominant driving force during this decade, demonstrating substantial and consistent growth in built-up areas. This urban growth reflects both horizontal city expansion and population growth, accompanied by developments in infrastructure and economic activities. The conversion of substantial land areas to urban uses represents one of the most notable changes in the territorial organization.

The agricultural sector underwent a qualitative transformation marked by distinct trends. Traditional agricultural areas experienced a sharp decline, primarily due to urban encroachment and land conversion. Simultaneously, intensive agriculture emerged as an alternative production model, demonstrating rapid expansion since 2018. This shift indicates a transition from seasonal rain-fed agriculture to protected year-round cultivation systems. The move toward agricultural intensification reflects efforts to achieve higher production efficiency and respond to market demands for consistent supply.

Natural resources exhibited contrasting trajectories during this period. Forest areas showed noticeable degradation, resulting from intensive human exploitation and land use changes. In contrast, water bodies maintained relative stability in their surface extent. Sandy areas and bare lands remained largely unchanged, suggesting either limited economic value or challenges in their utilization for development purposes.

This transformation highlights the inherent tension between urban and economic development requirements on one hand, and the necessity of preserving natural heritage on the other. The reduction in forested areas presents clear environmental challenges, including biodiversity loss and ecosystem degradation. Meanwhile, the shift toward more intensive agricultural practices represents an adaptive response to development pressures, aiming to maintain agricultural productivity within a changing land use context. These changes reflect broader patterns of economic development and environmental adaptation occurring across various regions.

*Figure 4. Vegetation Dynamics*

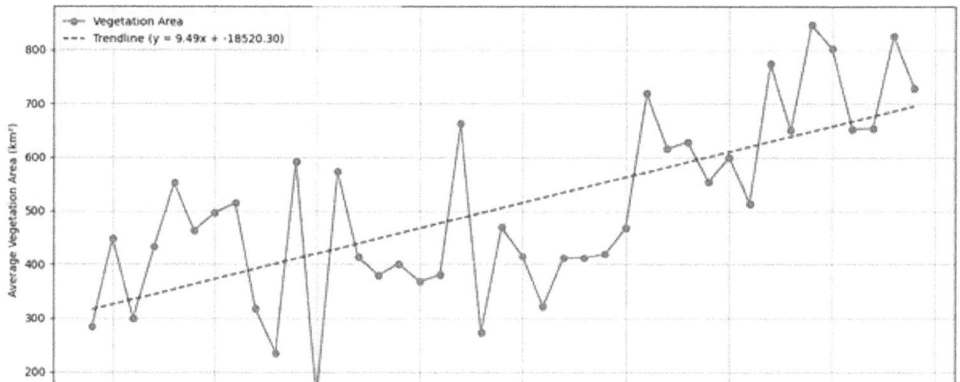

*Figure 5. NDVI distribution in the Moulay Bousselham–Merja Zerga region (2014, 2020 and 2024)*

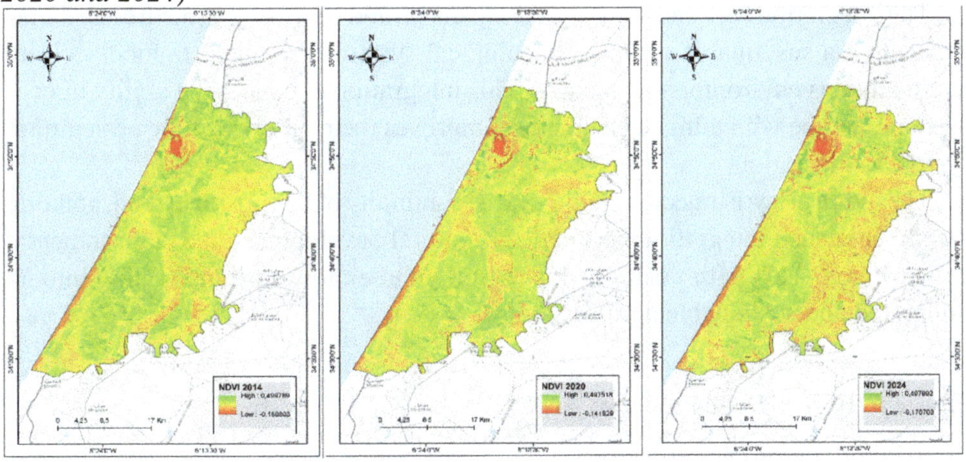


The vegetation cover in the coastal strip of the West has witnessed a remarkable transformation during the period from 1985 to 2025, revealing an interesting dynamic in the evolution of green spaces alongside the expansion of agricultural areas. The general trend curve (y = 9.49x - 18520.30) shows a steady upward pattern with an annual increase rate of approximately 9.5 square kilometers, indicating a noticeable improvement in the plant ecosystem.

This vegetative expansion is organically linked to the development of the agricultural sector in the region, where modern agricultural practices have contributed to creating an integrated ecological system. The transition to intensive farming and localized irrigation systems has increased cultivated areas and improved production efficiency, positively impacting the surrounding environment.

Integrated agriculture has enhanced vegetation cover through the adoption of practices such as intercropping, crop rotation, and the establishment of windbreaks. These methods have not only improved productivity but have also enhanced biodiversity and contributed to the formation of a denser and more sustainable plant structure.

Organized agricultural expansion plays a pivotal role in combating desertification and the encroachment of sand toward coastal areas. Agricultural projects have contributed to soil stabilization and reduced erosion, providing a suitable environment for the spread of natural plants and the increase of pasture areas.

Temporal analysis shows that the acceleration in vegetation growth since the beginning of the third millennium coincides with agricultural support policies and rural development programs. Improvements in agricultural infrastructure and the adoption of modern irrigation techniques have enhanced environmental conditions supporting plant growth.

The integration between agricultural expansion and vegetative growth contributes to creating a sustainable economic-ecological system that supports food security and maintains environmental balance. This integration enhances the ability of ecosystems to cope with climate change and improves the resilience of the agricultural system in the region.

The future growth trajectory indicates the continuity of this positive trend, pointing toward enhanced integration between agricultural development and environmental conservation. This path positions the region on track toward transitioning into an integrated and sustainable agro-ecological system.

*Figure 6. Waters Dynamics (1984-2024)*

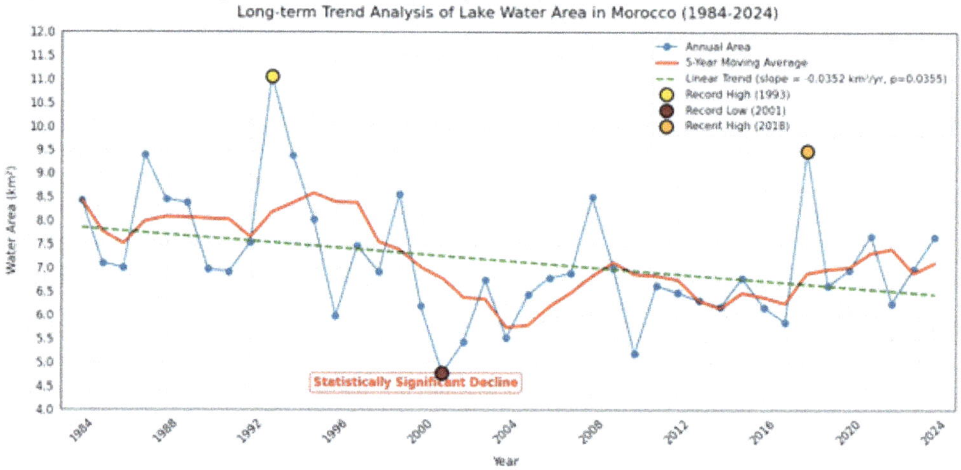

*Figure 7. NDWI of Lake Merja Zerga (2014, 2020 and 2024)*

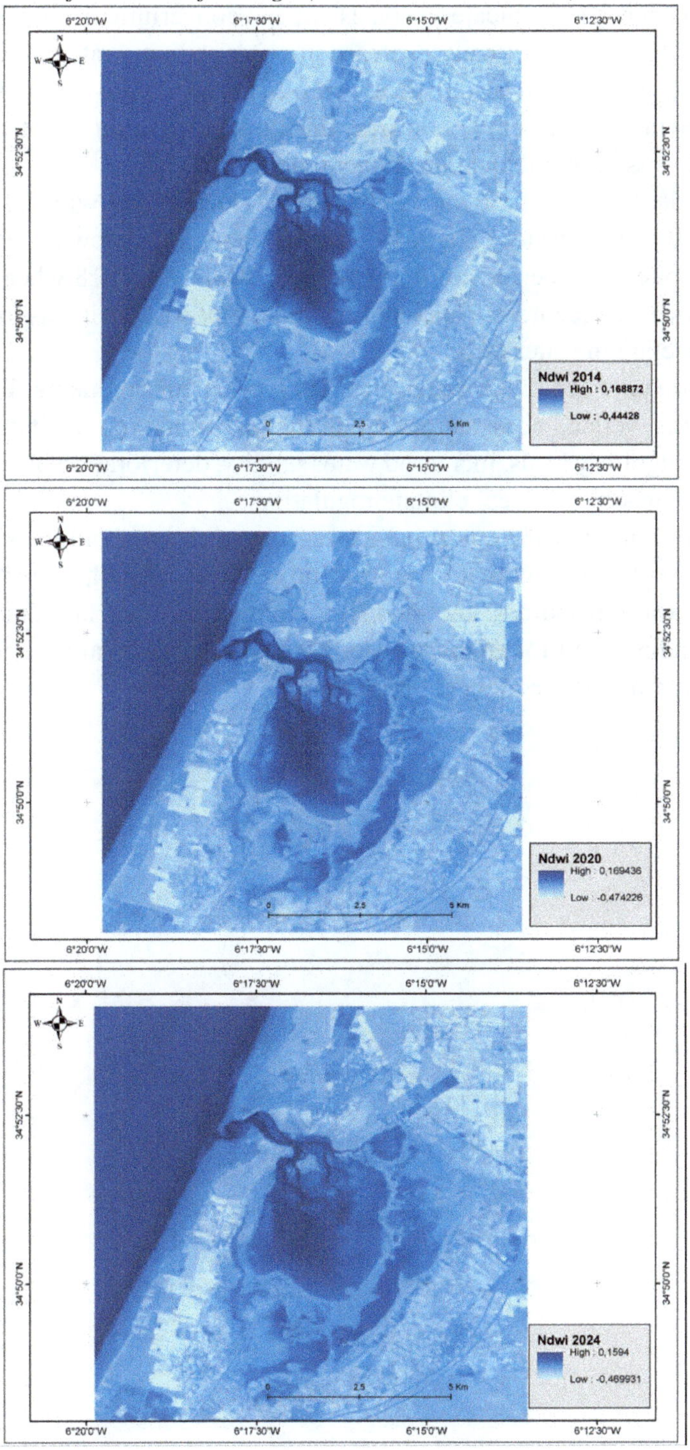

The data extracted from the analysis of the water surface area of Lake Merja Zerga between 1984 and 2024 indicates a consistent and concerning declining trend. The general trend line shows a steady decrease in water area at a rate of 0.0352 square kilometers per year, which is statistically significant with a p-value of 0.0355. The five-year moving average curve confirms this downward trend over the long term, although it is less severe than annual fluctuations.

The four-decade period was marked by notable fluctuations, with the lake recording its highest water surface area in 1993, while reaching its lowest recorded level in 2001. Despite some periods of improvement, such as in 2018 when a relatively high water area was recorded, this recovery did not persist, as it was followed by a noticeable decline in water area.

This continuous decline in water area has serious environmental implications, threatening the wetland ecosystem through the loss of natural habitats for aquatic species and migratory birds, increased water salinity, deterioration of water quality, and reduced capacity for local climate regulation.

This trend aligns with the water challenges facing many wetland areas in Morocco, resulting from changing climate patterns and reduced rainfall, increasing human pressures on water resources, and intensive exploitation of surface and groundwater. These findings call for enhanced monitoring and conservation efforts for this wetland of international ecological importance.

## III. RESULTS

### 3.1 Land Use and Land Cover 2014

*Figure 8. Land Use and Land Cover 2014*

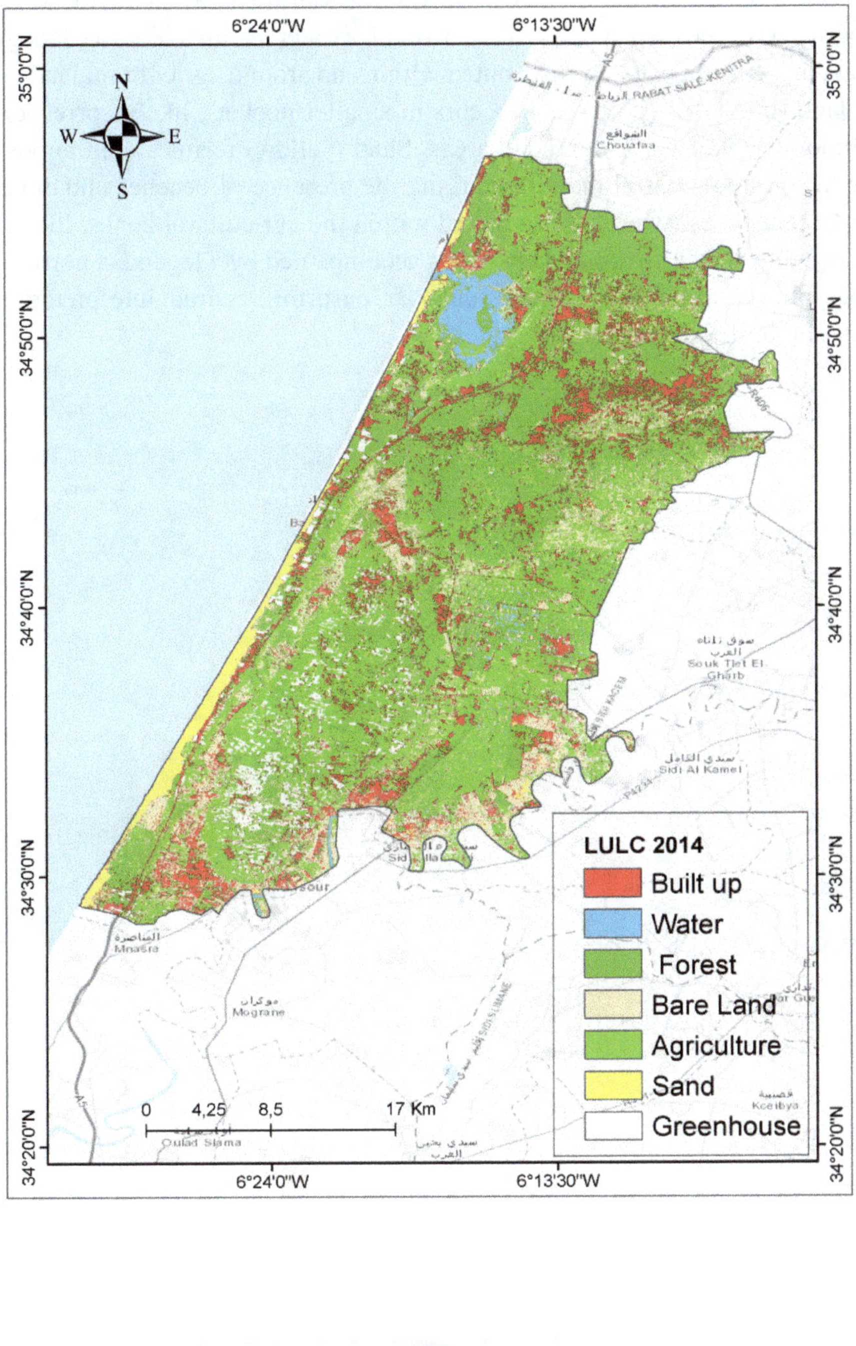

The Land Use/Land Cover (LULC) map of 2014 provides a detailed representation of the spatial distribution of land cover types in the Kénitra–Merja Zerga coastal region. The dominant class is agriculture (green), which extends widely across the inland areas and covers the majority of the mapped surface. Built-up zones (red) are visible around towns, villages, and transport corridors, particularly along the coastal strip, indicating areas occupied by residential, industrial, and service infrastructures. Water bodies (blue) are represented mainly by the Merja Zerga lagoon, located in the northern part of the map, as well as other smaller water features. Forest (dark green) appears in localized patches, distributed within and around agricultural landscapes. Bare land (brown) is limited and occurs in smaller pockets, likely corresponding to temporarily unused or degraded areas. Sand (yellow) forms a continuous strip along the western coastal margin, marking the presence of beaches and dune systems. Greenhouses (white) are scattered within the agricultural zones, illustrating areas of protected cultivation. The map is accompanied by a legend, a north arrow, a geographic coordinate grid, and a scale bar, ensuring accurate interpretation and orientation of the spatial data.

## 3.2 Land Use and Land Cover 2020

*Figure 9. Land Use and Land Cover 2020*

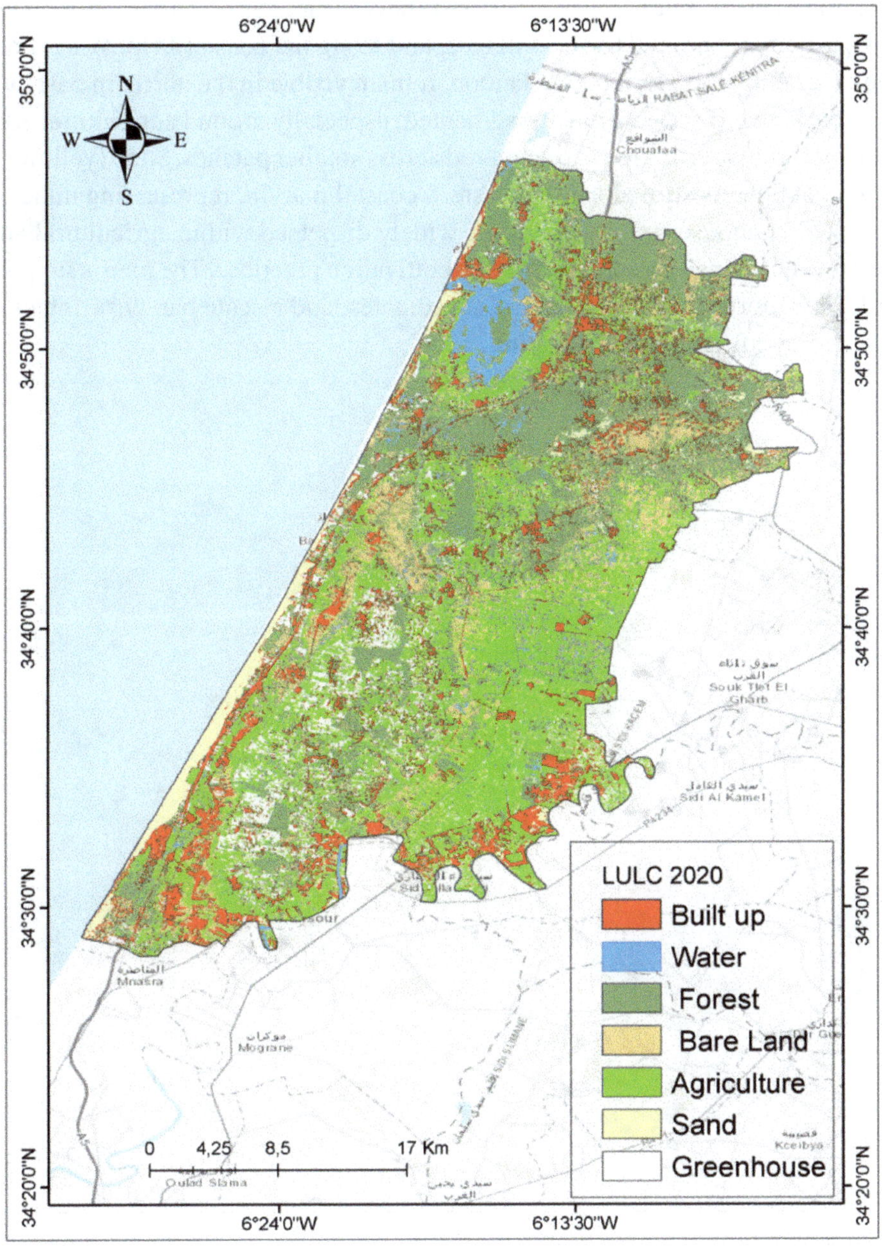

The Land Use/Land Cover (LULC) map for the year 2020 illustrates the spatial distribution of different land cover classes in the Kénitra–Merja Zerga coastal region. Agriculture (green) continues to dominate the landscape, covering large portions of the inland areas, but built-up areas (red) have noticeably expanded compared to 2014, particularly around towns, villages, and along the coastal strip. Water bodies (blue), including the Merja Zerga lagoon, remain visible in the northern part, while forest (dark green) appears more fragmented, especially around agricultural zones. Bare land (brown) is limited and scattered across smaller patches. Sand (yellow) still forms a continuous strip along the western coastal margin, representing dunes and beaches. Greenhouses (white) are more widely dispersed within agricultural areas, reflecting the intensification of protected cultivation practices. The map is supported by a legend, north arrow, geographic coordinates, and a scale bar, ensuring clarity and accuracy in spatial interpretation.

## 3.3 Land Use and Land Cover 2024

*Figure 10. Land Use and Land Cover 2024*

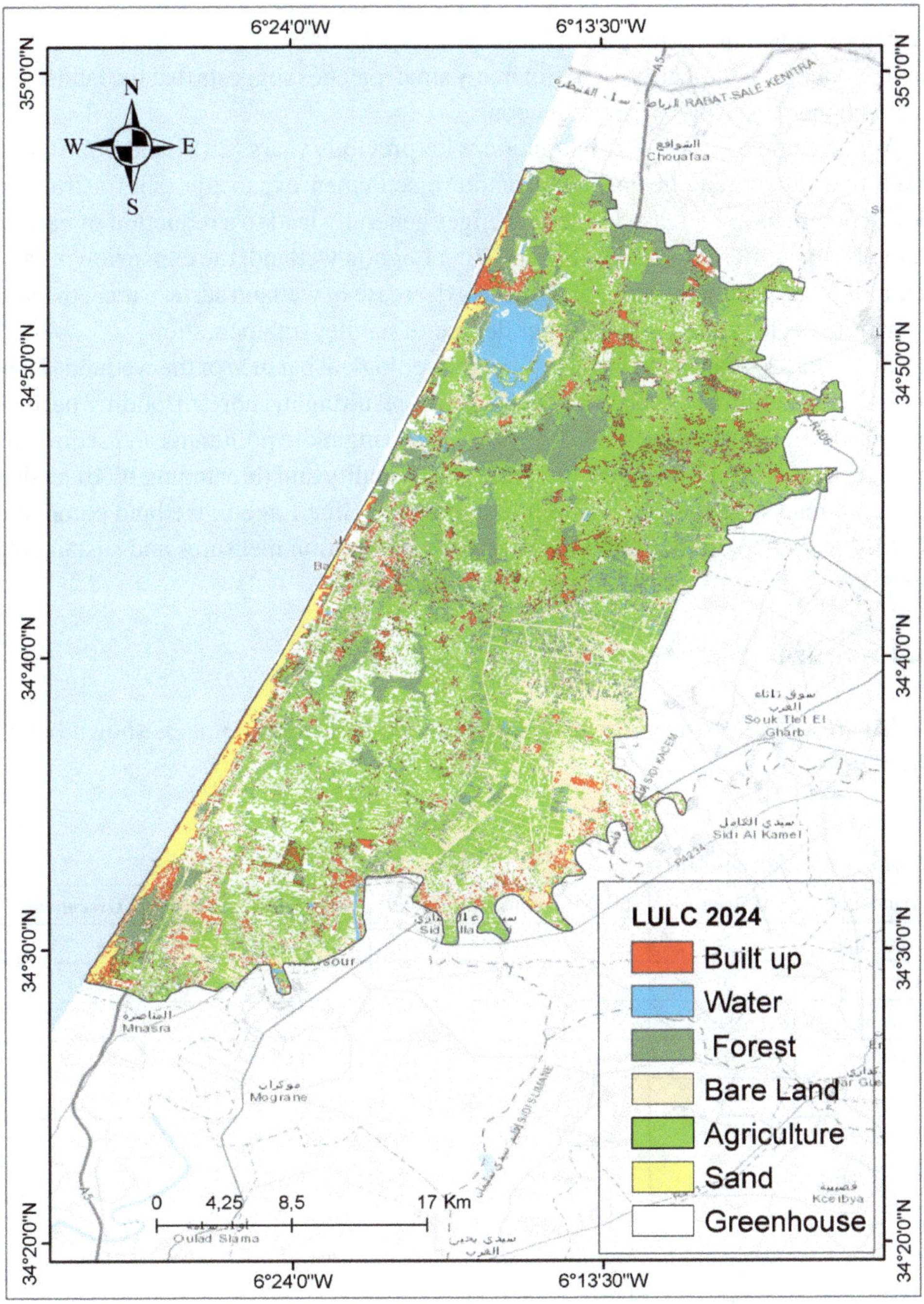

The land use map of 2024 shows a landscape dominated by several large classes, with one major class covering more than one fifth of the total area as a single large block, and another class that is widespread but fragmented into many small patches. This pattern indicates a mix of extensive land cover types, possibly agriculture or natural vegetation, along with more scattered features such as built-up areas or smaller land parcels. The presence of many small patches suggests that the landscape is fragmented and highly heterogeneous.

When comparing the 2024 situation with previous years (2014 and 2020), it is likely that human activities such as agriculture, settlement expansion, or infrastructure development have increased. These changes generally lead to a reduction of natural habitats, particularly wetlands. For the Blue Lagoon wetland, the expansion of land use around its perimeter probably caused a decrease in wetland surface area, the loss of natural buffer zones, and a higher degree of habitat fragmentation.

Such changes can significantly affect the ecological balance of the wetland. They may alter water inflows and drainage patterns, resulting in shorter flooding periods or even drying of some sections. They can also bring more pollutants and sediments from nearby agricultural lands, reducing water quality and threatening biodiversity, the 2024 map suggests increasing pressure on the Blue Lagoon wetland compared to 2014 and 2020, highlighting the need for conservation measures and sustainable land management to protect its ecological functions.

## 3.4 Accuracy Assessment

The resulting confusion matrix based on 200 validation points is shown below (Table 1):

*Table 1. Confusion Matrix*

| True Class | Agriculture | Forest | Built up | Sand | Bare Land | Water | Greenhouse |
|---|---|---|---|---|---|---|---|
| Agriculture | 56 | 1 | 2 | 3 | 2 | 4 | 2 |
| Forest | 0 | 32 | 2 | 2 | 3 | 0 | 1 |
| Built up | 1 | 1 | 32 | 0 | 0 | 4 | 2 |
| Sand | 0 | 0 | 1 | 16 | 1 | 2 | 0 |
| Bare Land | 2 | 0 | 0 | 0 | 11 | 0 | 1 |
| Water | 0 | 0 | 2 | 0 | 0 | 8 | 0 |
| Greenhouse | 0 | 0 | 1 | 0 | 1 | 0 | 4 |

- Overall Accuracy: 79.5%

- Kappa Coefficient: 0.74 (substantial agreement)

*Table 2. Class-wise Accuracy*

| Class | Producer's Accuracy | User's Accuracy |
|---|---|---|
| Agriculture | 0.800 | 0.949 |
| Forest | 0.800 | 0.941 |
| Built up | 0.800 | 0.800 |
| Sand | 0.800 | 0.762 |
| Bare Land | 0.786 | 0.611 |
| Water | 0.800 | 0.444 |
| Greenhouse | 0.667 | 0.400 |

The classification demonstrates high accuracy with a substantial Kappa value, confirming the robustness of the LULC map. Agriculture and forest show the strongest reliability (user's accuracy >94%), while built-up and sand are moderately accurate but subject to confusions with agriculture and bare land (Table 2). Water and greenhouse have the lowest accuracies, reflecting their smaller spatial extent and spectral overlap with adjacent land covers

# IV. DISCUSSION

## 4.1. Agricultural Intensification and Vegetation Change

Over recent decades, the Moulay Bousselham–Merja Zerga region has witnessed substantial increases in vegetation cover, largely driven by intensified agricultural practices such as modern irrigation, intercropping, and protected cultivation. These changes have boosted productivity and supported rural development programs, contributing to efforts against desertification in the coastal strip. Studies of land-use/ land-cover change in North African coastal wetlands, including Moulay Bouselham, show an increase in carbon storage associated with increased vegetation cover as agricultural lands expanded(Aitali et al. 2022) However, much of this vegetation growth reflects conversion of natural and semi-natural habitats into cultivated areas rather than recovery of native ecosystems, raising concerns about biodiversity loss. For example, bryophyte surveys in Merja Zerga note that anthropic disturbance (e.g. overgrazing, reed cutting) has altered species composition and reduced native flora richness. (Chtaibi et al. 2017)

*Figure 11. The Interaction Between Agricultural Activity and the Wetland Ecosystem in Merja Zerga – Moulay Bousselham*

## 4.2. Wetland Decline and Hydrological Stress

While vegetation increases are apparent, the wetland water surface of Merja Zerga has shown significant long-term decline, consistent with regional patterns of wetland shrinkage. Drivers include reductions in rainfall, enhanced evapotranspiration, and substantial human extraction of groundwater. Although no published study yet reports exactly a –0.0352 km$^2$ per year rate over 1985-2025, remote sensing work (e.g. Ramdani et al. 2009) has documented decreasing freshwater inflows and changes in channel morphology associated with environmental pressure around Merja Zerga. (Ahmed et al. 2009) Also, monitoring of submerged vegetation (such as *Zostera noltei*) indicates that fluctuations in water levels and aquatic habitat areas are sensitive to hydrological stressors. (Benmokhtar et al. 2023) The ecological risks include threats to aquatic biodiversity, migratory birds, and the loss of wetland services such as groundwater recharge and climate buffering.

## 4.3. Urban Expansion and Land-Use Conflicts

Since around 2013, urbanization and infrastructural expansion have increasingly driven land-use change in areas surrounding Merja Zerga. Built-up areas encroach on traditional agriculture, fragmenting natural habitats and increasing pollution risks via runoff of fertilizers and pesticides into the wetland. The shift after 2018 toward more commercialized and intensive agriculture in Morocco also intensifies these tensions. While there is not a single longitudinal study that quantifies all these shifts exactly in the 2013-2024 period for Merja Zerga, socio-ecological assessments of the lagoon show that economic development (agriculture, fisheries, urban growth) is a major driver of pressures on the ecosystem.

## 4.4. Ecological and Policy Implications

The combination of increasing cultivated vegetation, shrinking wetland surfaces, and expanding urbanization suggests a fragile balance where economic gains may come at the cost of ecological integrity. Documented ecological costs include decline in native species richness (for example from bryophytes and halophytic plants), degradation in sediment quality due to metal contamination, and changes in hydrological regimes (Anhichem et al. 2021). To manage these challenges, integrated land and water management is essential. Strategies could include controlling groundwater abstraction, establishing buffer zones around the wetland, promoting more eco-friendly irrigation practices, and reinforcing the Ramsar site designation with stronger enforcement mechanisms and community-based conservation. Several recent studies in Moulay Bousselham emphasize exactly such responses, including through DPSIR-framework assessments, to reconcile conservation with livelihoods. (Erraj Chahid et al. 2023)

## 4.5. Contribution to Broader Debates

This case contributes to wider debates about the sustainability of wetland-agriculture systems in arid and semi-arid regions. By using remote sensing and long-term monitoring tools, studies have quantified both vegetation gains (especially in agricultural zones) and wetland decline, highlighting the paradox that efforts to combat desertification and improve food security can simultaneously undermine the ecosystems essential for long-term resilience. The Merja Zerga example aligns with findings in other Mediterranean and North African systems, reinforcing the need for balanced policy that values both economic development and ecological health.

## V. PROPOSALS FOR THE CONSERVATION OF THE MERJA ZERGA WETLAND

Merja Zerga is a natural area of significant ecological, environmental, touristic, and economic value, not only at the regional and national levels but also internationally. However, this site faces major challenges linked to the decline of groundwater reserves, the expansion of agricultural activities, and unregulated urban growth, which collectively reduce its value and threaten its sustainability. To overcome these constraints, we propose a set of urgent and practical measures:

- Control urban and agricultural expansion by putting an end to drainage practices and the overexploitation of wetland areas, while adopting land-use and agricultural plans that respect the ecological balance.
- Rationalize the use of groundwater by promoting efficient irrigation techniques (such as drip irrigation) and prohibiting unauthorized and uncontrolled well drilling.
- Implement programs to recharge groundwater reserves through the use of treated wastewater or rainwater harvesting and storage.
- Raise awareness among civil society and local communities about the importance of Merja Zerga and their role in its protection, through workshops and environmental campaigns.
- Establish new nature reserves and national parks to safeguard biodiversity and the ecosystems associated with the wetland.
- Strengthen monitoring of hunting and fishing activities, prohibiting the targeting of endangered species and enforcing strict penalties on violations.
- Address the problem of sand encroachment through reforestation and soil stabilization programs along the adjacent coastline.
- Adopt and effectively enforce strict regulations to protect the wetland from all forms of unsustainable exploitation.
- Encourage scientific research and environmental studies to better understand the ecological dynamics of the wetland and guide public policies.
- Engage farmers in a participatory approach that provides them with sustainable economic alternatives to reduce pressure on natural resources.

## VI. CONCLUSION

This study has provided a comprehensive assessment of land use, vegetation dynamics, and hydrological changes in the Moulay Bousselham–Merja Zerga region using multi-source remote sensing data and Google Earth Engine analysis. The findings reveal a dual trajectory: on the one hand, intensive agricultural practices and rural development programs have supported a marked increase in vegetation cover, reflecting Morocco's commitment to enhancing productivity and combating desertification. On the other hand, this progress has come at the expense of natural ecosystems, with the Merja Zerga wetland showing a persistent and statistically significant decline in its water surface area.

The juxtaposition of expanding cultivated vegetation and shrinking wetland habitats underscores the fragile balance between economic development and ecological sustainability. Urban expansion, hydrological stress, and agricultural intensification threaten biodiversity, water quality, and the long-term ecological functions of this Ramsar-designated wetland. These results highlight the urgent need for integrated land and water management strategies that reconcile agricultural growth with the conservation of critical ecosystems.

Future management of the Merja Zerga wetland should prioritize sustainable irrigation practices, strict regulation of groundwater abstraction, establishment of protective buffer zones, and reinforcement of community-based conservation initiatives. More broadly, the case of Moulay Bousselham contributes to regional and global debates on wetland sustainability, demonstrating how modern agricultural gains can paradoxically undermine the very ecosystems needed to ensure long-term resilience and food security.

## REFERENCES


Ahmed, M. H., El Leithy, B. M., Thompson, J. R., Flower, R. J., Ramdani, M., Ayache, F., & Hassan, S. M. (2009). Application of remote sensing to site characterisation and environmental change analysis of North African coastal lagoons. *Hydrobiologia, 622*(1), 147–171. DOI: 10.1007/s10750-008-9682-8

Aitali, R., Snoussi, M., Kolker, A. S., Oujidi, B., & Mhammdi, N. (2022). Effects of Land Use/Land Cover Changes on Carbon Storage in North African Coastal Wetlands. *Journal of Marine Science and Engineering*, *10*(3), 364. DOI: 10.3390/jmse10030364

Alaoui, A. M., Choura, M., Maanan, M., Zourarah, B., Robin, M., Conceição, M. F., Andrade, C., Khalid, M., & Carruesco, C. (2010). Metal fluxes to the sediments of the Moulay Bousselham lagoon, Morocco. *Environmental Earth Sciences*, *61*(2), 275–286. DOI: 10.1007/s12665-009-0341-9

Anhichem, M., Benbrahim, S., Idrissi, M., Chfiri, R., & Yahyaoui, A. (2021). Environmental state of the Merja Zerga lagoon in Moulay Bousselham, Morocco: Metallic contamination levels and origins in sediments. *Aquaculture, Aquarium, Conservation & Legislation*, *14*(2), 774–785.

Arabi, M., Mechkirrou, L., El Malki, M., Alaoui, K., Chaieb, A., Maaroufi, F., & Karmich, S. (2024). Overview of Ecological Dynamics in Morocco – Biodiversity, Water Scarcity, Climate Change, Anthropogenic Pressures, and Energy Resources – Navigating Towards Ecosolutions and Sustainable Development. *E3S Web of Conferences, 527*, 01001. DOI: 10.1051/e3sconf/202452701001

Asawra, K., & Koli, V. K. (2025). Foraging strategies of three co-occurring ibis species (Black-headed, Glossy and Red-naped) in a semi-arid landscape of western India. *Avian Biology Research*, *18*(1), 12–20. DOI: 10.1177/17581559251314345

Ayache, F., Thompson, J. R., Flower, R. J., Boujarra, A., Rouatbi, F., & Makina, H. (2009). Environmental characteristics, landscape history and pressures on three coastal lagoons in the Southern Mediterranean Region: Merja Zerga (Morocco), Ghar El Melh (Tunisia) and Lake Manzala (Egypt). *Hydrobiologia*, *622*(1), 15–43. DOI: 10.1007/s10750-008-9676-6

Bahouq, T., Moumane, A., & Touhami, N. (2024). Monitoring Agricultural Dynamics and Its Impact on Water Resources Using GIS and Remote Sensing Data: A Case Study in the Ben Mansour-Mnasra Region, Kenitra, Morocco. In Batchi, M., & Moumane, A. (Eds.), (pp. 183–210). Advances in Geospatial Technologies. IGI Global., DOI: 10.4018/979-8-3693-9651-3.ch007

Benmokhtar, S., Robin, M., Maanan, M., Boutoumit, S., Badaoui, B., & Bazairi, H. (2023). Monitoring the Spatial and Interannual Dynamic of Zostera noltei. *Wetlands*, *43*(5), 43. DOI: 10.1007/s13157-023-01690-7 PMID: 37153812

Boyer, M. (2025). Historicizing The Atlantic Morocco From the Horizon of the New Worlds to the Globalized World: *Afrique(s) en mouvement, n° 9*(3), 32–43. DOI: 10.3917/aem.009.0032

Chafiq, T., Hmamou, M., Ouhammou, I., Azmi, R., & Kumar, M. (2024). Modelling change detection for unveiling urban transitions: Using machine learning algorithms and Sentinel-2 data in Larache City, Morocco. *Modeling Earth Systems and Environment, 10*(2), 1711–1725. DOI: 10.1007/s40808-023-01860-w

Chtaibi, N., Magri, N., Belahbib, N., Zidane, L., Douira, A., & Dahmani, J. (2017). Study of Bryophytic Flora in the Ramsar Wetland of Merja Zerga (North-west of Morocco). *International Journal of Environment, Agriculture and Biotechnology*, *2*(3), 1032–1037. DOI: 10.22161/ijeab/2.3.4

Elmotawakkil, A., & Enneya, N. (2024). Advanced machine learning for predicting groundwater decline and drought in the Rabat–Salé–Kénitra region, Morocco. *Journal of Hydroinformatics*, *26*(11), 2980–3007. DOI: 10.2166/hydro.2024.328

Ennabili, A., Libiad, M., El Haissoufi, M., Khabbach, A., & Bennas, N. (2025). Biodiversity, Change, and Use of Moroccan Wetlands. In Pullaiah, T. (Ed.), *Wetlands of Tropical and Subtropical Asia and Africa* (1st ed., pp. 265–296). Wiley., DOI: 10.1002/9781394235278.ch12

Erraj Chahid, N., Bouchkara, M., Canu, D. M., El Khalidi, K., & Zourarah, B. (2023). Assessing the Dynamics of the Moulay Bousselham Lagoon: An Integrated DPSIR Approach to Understand Drivers, Impacts, and Human Response. In *IECG 2022* (p. 31). Presented at the IECG 2022, MDPI. DOI: 10.3390/IECG2022-14553

He, H., Li, X., & Li, T. (2025). The Sustainable Development of Wetlands and Agriculture: A Literature Review. *Agronomy (Basel)*, *15*(3), 746. DOI: 10.3390/agronomy15030746

McFEETERS, S. K.McFEETERS. (1996). The use of the Normalized Difference Water Index (NDWI) in the delineation of open water features. *International Journal of Remote Sensing*, *17*(7), 1425–1432. DOI: 10.1080/01431169608948714

Moumane, A., Bahouq, T., & Ghazi, A. K. (2025). Integrating GIS, remote sensing, and analytic hierarchical process (AHP) for mapping of groundwater-recharge potential zones in Kenitra Province, Morocco: Toward sustainable development goals achievement. *Euro-Mediterranean Journal for Environmental Integration, 10*(2), 831–851. DOI: 10.1007/s41207-024-00706-w

Moumane, A., Bahouq, T., Karmaoui, A., Laghfiri, D., Yassine, M., Karkouri, J. A., Batchi, M., Faouzi, M., Boulakhbar, M., & Youssef, A. A. (2025). Lake Iriqui's Remarkable Revival: Field Observations and a Google Earth Engine Analysis of Its Recovery After over Half a Century of Desiccation. *Land (Basel), 14*(1), 104. DOI: 10.3390/land14010104

Moumane, A., El Ghazali, F. E., Al Karkouri, J., Delorme, J., Batchi, M., Chafiki, D., & Karmaoui, A. (2021). Monitoring spatiotemporal variation of groundwater level and salinity under land use change using integrated field measurements, GIS, geostatistical, and remote-sensing approach: Case study of the Feija aquifer, Middle Draa watershed, Moroccan Sahara. *Environmental Monitoring and Assessment, 193*(12), 769. DOI: 10.1007/s10661-021-09581-2 PMID: 34735624

Purba, H. J., Azahari, D. H., Dani, F. Z. D. P., Alsyouf, I., Masmoudi, M., & Obaideen, K. (2024). Enhancing rice resilience and sustainability: A bibliometric analysis of innovations for food security and environmental conservation. *BIO Web of Conferences, 119*, 05003. DOI: 10.1051/bioconf/202411905003

Rad, S. M., Ray, A. K., & Barghi, S. (2022). Water Pollution and Agriculture Pesticide. *Cleanroom Technology, 4*(4), 1088–1102. DOI: 10.3390/cleantechnol4040066

Rehman, S. U., Bhatti, A., Kraus, S., & Ferreira, J. J. M. (2021). The role of environmental management control systems for ecological sustainability and sustainable performance. *Management Decision, 59*(9), 2217–2237. DOI: 10.1108/MD-06-2020-0800

Sharma, S., Sharma, U., Sharma, S., Sharma, S., Likhita, J., Thakur, S., & Sharma, N. (2025). Strategies for Sustainable Land Management Through Climate-Resilient Horticulture: A Systematic Review. *Land Degradation & Development*, ldr.70082. DOI: 10.1002/ldr.70082

Thompson, J. R., Flower, R. J., Ramdani, M., Ayache, F., Ahmed, M. H., Rasmussen, E. K., & Petersen, O. S. (2009). Hydrological characteristics of three North African coastal lagoons: Insights from the MELMARINA project. *Hydrobiologia, 622*(1), 45–84. DOI: 10.1007/s10750-008-9680-x

# Chapter 5
# Harnessing Precision Agriculture and Artificial Intelligence in Sustainable Farming

**Muhammad Safdar**
https://orcid.org/0009-0006-1779-6967
*Agricultural Remote Sensing Lab, University of Agriculture, Faisalabad, Pakistan*

**Wasiq Farooq**
*Agricultural Remote Sensing Lab, University of Agriculture, Faisalabad, Pakistan*

**Muhammad Sajid Mehmood**
https://orcid.org/0000-0002-2055-3505
*School of Tourism and Planning, Pingdingshan University, Pingdingshan, China*

**Abdul Rauf**
*Department of Irrigation and Drainage, University of Agriculture, Faisalabad, Pakistan*

**Amina Rashid**
https://orcid.org/0000-0001-9772-1063
*Department of Agronomy, University of Agriculture, Faisalabad, Pakistan*

**Kashif Mehmood**
*Department of Irrigation and Drainage, University of Agriculture, Faisalabad, Pakistan*

**Muntaha Munir**
https://orcid.org/0009-0005-3404-3636
*Institute of Botany, University of the Punjab, Lahore, Pakistan*

**Nalain E. Muhammad**
*Department of Irrigation and Drainage, University of Agriculture, Faisalabad, Pakistan*

**Hafiz Muhammad Bilawal Akram**
https://orcid.org/0000-0001-6720-7266
*Department of Agronomy, University of Agriculture, Faisalabad, Pakistan*

**Hafiz Muhammad Mohsin Raza**
*Department of Agricultural Sciences, Allama Iqbal Open University, Islamabad, Pakistan*

DOI: 10.4018/979-8-3373-6608-1.ch005

**Aisha Nazir**
*Institute of Botany, University of the Punjab, Lahore, Pakistan*

**Hafiz Muhammad Awais**
https://orcid.org/0000-0002-1757-9747
*Agricultural Remote Sensing Lab, University of Agriculture, Faisalabad, Pakistan*

## ABSTRACT

*The chapter explores the role of Artificial Intelligence (AI) in revolutionizing precision agriculture, focusing on data-driven, adaptive, and efficient practices. It traces the evolution of AI-integrated digital farming systems, including GNSS, drones, sensors, and IoT technologies. Key AI applications include crop classification, yield forecasting, pest detection, irrigation scheduling, and nutrient management. The chapter also discusses technologies like UAVs, hyperspectral imaging, edge computing, and cloud-based platforms for improved farm-level decision-making. The chapter discusses the potential of AI in large-scale and smallholder farming systems but also highlights challenges like heterogeneous data, economic barriers, limited digital literacy, and low adoption among smallholder farmers. The chapter discusses future opportunities in AI-integrated robotics, blockchain-based supply chains, and big data analytics for climate-resilient agriculture and calls for inclusive policy reforms, capacity-building initiatives, and collaborative innovation for equitable scale.*

## I. INTRODUCTION

Agriculture is a practice of cultivating the soil, planting and harvesting crops. Agriculture is the backbone of human civilization which is facing severe challenge. The world is entering a phase of food security with increasing population the demand of food is expected to rise significantly placing immense pressure on the agriculture system. Increasing population, changing land and climate are putting drastic stress on agricultural land. Urbanization, rapid soil degradation and unstable farming practices are reducing cultivable land. World full demand is anticipated to increase by 50% till 2050. All the challenges underline the critical need for better and sustainable farming practices. Sustainable practices are no longer optional but are the need for meeting future food requirements.

Till now agriculture has relied on traditional farming practices. In the face of modern challenges these traditional farming practices are proving to be unsustainable as they require high level of water and chemical inputs which not only escalate production but also contribute to soil degradation, water wastage and long term environmental pollution. Climate change continue to challenge crop productivity.

Developing countries, the backbone of global food production are facing the adverse effect of climate change due to lack of technology, financial resources and infrastructure. All these problems are widening the gap between rising food demand and availability of existing farm system it cannot be resolved without innovative technological development solutions.

Precision agriculture has arisen as a solution to these difficulties. Combining geographic information system, remote sensing and Internet of Things; precision agriculture now allows farmers to monitor and manage crops. Farmers can apply water, fertilizer, and insecticides precisely where and when they are needed (Gebbers & Adamchuk, 2010; Liakos et al., 2018). Resource wastage and environmental pollution are reduced due to precision agriculture. Research revealed that precision irrigation can reduce water use by up to 30% while maintaining crop yields (Fountas et al., 2020). Similarly, precision nutrient management helps prevent fertilizer overuse, reducing the production costs (Gebbers & Adamchuk, 2010).

Precision agriculture not only helps to increase the production but also enhances the resilience. Integrating sensors, drones, and satellite data, farmers can now monitor crop health, detect stress earlier, and take timely actions. Precision agriculture reduces prop loss due to pest disease and extreme weather which are the leading cause of food shortages (Liakos et al., 2018). Moreover precision agriculture can empower small older farmers who are the backbone of developing countries. By adopting low cost precision tool and mobile based system farmers can increase their productivity while contributing to local food security (Wolfert et al., 2017).

Precision agriculture provides the foundation for site suitable crops and source management. The integration of Artificial Intelligence significantly amplifies precision agriculture potential. Traditional farming technologies mainly collect raw data. However, the complexity of agricultural systems often requires deeper insights than raw numbers alone. AI tools like machine learning, and predictive analytics to process real-time data. Algorithms can analyze imagery to differentiate between healthy and stressed plants. Machine learning models can forecast future yield outcomes based on soil, weather, and historical yield data. AI enhances precision farming by enabling real-time, data-driven decision-making. (Chlingaryan et al., 2018; Liakos et al., 2018).

Artificial intelligence has become a keystone for precision agriculture which enabled predictive analysis automation and adaptive decision making. Artificial intelligence is analyzing large data set from satellite drones helping farmers anticipate challenges. AI models can predict weather radiation identify Crohn's disease and recommend time the intervention increasing the overall yield (Kamilaris & Prenafeta-Boldu, 2018). Nowadays, AI is being widely used today weeds best and disease diagnosis (Singh et al., 2016).

AI technologies like machine learning deep learning and predictive analysis can help in dealing with massive amount of data providing unavailable actionable insights (Chlingaryan et al., 2018; Kamilaris & Prenafeta-Boldu, 2018). Combining precision agriculture and artificial intelligence can boost productivity while aligning with sustainable goals (Wolfert et al., 2017). The motivation behind integrating artificial intelligence into precision farming is decision making capacity (Kamilaris & Prenafeta-Boldu, 2018) and sustainability by enabling precise input management and early detection of threats (Shamshiri et al., 2018).

The integration of precision agriculture with artificial intelligence is the future of sustainable farming. By combining the satellite image, drones, Internet of Things and machine learning algorithms farmers can monitor crops at micro level. Satellite data provides large scale monitoring, while drones capture high resolution image, IoT based soil and climate sensors provide continues data to AI models which then analyze patterns to predict irrigation needs, disease outbreak and yield outcomes. The interaction between these technologies not only improves efficiency but also strengthens resilience. Machine learning candidate early signs of crop stress (Liakos et al., 2018) while AI driven irrigation scheduling has optimized water use in drought prone regions (Alvim et al., 2022).

The chapter is structured around several objectives to highlight the transformative role of precision agriculture and RPG intelligence. Firstly the chapter will explain the transformation of precision agriculture from the era of mechanization to modern precision farming. The chapter will address global food security posed by population growth shrinking land and climate change. Secondly the chapter will introduce a technology in agriculture including predictive analysis crop management and resource optimization. Thirdly, the chapter will analyze how AI enhances efficiency, productivity and sustainability by enabling data-driven decisions. Additionally the chapter will explore real world examples and case studies that demonstrate successful integration of precision agriculture and art vision intelligence finally it highlights the challenges values and limitations associated with implementing artificial intelligence in precision agriculture.

In summary, AI-driven precision agriculture represents a paradigm shift. It not only offers solutions that are not only technologically advanced but also environmentally sustainable and economically viable. In this new digital era it is crucial to understand how this advancements have evolved overtime.

## II. EVOLUTION OF PRECISION AGRICULTURE

With the passage of time agriculture has evolved through various waves of innovation. These innovations were driven by need to increase productivity. Precision agriculture is the product of this long innovative trajectory, evolving from early mechanization to modern digital farming systems.

1) **The era of mechanization**

It began with adoption of tractors, mechanical harvesters, and irrigation systems in the late nineteenth centuries. These innovations replaced animal and human labor, significantly improving farm productivity and reducing the time required for cultivation (Pingali, 2007). Mechanization placed the foundation for input-intensive agriculture, which later defined the Green Revolution.

2) **The Green Revolution**

It started with improved crop varieties, chemical fertilizers, and pesticides to further increase the yield. Green evolution successfully reduced hunger in many parts of the world but reliance completely on fertilizer and pesticides led to environmental degradation (Evenson & Gollin, 2003).

3) **The digitization of agriculture**

Satellite positioning systems, remote sensing technologies, and geographic information systems made foundation of Digitization in Agriculture. It enabled farmers to collect and analyze spatial data. Digitization allowed site-specific management of crops, reducing input waste and improving efficiency (Mulla, 2013).

*Figure 1. The evolution of Agriculture*

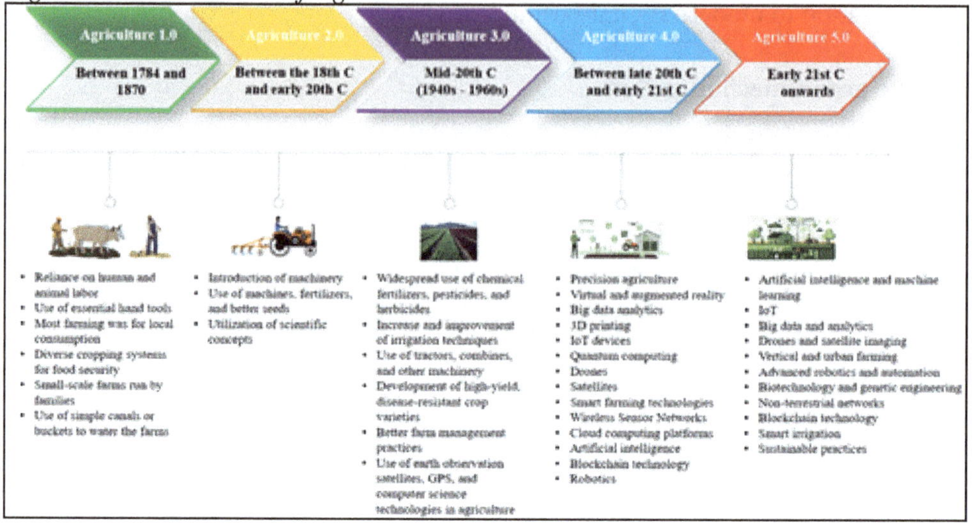


Nowadays, precision agriculture is going through new revolution driven by digitization and artificial intelligence. Machine learning, and cloud computing have transformed farming into a data-rich initiative, now decisions are based not just on observations but on predictive models and simulations.

## 1. Role of GNSS, Sensors, Drones, and IoT

GNSS, sensors, drones and IoT are the technologies which allowed farmers to monitor, analyze and manage field variability with accuracy. All these technologies form the foundation of precision agriculture playing a crucial role in collecting data and enabling management strategies.

### 1.1 Global Navigation Satellite Systems (GNSS)

Global navigation satellite system provide highly precise positioning which is important for guiding agriculture machinery with high level accuracy. GNSS technologies provide accuracy required for precision field operations. Farmers require GNSS to guide tractors, harvesters and sprayers (Alonge & Isreal, 2025). GNSS supports yield mapping through data collected during harvesting, enabling farmers to analyze spatial yield variability.

*Figure 2. GNSS*

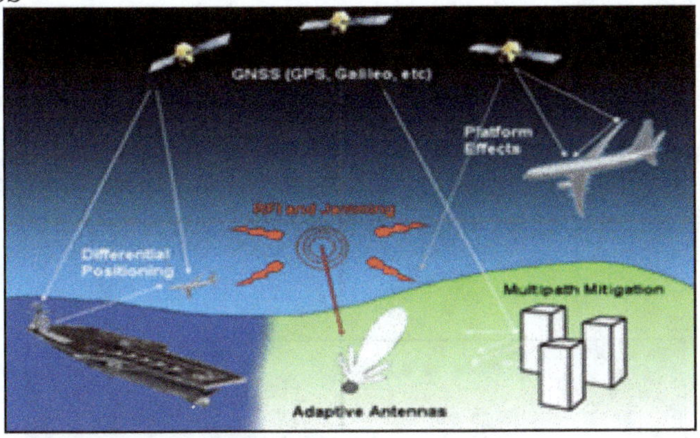


## 1.2 Sensors

Various sensors used to collect real time data about crops, soil and environmental conditions. Soil sensors collect data of soil moisture content, temperature and neutral levels while crop sensors detect chlorophyll concentration, crop health and stress level (Jurišić et al., 2021). On the other hand environmental sensors provide data on temperature rainfall humidity and wind speed. All this data lays the foundation play mission scheduling fertilizer application and various other aspects (Saleem et al., 2024).

*Figure 3. Sensors*

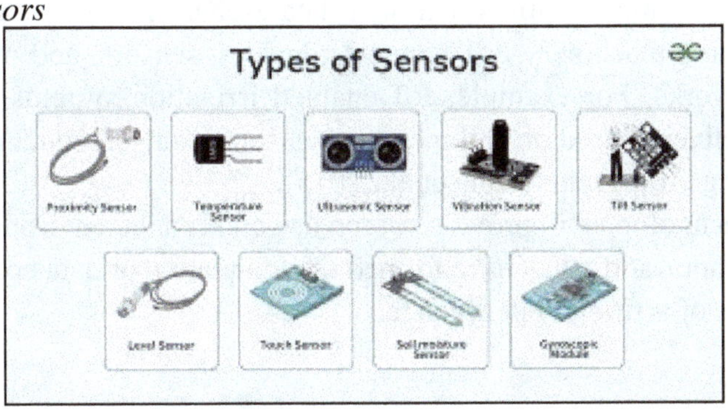

## 1.3 Drones (Unmanned Aerial Vehicles, UAVs)

Precision agriculture require high resolution imagery to monitor crop health, detect pest infestations, and assess plant stress. Drones equipped with multispectral, thermal and RGB cameras are utilized for this purpose (Tsouros et al., 2019). Unlike satellites, drones can operate under cloud cover. Drones also facilitate tasks such as targeted spraying of pesticides and fertilizers.

*Figure 4. Drones*

## 1.4 Internet of Things (IoT)

Precision agriculture requires real time data exchange between agriculture devices, sensors and machinery. IoT connects devices, sensors, and machinery into integrated networks. For example, IoT-enabled irrigation systems automatically adjust water delivery based on soil moisture sensor readings, reducing water waste while improving crop health (Ojha et al., 2015).

All these technologies improve resource use and reduce environmental degradation. More importantly they have formed the foundation of data collection which is the backbone of artificial intelligence.

## III. AI APPLICATIONS IN PRECISION FARMING

Agriculture farming has become more productive efficient and sustainable due to the integration of artificial intelligence add automation. Smart irrigation systems are being used to automatically adjust water supply which helps reduce water wastage and ensures crops get exactly what they require (Miller et al., 2025). Artificial intelligence process data obtain from IoT devices, drones, GNSS and satellite imagery to monitor crop health, predict weather and event forecast pest attacks. Machine learning provides farmers with localized, practical insights on which they can act immediately by analyzing these data streams. On the ground, AI powered machinery has taken over labor extensive task life planting, spraying and harvesting. In the air, drones equipped with AI-driven vision systems detect crop stress, pests, or nutrient deficiencies instantly (Ahmad et al., 2025). All these innovations reduce labor cost, tackle workforce shortages and increase operational productivity. Robotic automation also supports better timing and consistency of farming. Together, these innovations implies that AI does not simply support precision farming but actively drives automation, making farming systems more resilient and sustainable.

## 1. Crop Classification and Yield Prediction

Nowadays, artificial intelligence is being used for crop classification and yield prediction. Crop classification is the identification and categorization of different crop types in agriculture field using data-driven algorithm while yield prediction is the analysis of crop outcome after harvesting. Crop classification and yield prediction traditional methods relied on manual surveys. Yield prediction models are used to forecast crop harvest quantities before actual harvesting (Pang et al., 2022). AI techniques now integrate satellite imagery, drone data, and ground-based sensors to classify crop types and forecast yields well before harvest (Khaki & Wang, 2019). For example, convolutional neural networks have been applied to multispectral images from Sentinel-2 and Landsat platforms to distinguish between crop varieties while recurrent neural networks are increasingly used to model temporal growth patterns and predict yield outcomes (Lu et al., 2024). All these approaches not only improve planning for farmers but also support policymakers by providing reliable production forecasts.

*Figure 5. Crop Classification*

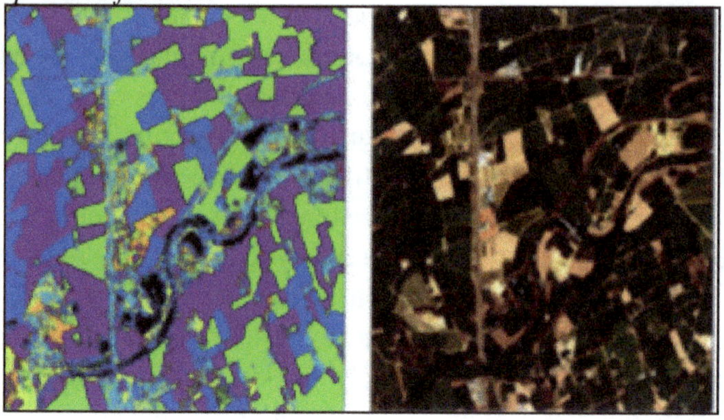

## 2. Disease and Pest Detection Using Image Processing

Nowadays, Artificial intelligence is being used for the detection of Crop diseases and pests. Conventional methods for disease and test detection were often slow and were dependent on human expertise which usually get delay due to interventions. Nowadays digital image processing methods are used to analyze photographs captured from crops to detect visual symptoms of disease and pests. Primarily image acquisition is done using camera or drones followed by reprocessing steps which includes noise reduction in segmentation to separate affected areas. Feature extraction techniques are used to identify characteristics such as color, shape, and texture changes indicative of disease or pest damage (Lu et al., 2017). However, challenges such as varying light conditions, hidden plant parts, and subtle disease symptoms can affect accuracy, making continuous model training and image enhancement crucial. Convolutional Neural Networks can differentiate between healthy and diseased plant tissues with remarkable accuracy, even when symptoms are not visible to the naked eye (Barbedo, 2019). Similarly, deep learning models have been trained to detect specific pests enabling rapid response (Obasekore et al., 2023)(Brahimi et al., 2023). Overall, combining image processing with AI offers a practical and scalable solution for timely crop health monitoring, contributing to higher yields and more sustainable farming.

*Figure 6. Disease and Pest Identification*

## 3. Irrigation Scheduling and Water Use Efficiency

Sustainable agriculture requires efficient water management. Traditionally, farmers relied on experience or simple calendars to decide water requirements which often lead to overuse or underuse of water (Mushtaq et al., 2024). AI has significantly advanced irrigation scheduling by moving beyond fixed timetables to data-driven systems (Kaur et al., 2024). Machine learning algorithms can predict evapotranspiration rates, thereby automating irrigation decisions that improve yields while reducing water losses (Krishnan et al., 2022). Moreover IoT enable smart irrigation systems with AI platforms have demonstrated water savings up to 40 to 60% highlighting their potential in effective water management (Miller et al., 2025). AI driven irrigation is proving to be crucial tool by enhancing both water use efficiency and crop productivity.

*Figure 7. Irrigation scheduling and water use system (Miller et al., 2025)*

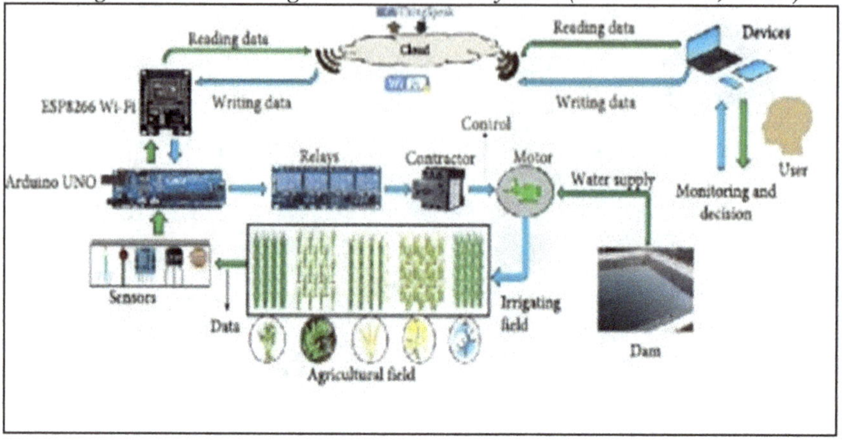


## 4. AI for Fertilizer Application and Soil Health Monitoring

Traditional fertilizer application did not follow any specific recommendations, which can lead to nutrient overuse, soil degradation, and environmental pollution. AI-powered decision support systems analyze soil properties, crop nutrient requirements, and weather conditions to determine precise fertilizer required to specific field (Alreshidi, 2023). Deficiencies in nutrients can be predicted using soil nutrient data and yield histories enabling farmers to apply the right amount of fertilizer at the right time (Bondre & Mahagaonkar, 2019). Moreover, AI-Models integrates data from remote sensing, sensors, and laboratory analyses to track soil organic matter, pH, and microbial activity (Sabir et al., 2024). Studies revealed that AI-assisted nutrient management can increase crop yields while cutting fertilizer costs improving both farm profitability and environmental outcomes (Chlingaryan et al., 2018).

*Figure 8. AI for fertilizer application and soil health monitoring(Bondre & Mahagaonkar, 2019)*

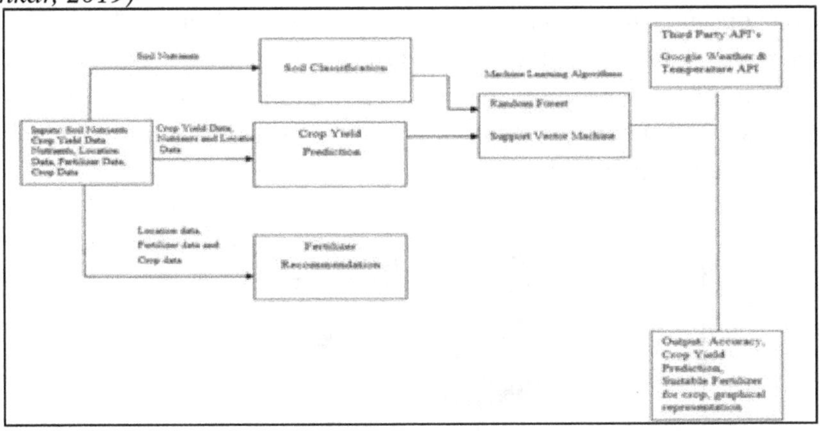


## IV. TECHNOLOGIES ENABLING AI IN PRECISION AGRICULTURE

### 1. UAVs, Hyperspectral and Multispectral Imaging

Unmanned Aerial Vehicles (UAVs) are transforming modern agriculture by providing farmers with a bird's-eye view of their fields. Unlike satellite imagery, which may be limited by cloud cover or low resolution, drones can fly at low altitudes and capture high-resolution images at specific times determined by the farmer. UAVs equipped with various sensors and imaging system provide high resolution real time data of crop fields enabling AI application in precision agriculture. UAVs combined with hyperspectral and multi spectral imaging performs various tasks such as crop health monitoring, nutrient assessment and disease detection. Unlike traditional field surveys UAVs provide high resolution data over large areas (Sishodia et al., 2020). Hyperspectral imaging goes ahead selecting hundreds of narrow spectral bands enabling the identification of physiological changes in plant early stage diseases and even soil nutrient status (García-Vera et al., 2024). When the data collected from UAB hyperspectral and multi spectral images is processed with AI algorithm a massive data is generated which can be transformed into actionable insights. Recent research revealed dad UAB base hyperspectral imaging integrated with AI models outperform traditional remote sensing methods in both disease detection and yield forecasting (Maes & Steppe, 2019). Together, UAVs and advanced imaging technologies bridge the gap between data collection and AI-driven decision-making, making them indispensable tools in modern precision farming. They help farmers detect variability within fields, apply fertilizers or pesticides only where needed, and optimize irrigation. This not only boosts productivity but also reduces unnecessary input use, lowering costs and minimizing environmental impacts. As these technologies become more affordable and accessible, they are opening new opportunities for farmers to make informed, data-driven decisions that support both profitability and sustainability.

*Figure 9. hyperspectral and multispectral imaging*

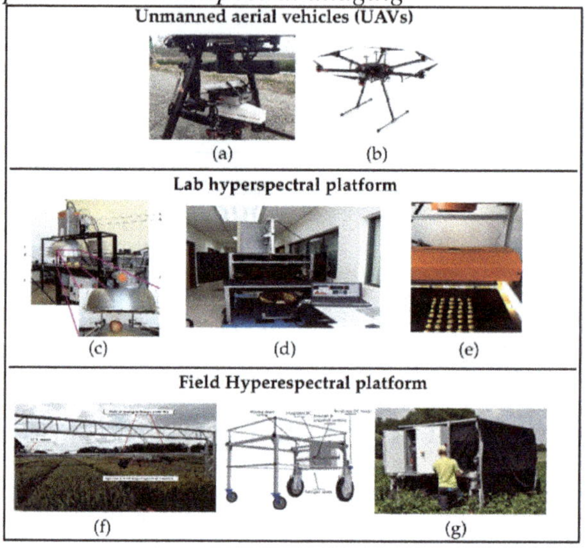


## 2. Edge Computing, IoT, and Wireless Sensor Networks

Edge computing, Internet of Things and wireless sensor networks are transforming the techniques of data collection, data processing and data application in precision agriculture. IoT devices create a dense web of real-time field observations. Traditionally data collected had to be transmitted to cloud service for analysis which open cause delays and requires stable Internet connectivity but edge computing address this challenge by processing data closer to the source. Edge computing allows AI algorithms to generate rapid insights directly in the field which reduces the noise ensures reliability in the areas poor connectivity (Wolfert et al., 2017). Wireless sensor networks enhance the system by connecting multiple nodes across large farm areas, enabling continuous monitoring of various parameters (Alvim et al., 2022). Together, IoT, WSNs, and edge computing create a responsive digital infrastructure that empowers AI applications to operate efficiently, even in regions with limited connectivity. For instance, a soil sensor connected to an edge device can immediately analyze moisture data and trigger irrigation only when and where it is needed, preventing both under- and over-watering. Similarly, these systems can support real-time pest or disease alerts, allowing farmers to take corrective action quickly. By enabling faster decision-making, reducing dependency on cloud services, and making efficient use of resources, these technologies represent a crucial step toward more climate-resilient and sustainable agriculture.

*Figure 10. Edge computing, IoT, and wireless sensor networks*

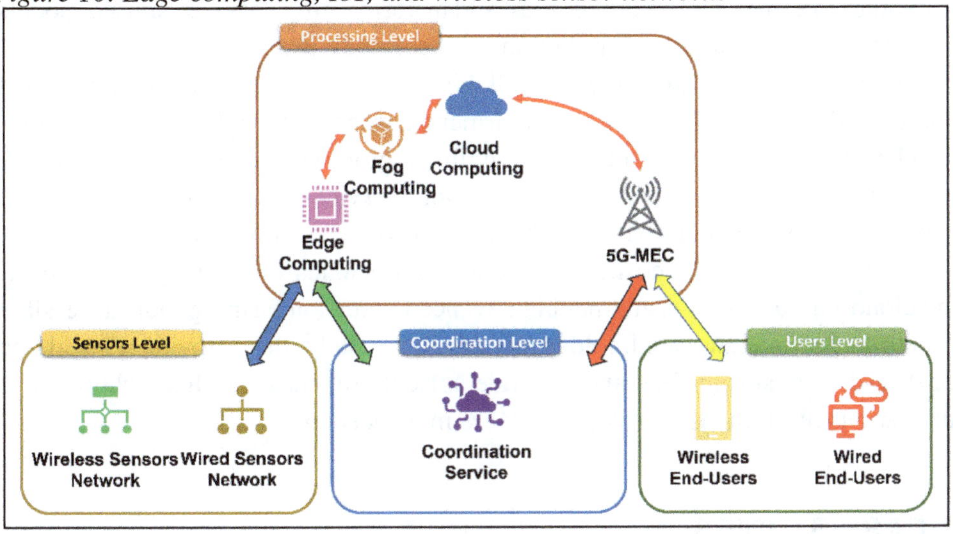


## 3. AI Platforms and Cloud-Based Solutions

AI platforms and cloud based solution play a crucial role in scaling precision agriculture. They provide the computational power and storage required to process agriculture data. Farmers and researchers now depend on cloud services to integrate data from UAVs IoT sensors and weather station to centralized platforms AAI can perform advanced analytics (Wolfert et al., 2017).

Cloud computing enable centralized data storage allowing farmers to access data analytics from anywhere. On the other hand, cloud base a models enables crop classification, disease detection and yield prediction without any need of expensive infrastructure. AI Platforms and cloud Computing makes precision farming accessible even to resource-limited regions (Senoo et al., 2024). Cloud infrastructure supports scalability and flexibility for deploying AI applications. Microsoft Azure Farm Beats and Google Earth Engine exemplify how cloud solutions support large-scale data management and decision-making by combining geospatial analytics with machine learning. Moreover, these systems provide user-friendly dashboards and mobile interfaces, allowing farmers to visualize field variability (Ahmad et al., 2025).

Cloud-based platforms also enable scalability and accessibility. Instead of requiring expensive on-site servers or technical expertise, farmers can access AI-driven recommendations through user-friendly dashboards or mobile applications. For example, a farmer can receive daily irrigation advice, crop stress alerts, or fertilizer recommendations directly on a smartphone, making decision support more practical and inclusive. These platforms further allow for integration of diverse datasets such as weather forecasts, soil health records, and historical crop data into unified models that enhance the accuracy of predictions.

Another important advantage is collaboration and knowledge sharing. Cloud solutions allow multiple stakeholders farmers, agronomists, policymakers, and researchers to access shared data pools, enabling region-wide analysis of crop health, pest outbreaks, or climate risks. In addition, cloud services support continuous model improvement by updating AI algorithms as new data becomes available.

By reducing barriers of cost, infrastructure, and technical complexity, AI platforms and cloud-based solutions are making advanced precision farming tools accessible to both large-scale and smallholder farmers. Ultimately, they provide a backbone for data-driven, sustainable, and adaptive agriculture, ensuring that technological innovations reach the fields where they are most needed.

## V. CASE STUDIES

### 1. AI-Based Smart Farming Systems in India

The adoption of AI precision farming has expanded across different countries which reflects agriculture priorities. In India, AI based mobile application are increasingly used for crop disease detection, weather forecasting and advisory services to small farmers. (Pandey et al., 2024) conducted a field study at Amity University, Lucknow, to test how IoT-based technologies could improve banana crop production. The work focused on building a smart farming system that combined sensors, drones, and mobile communication to give farmers real-time insights into crop and soil conditions. The experimental farm included 300 banana plants equipped with different sensors soil moisture probes, NPK nutrient sensors, pH meters, humidity and temperature sensors, a rain detection module, and PIR motion sensors. These devices were connected to an Arduino Uno board, supported by a GSM module for wireless data transmission to farmers' mobile phones. Drones with cameras were also deployed to monitor crop density, detect infection. The system continuously collected data on key growth parameters soil pH (6.5), NPK levels (120 ppm nitrogen, 50 ppm phosphorus, 180 ppm potassium), soil moisture (70%), rainfall (1500 mm annually), temperature (28.5 °C), and humidity (65%). While these values were within acceptable ranges, the system highlighted where slight adjustments could further optimize yield. Farmers receive automatic alerts via their phones allowing them the time actions. Outcome revealed that this IoT based monitoring system offer a post effective and efficient way to manage crops. It also reduce resources by aligning irrigation and fertilization with actual crop requirements. The addition of drones further enhanced system capabilities. Overall the study demonstrated that IoT enabled precision farming can make banana cultivation more sustainable and profitable (Pandey et al., 2024).

## 2. AI-Based Crop Mapping in Jilin Province, China

In China, AI technologies combined with IoT sensors are widely employed to monitor crop growth, pest detection and optimize irrigation. (Jiang et al., 2022) study in Jilin Province demonstrated how artificial intelligence can significantly improve agricultural monitoring. They developed a crop mapping framework that combined deep neural networks with historical crop data and Sentinel-2 satellite imagery. The objective was to avoid costly and time consuming field surveys and create reliable crop classification maps. The researchers processed multi-temporal Sentinel-2 images in Google Earth Engine to capture seasonal variations. An ensemble learning approach was used to refine the historical crop data and generate high-quality training samples. Several machine learning algorithms support vector machine, decision tree, random forest and DNN were tested. Among these, the deep neural network consistently outperformed the others, producing classification maps with a kappa coefficient of up to 0.95. Rice and corn were identified with high accuracy, while soybeans remained more challenging due to spectral similarity with other crops. Some confusion also occurred between wetlands and rice paddies, highlighting the limitations of spectral-based approaches.

The study revealed that integrating AI with remote sensing enables large-scale, accurate, and cost-effective crop monitoring. Such systems not only support farm-level management but also contribute to national-level goals of food security, yield forecasting, and disaster assessment. By reducing reliance on extensive fieldwork, the framework provides a practical pathway for countries like China to modernize agricultural monitoring and decision-making (Jiang et al., 2022).

## 3. AI-Based Weed Detection and Precision Spraying in the United States

In the United States, artificial intelligence is being increasingly integrated into precision agriculture to address the challenge of weed management, which traditionally relies on heavy herbicide use. (Azghadi et al., 2024) study tested the application of YOLOv7, a deep learning object detection model, for identifying Palmer amaranth, one of the most aggressive weeds in U.S. cotton fields. Researchers joined this model with a robotic sprayer that selectively applied herbicide only where weeds were detected, rather than across entire fields. The system achieved a detection precision of about 60% and a recall of 62%, demonstrating the potential of AI to reduce unnecessary chemical application. The case study revealed how robotics and AI can jointly reduce chemical dependency, lower production costs, and limit environmental impacts. Building on this, ongoing projects at Pennsylvania State University are developing robotic precision sprayers for vineyards and berry crops. These autonomous ground robots adjust spray rates according to plant density. Together, these innovations highlight how AI-driven solutions in the U.S. are transforming weed and pesticide management into more sustainable, efficient, and environmentally responsible practices(Azghadi et al., 2024).

The survey further emphasizes that deep learning often outperforms traditional machine learning methods such as support vector machines or random forests, especially in tasks involving large and complex datasets. This means that farmers and researchers can rely on AI systems for faster, more reliable insights that were previously difficult to obtain. However, the authors also point out some challenges, including the need for large, well-annotated datasets and the computational demands of training deep models. In the context of precision agriculture, this study is particularly valuable because it demonstrates how AI can move beyond supportive roles and become central to automation.

## 4. Smart Irrigation System for Wheat in Southern Punjab, Pakistan

Wheat is one of the most important crops in Southern Punjab, but traditional farming practices often lead to significant yield losses due to climate fluctuations. Farmers usually depend on fixed irrigation schedules which fail to capture real-time changes in field conditions. To address this challenge, (Mustafa et al., 2024) developed an IoT-based monitoring and alerting system that helps farmers maintain an optimal conditions for wheat production. The system integrates simple yet effective components which includes sensor to measure temperature and humidity, an ESP8266 microcontroller for wireless monitoring, and a GSM module to send timely alerts directly to farmers via SMS when conditions are not ideal. The study demonstrated that low-cost sensors and mobile-based alerts can provide farmers with real-time guidance, enabling them to adjust irrigation. Findings revealed that such interventions reduce crop stress, optimize resource use and enhance wheat productivity. The case study underscores that precision agriculture does not always require high-end technologies simple, context-appropriate IoT solutions can make farming more resilient and productive when tailored to local farmer needs (Mustafa et al., 2024).

Pakistan has launched several other initiatives to integrate AI and smart technologies into agriculture. Smart farming systems are introduced in Lahore, Vihari, Sargodha and layyah by the collaboration of agriculture department and private partners. The project employs sensors along with gateway devices to collect real time crop data which is analyzed using AI Models. Similarly, the University of Agriculture has established a Pak-China joint lab under this initiative. The "High-Throughput Crop Monitoring Using Computer Vision for Climate Smart Agriculture" pilot has engaged around 600 farmers, with drone-based spraying. Complementing these efforts, the China-Pakistan developed mobile application Kisan360. Kisan360 provides farmers with AI-powered recommendations on crop health, irrigation, and fertilizer usage. Training sessions have already reached about 1,000 farmers across Punjab and Islamabad. In parallel, Lahore Garrison University developed an IoT-based system capable of predicting rice leaf diseases with an accuracy of nearly 97%.

Collectively, these initiatives represent Pakistan's growing commitment to adopting AI-driven smart farming technologies for improving productivity, sustainability, and resilience in agriculture.

*Figure 11. Kisan 360 App*

## VI. CHALLENGES AND BOTTLENECKS

The application of artificial intelligence in precision agriculture faces a number of challenges. Data heterogeneity is the most serious, which occurs when agriculture data is collected from many sources such as satellites UAVs IoT sensors weather station and farm records. All these data sets differ greatly in size, resolution and quality making it challenging for successful model training (Kamilaris & Prenafeta-Boldu, 2018). Agriculture environment differ significantly across regions, seasons and crop variabilities which makes it difficult to gather label datasets which are required for AI modeling, resulting in models that often lack transferability (Liakos et al., 2018). In many cases, models trained in one region under specific climatic and soil conditions fail when applied to another. This highlights the need for localized datasets and improved data-sharing frameworks to overcome heterogeneity and enhance reliability (Sarker et al., 2020).

Infrastructure deficits remain one of the most pressing barriers to modernizing agriculture in developing regions. Poor road networks and inadequate transport systems limit farmers' ability to access markets efficiently, often forcing them to sell produce at low farm-gate prices. Limited irrigation coverage further compounds the issue, as many farmers are still heavily dependent on rain fed agriculture, leaving them highly vulnerable.

Beyond physical infrastructure, digital literacy poses another challenge. Many farmers are unfamiliar with operating AI-based systems, drones, or IoT-enabled platforms, which makes it difficult for them to benefit from precision agriculture tools even when they are available. The high initial cost of these technologies including sensors, UAVs, and data platforms creates an additional barrier, especially for small-scale farmers with limited access to credit or financial services. Overcoming all these will require integrated strategies investments in rural infrastructure, financial inclusion, localized and affordable technologies, as well as farmer training and policy support to ensure that digital agriculture benefits even the most resource-constrained farmers.

## VII. FUTURE PROSPECTS

The convergence of big datasets, blockchain and robotics are the future of precision agriculture. Big data analytics enables the integration of data set ranging from local to satellite data align the algorithms to provide increasingly precise recommendations (Wolfert et al., 2017). Blockchain technologies can address trust challenges by ensuring transparency of agricultural products across supply chain, improving food safety and empowering farmers (Tripoli & Schmidhuber, 2020).

Meanwhile, robotics promise to reduce dependence on manual labor and enhance efficiency in large- and small-scale farms (Shamshiri et al., 2018). The combination of these technologies with AI is expected to create self-learning, automated farms capable of continuously optimizing resource use and productivity.

The integration of artificial intelligence (AI) in precision agriculture does not rely on technology alone but also on the presence of supportive policies and strong educational frameworks. Policy interventions must go beyond promoting innovation; they should directly address infrastructure gaps such as poor connectivity and limited energy supply. At the same time, policies need to ensure that AI tools are affordable, accessible, and designed with the needs of smallholder farmers. The establishment of ethical standards for data ownership and privacy is also important, as farmers should feel secure in sharing information (Bronson, 2019). Equally important is the role of education and training. Without adequate digital literacy and technical knowledge, even the most advanced systems may remain underutilized. Therefore, farmer-focused educational programs, community-based training sessions, and vocational courses are essential to effectively operate AI technologies (Klerkx et al., 2019).

Ultimately the future of agriculture hinge on multi stakeholder approach, combination of technology development with farmers training, policy support and global collaboration.

## VIII. CONCLUSION

Artificial intelligence has emerged as a transformative force in precision agriculture. Artificial intelligence offer solutions that address the core challenges of modern farming practices. AI-powered applications have demonstrated significant potential from crop classification and yield prediction to disease detection, irrigation scheduling and soil health monitoring (Jha et al., 2019; Liakos et al., 2018). Technologies such as UAVs, IoT devices, hyperspectral imaging, and cloud-based platforms have created an ecosystem where vast and complex datasets can be transformed into actionable insights for farmers. By integrating AI with automation, blockchain, and robotics, agriculture is moving toward data-driven and climate-smart practices (Shamshiri et al., 2018; Wolfert et al., 2017).

AI offers immense promise but its benefits will only be realized through adoption strategies that covers the existing digital social and infrastructure gap. Looking ahead, widespread adoption of AI in agriculture will require multi-stakeholder collaboration in which governments, researchers, commercial firms, and farmers work together to develop solutions adapted to specific agricultural problems. In this approach, artificial intelligence can become a cornerstone of sustainable farming systems, assuring food security, climate resilience, and equitable growth for decades to come.

## REFERENCES


Ahmad, S., Wassay, M., & Hussain, M. A. (2025). Cloud Computing: Empowering the Next Generation of Agricultural Research. *Trends in Applied Sciences*. 10.62324/TAPS/2025.060

Alonge, M., & Isreal, O. (2025). GPS and GNSS-based guidance systems for agricultural vehicles.

Alvim, S. J., Guimarães, C. M., Sousa, E. F., Garcia, R. F., & Marciano, C. R. (2022). Application of artificial intelligence for irrigation management: A systematic review. *Engenharia Agrícola*, *42*(spe), e20210159. 10.1590/1809-4430-Eng.Agric.v42nepe20210159/2022. DOI: 10.1590/1809-4430-eng.agric.v42nepe20210159/2022

Azghadi, M. R., Olsen, A., Wood, J., Saleh, A., Calvert, B., Granshaw, T., Fillols, E., & Philippa, B. (2024). Precise robotic weed spot-spraying for reduced herbicide usage and improved environmental outcomes—a real-world case study. *arXiv preprint arXiv:2401.13931*. 10.1016/j.compag.2025.110365

Barbedo, J. G. A. (2019). Plant disease identification from individual lesions and spots using deep learning. *Biosystems Engineering*, *180*, 96–107. 10.1016/j.biosystemseng.2019.02.002. DOI: 10.1016/j.biosystemseng.2019.02.002

Bondre, D. A., & Mahagaonkar, S. (2019). Prediction of crop yield and fertilizer recommendation using machine learning algorithms. *International Journal of Engineering Applied Sciences and Technology*, *4*(5), 371–376. https://doi.org/http://www.ijeast.com/. DOI: 10.33564/IJEAST.2019.v04i05.055

Bronson, K. (2019). Looking through a responsible innovation lens at uneven engagements with digital farming. *NJAS Wageningen Journal of Life Sciences*, *90*(1), 100294. 10.1016/j.njas.2019.03.001. DOI: 10.1016/j.njas.2019.03.001

Chlingaryan, A., Sukkarieh, S., & Whelan, B. (2018). Machine learning approaches for crop yield prediction and nitrogen status estimation in precision agriculture: A review. *Computers and Electronics in Agriculture*, *151*, 61–69. 10.1016/j.compag.2018.05.012. DOI: 10.1016/j.compag.2018.05.012

Evenson, R. E., & Gollin, D. (2003). Assessing the impact of the Green Revolution, 1960 to 2000. *Science*, *300*(5620), 758–762. DOI: 10.1126/science.1078710 PMID: 12730592

Fountas, S., Mylonas, N., Malounas, I., Rodias, E., Hellmann Santos, C., & Pekkeriet, E. (2020). Agricultural robotics for field operations. *Sensors (Basel), 20*(9), 2672. 10.3390/s20092672. DOI: 10.3390/s20092672 PMID: 32392872

García-Vera, Y. E., Polochè-Arango, A., Mendivelso-Fajardo, C. A., & Gutiérrez-Bernal, F. J. (2024). Hyperspectral image analysis and machine learning techniques for crop disease detection and identification: A review. *Sustainability (Basel), 16*(14), 6064. 10.3390/su16146064. DOI: 10.3390/su16146064

Gebbers, R., & Adamchuk, V. I. (2010). Precision agriculture and food security. *Science, 327*(5967), 828–831. 10.1126/science.1183899. DOI: 10.1126/science.1183899 PMID: 20150492

Jha, K., Doshi, A., Patel, P., & Shah, M. (2019). A comprehensive review on automation in agriculture using artificial intelligence. *Artificial Intelligence in Agriculture, 2*, 1-12. 10.1016/j.aiia.2019.05.004

Jiang, D., Chen, S., Useya, J., Cao, L., & Lu, T. (2022). Crop mapping using the historical crop data layer and deep neural networks: A case study in jilin province, china. *Sensors (Basel), 22*(15), 5853. 10.3390/s22155853. DOI: 10.3390/s22155853 PMID: 35957410

Jurišić, M., Plaščak, I., Barač, Ž., Radočaj, D., & Zimmer, D. (2021). Sensors and their application in precision agriculture. *Tehnički glasnik, 15*(4), 529-533. 10.31803/tg-20201015132216

Kaur, A., Bhatt, D. P., & Raja, L. (2024). Developing a hybrid irrigation system for smart agriculture using IoT sensors and machine learning in Sri Ganganagar, Rajasthan. *Journal of Sensors, 2024*(1), 6676907. 10.1155/2024/6676907. DOI: 10.1155/2024/6676907

Khaki, S., & Wang, L. (2019). Crop yield prediction using deep neural networks. *Frontiers in Plant Science, 10*, 621. 10.3389/fpls.2019.00621. DOI: 10.3389/fpls.2019.00621 PMID: 31191564

Klerkx, L., Jakku, E., & Labarthe, P. (2019). A review of social science on digital agriculture, smart farming and agriculture 4.0: New contributions and a future research agenda. *NJAS Wageningen Journal of Life Sciences, 90*(1), 100315. 10.1016/j.njas.2019.100315. DOI: 10.1016/j.njas.2019.100315

Krishnan, S. R., Nallakaruppan, M., Chengoden, R., Koppu, S., Iyapparaja, M., Sadhasivam, J., & Sethuraman, S. (2022). Smart water resource management using Artificial Intelligence—A review. *Sustainability (Basel), 14*(20), 13384. 10.3390/su142013384. DOI: 10.3390/su142013384

Liakos, K. G., Busato, P., Moshou, D., Pearson, S., & Bochtis, D. (2018). Machine learning in agriculture: A review. *Sensors (Basel)*, *18*(8), 2674. 10.3390/s18082674. DOI: 10.3390/s18082674 PMID: 30110960

Lu, J., Hu, J., Zhao, G., Mei, F., & Zhang, C. (2017). An in-field automatic wheat disease diagnosis system. *Computers and Electronics in Agriculture*, *142*, 369–379. 10.1016/j.compag.2017.09.012. DOI: 10.1016/j.compag.2017.09.012

Lu, J., Li, J., Fu, H., Tang, X., Liu, Z., Chen, H., Sun, Y., & Ning, X. (2024). Deep learning for multi-source data-driven crop yield prediction in northeast China. *Agriculture*, *14*(6), 794. 10.3390/agriculture14060794. DOI: 10.3390/agriculture14060794

Maes, W. H., & Steppe, K. (2019). Perspectives for remote sensing with unmanned aerial vehicles in precision agriculture. *Trends in Plant Science*, *24*(2), 152–164. 10.1016/j.tplants.2018.11.007. DOI: 10.1016/j.tplants.2018.11.007 PMID: 30558964

Miller, T., Mikiciuk, G., Durlik, I., Mikiciuk, M., Łobodzińska, A., & Śnieg, M. (2025). The IoT and AI in Agriculture: The Time Is Now—A Systematic Review of Smart Sensing Technologies. *Sensors (Basel)*, *25*(12), 3583. 10.3390/s25123583. DOI: 10.3390/s25123583 PMID: 40573470

Mulla, D. J. (2013). Twenty five years of remote sensing in precision agriculture: Key advances and remaining knowledge gaps. *Biosystems Engineering*, *114*(4), 358–371. 10.1016/j.biosystemseng.2012.08.009. DOI: 10.1016/j.biosystemseng.2012.08.009

Mushtaq, M., Ali, H., Raza, A., Maqbool, S., Safdar, M., Ahmed, M., & Sattar, J. (2024). Precision irrigation for sustainable agricultural productivity. In *Emerging Technologies and Marketing Strategies for Sustainable Agriculture* (pp. 184–208). IGI Global Scientific Publishing., DOI: 10.4018/979-8-3693-4864-2.ch010

Mustafa, S., Salahuddin, H., Ashraf, A., Shafique, M., Salahuddin, S., & Akbar, M. (2024). Smart Irrigation System for Wheat in Southern Punjab Using IOT-based Wireless Technology. *Journal of Computing & Biomedical Informatics*, *8*(01). 10.56979/801/2024

Obasekore, H., Fanni, M., Ahmed, S. M., Parque, V., & Kang, B.-Y. (2023). Agricultural robot-centered recognition of early-developmental pest stage based on deep learning: A case study on fall armyworm (Spodoptera frugiperda). *Sensors (Basel)*, *23*(6), 3147. 10.3390/s23063147. DOI: 10.3390/s23063147 PMID: 36991858

Ojha, T., Misra, S., & Raghuwanshi, N. S. (2015). Wireless sensor networks for agriculture: The state-of-the-art in practice and future challenges. *Computers and Electronics in Agriculture*, *118*, 66–84. 10.1016/j.compag.2015.08.011. DOI: 10.1016/j.compag.2015.08.011

Pandey, P., Kumar, S., Khanna, P., & Pragya, . (2024). Enhancing Agricultural Practices Through IoT-Based Smart Farming Technologies: A Case Study on Banana Crop Yield Optimization. *Asian Journal of Research in Computer Science, 17*(7), 55–66. DOI: 10.9734/ajrcos/2024/v17i7477

Pang, A., Chang, M. W., & Chen, Y. (2022). Evaluation of random forests (RF) for regional and local-scale wheat yield prediction in southeast Australia. *Sensors (Basel), 22*(3), 717. 10.3390/s22030717. DOI: 10.3390/s22030717 PMID: 35161467

Pingali, P. (2007). Agricultural mechanization: adoption patterns and economic impact. *Handbook of agricultural economics, 3*, 2779-2805. 10.1016/S1574-0072(06)03054-4

Sabir, R. M., Mehmood, K., Sarwar, A., Safdar, M., Muhammad, N. E., Gul, N., Athar, F., Majeed, M. D., Sattar, J., & Khan, Z. (2024). Remote sensing and precision agriculture: a sustainable future. In *Transforming agricultural management for a sustainable future: climate change and machine learning perspectives* (pp. 75–103). Springer., 10.1007/978-3-031-63430-7_4 DOI: 10.1007/978-3-031-63430-7_4

Saleem, M. F., Raza, A., Sabir, R. M., Safdar, M., Faheem, M., Al Ansari, M. S., & Hussain, S. (2024). Applications of sensors in precision agriculture for a sustainable future. In *Agriculture and Aquaculture Applications of Biosensors and Bioelectronics* (pp. 109–137). IGI Global Scientific Publishing., DOI: 10.4018/979-8-3693-2069-3.ch006

Sarker, I. H., Kayes, A., Badsha, S., Alqahtani, H., Watters, P., & Ng, A. (2020). Cybersecurity data science: An overview from machine learning perspective. *Journal of Big Data, 7*(1), 41. 10.1186/s40537-020-00318-5. DOI: 10.1186/s40537-020-00318-5

Senoo, E. E. K., Anggraini, L., Kumi, J. A., Karolina, L. B., Akansah, E., Sulyman, H. A., Mendonça, I., & Aritsugi, M. (2024). IoT solutions with artificial intelligence technologies for precision agriculture: Definitions, applications, challenges, and opportunities. *Electronics (Basel), 13*(10), 1894. 10.3390/electronics13101894. DOI: 10.3390/electronics13101894

Shamshiri, R., Kalantari, F., Ting, K., Thorp, K. R., Hameed, I. A., Weltzien, C., Ahmad, D., & Shad, Z. M. (2018). Advances in greenhouse automation and controlled environment agriculture: A transition to plant factories and urban agriculture. *International Journal of Agricultural and Biological Engineering, 11*(1), 1–22. https://doi.org/https://dx.doi.org/10.25165/j.ijabe.20181101.3210. DOI: 10.25165/j.ijabe.20181101.3210

Singh, A., Ganapathysubramanian, B., Singh, A. K., & Sarkar, S. (2016). Machine learning for high-throughput stress phenotyping in plants. *Trends in Plant Science*, *21*(2), 110–124. 10.1016/j.tplants.2015.10.015. DOI: 10.1016/j.tplants.2015.10.015 PMID: 26651918

Sishodia, R. P., Ray, R. L., & Singh, S. K. (2020). Applications of remote sensing in precision agriculture: A review. *Remote Sensing (Basel)*, *12*(19), 3136. 10.3390/rs12193136. DOI: 10.3390/rs12193136

Tripoli, M., & Schmidhuber, J. (2020). Emerging opportunities for the application of blockchain in the agri-food industry.

Tsouros, D. C., Bibi, S., & Sarigiannidis, P. G. (2019). A review on UAV-based applications for precision agriculture. *Information (Basel)*, *10*(11), 349. 10.3390/info10110349. DOI: 10.3390/info10110349

# Chapter 6
# Applications of Artificial Intelligence and Remote Sensing in Water Resources Management


**Wasiq Farooq**
*Agricultural Remote Sensing Lab, University of Agriculture, Faisalabad, Pakistan*

**Muhammad Safdar**
https://orcid.org/0009-0006-1779-6967
*Agricultural Remote Sensing Lab, University of Agriculture, Faisalabad, Pakistan*

**Hamza Anjum**
*Department of Computer Science, Ghulam Ishaq Khan Institute of Engineering Sciences and Technology, Khyber Pakhtunkhwa, Pakistan*

**Muhammad Adnan Shahid**
https://orcid.org/0000-0001-6403-0290
*Agricultural Remote Sensing Lab, University of Agriculture, Faisalabad, Pakistan*

**Muhammad Sajid Mehmood**
https://orcid.org/0000-0002-2055-3505
*School of Tourism and Planning, Pingdingshan University, Pingdingshan, China*

**Amina Rashid**
https://orcid.org/0000-0001-9772-1063
*Department of Agronomy, University of Agriculture, Faisalabad, Pakistan*

**Abdul Rauf**
*Department of Irrigation and Drainage, University of Agriculture, Faisalabad, Pakistan*

**Fahd Rasul**
*Department of Agronomy, University of Agriculture, Faisalabad, Pakistan*

**Imran Shauket**
https://orcid.org/0000-0001-8113-2248
*Department of Structures and Environmental Engineering, University of Agriculture, Faisalabad, Pakistan*


DOI: 10.4018/979-8-3373-6608-1.ch006

**Naeem Saddique**
*Department of Irrigation and Drainage, University of Agriculture, Faisalabad, Pakistan*

**Muntaha Munir**
https://orcid.org/0009-0005-3404-3636
*Institute of Botany, University of the Punjab, Lahore, Pakistan*

**Hafiz Muhammad Bilawal Akram**
https://orcid.org/0000-0001-6720-7266
*Department of Agronomy, University of Agriculture, Faisalabad, Pakistan*

## ABSTRACT

*The chapter discusses the need for improved water resource management due to climate change, population growth, and freshwater issues. The potential of artificial intelligence, including machine learning, deep learning, and hybrid models, in reshaping water management systems. The chapter explores AI-powered solutions for improved water assessment practices, including rainfall-runoff modeling, groundwater prediction, drought/flood forecasting, streamflow forecasting, aquifer recharge analysis, and hydrological anomaly detection. The integration of AI with remote sensing and geographic information systems is particularly important for enabling spatial-temporal watershed monitoring. Pakistan's AI deployments in early warning systems and water governance have made significant progress, but challenges like data scarcity, model overfitting, computational complexity, and skill gaps persist. The chapter discusses the need for scalable, transferable, and policy-supported AI solutions for water management in both developed and developing regions, emphasizing the adoption of explainable AI.*

## 1. INTRODUCTION

Water is the foundation of life, agriculture, industry, and human health. Although about three-quarters of Earth's surface is covered with water, only about 2.5% is fresh and about 1% is available for human use (Shiklomanov & Rodda, 2003). The available fresh water is facing immense pressure due to the growing global population, rapid urbanization, and industrial expansion. Meanwhile, climate change is altering rainfall patterns, which is complicating the availability and management of water resources (Parmesan et al., 2022).

Water resource challenges are becoming increasingly complex due to population growth, climate change, and stricter regulations. Managing water now requires long-term planning that considers both surface and groundwater (Muzammal et al., 2024). Such large-scale decisions involve multiple stakeholders, environmental concerns, and the need for fair distribution. At the same time, advances in computing power have transformed how information is processed, making technology an essential partner in water management. Finally, modern analytical tools, particularly systems analysis, have simplified decision-making, helping managers address the complexity of today's water issues more effectively (Simonovic, 2012). However, the complexity of hydrological systems often exceeds the predictive and adaptive capacities of traditional models.

*Figure 1. Relation between domain complexity, computer power, and complexity of modelling tool with time*

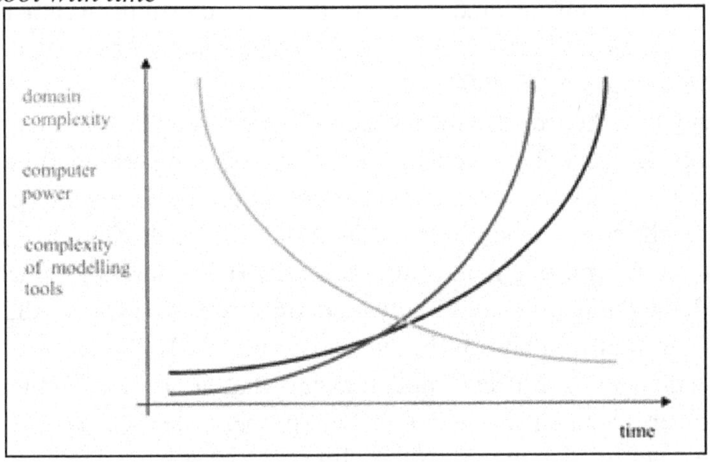


Water Resources Management is the process of planning, developing, and managing the optimum use of water resources (Loucks & Van Beek, 2017). Traditionally, water resource management focuses mainly on infrastructure developments, which include dams, canals, reservoirs, and levees, but the modern integrated water source management approach balances human and ecological necessities. Integrated water resource management promotes climate reduction and adaptation, and focuses on sustainable development. The goal of WRM is to ensure that water availability is ensured while also addressing challenges such as scarcity, pollution, and seasonal variability (van der Zaag & Savenije, 2004). WRM also addresses risks such as floods, droughts, and pollution. Modern WRM goes beyond infrastructure development; it incorporates resilience building, climate adaptation, and ecosystem sustainability (Xie, 2006).

Hydrological processes are not linear as they are influenced by numerous variables. Due to the uncertainty caused by climate change, traditional models often struggle to capture these variables. AI enables more accurate, timely, and data-driven decision-making; therefore, integrating Artificial Intelligence is crucial. AI has made us capable of analyzing large and complex datasets (Mosavi et al., 2018). Effective management requires both structural and non-structural approaches. Technological advancements have benefited WRM in recent years. Tools like Remote Sensing, Geographic Information Systems, and Artificial Intelligence have provided new capabilities to monitor, model, and predict hydrological processes (Mishra & Coulibaly, 2009).

Climate change has emerged as the greatest threat to sustainable water management in the current century. Climate change is impacting water availability, quality, and distribution through altered rainfall patterns, melting glaciers, changes in snowmelt, etc. (Sukanya & Joseph, 2023). Hydrological cycles have been disturbed by rising global temperatures, shifting precipitation patterns, and extreme weather events. All these extreme changes are also depressing the consistency of existing water supply systems (Kundzewicz et al., 2008).

Floods and droughts are also the most pressing challenges. Flash floods and riverine flooding are occurring frequently due to warmer temperatures and increasing rainfall. On the contrary, many regions experience prolonged dry spells, contributing to severe drought conditions (Hirabayashi et al., 2013). Agricultural production, food security, and rural livelihoods are facing enormous stress (Whitehead et al., 2009). Accelerated glacier melt initially increases water flow, but in the long term, it reduces freshwater availability (Immerzeel et al., 2010). Urban areas are facing additional challenges with heavy rainfall events that are overwhelming the drainage infrastructure, leading to urban flooding, property damage, and public health risks. On the other hand, hotter and drier climates increase the demand for water in growing cities (Boudreau et al., 2022).

All these challenges highlight the need for innovative approaches that integrate monitoring, prediction, and adaptive management. Remote sensing and Geographic Information Systems have already transformed hydrology by providing large-scale data. However, the exploration of available data requires equally advanced tools to process, interpret, and apply this information. Here, AI emerges as a transformative technology, capable of converting raw data into actionable insights.

AI techniques such as machine learning, deep learning, and expert systems are being used for flood forecasting, groundwater modelling, water quality monitoring, reservoir operation, and climate change adaptation. AI models trained on hydrological and meteorological data have improved short-term and long-term predictions. All these applications demonstrate that AI is not merely a sporting tool but a crucial enabler of modern water resource management. Unlike traditional tools, AI systems can continuously learn and improve with the amount of data available.

Despite significant advances in hydrological modeling and water management strategies, many areas still face increasing water-related risk due to climate change, population growth, and urbanization. Traditional models, while scientifically rigorous, are often constrained by data limitations, parameter uncertainties, and computational complexity. Therefore, they fail to provide reliable predictions under uncertain conditions.

On the other hand, today we have vast amounts of hydrological, meteorological, and remote sensing data, yet these are underutilized in many decision-making processes. The challenge lies in effectively handling these complex datasets to provide accurate predictions. Artificial Intelligence provides a set of powerful tools to address these gaps. However, the integration of AI into water resource engineering remains limited by challenges such as model interpretability, data quality, and the need for interdisciplinary collaboration.

Overall, all these insights outline the challenges of climate change to water resources, highlighting the need for integrated adaptive management and innovative technologies. Integrating AI into water resource management improves prediction accuracy, allows adaptive responses, optimizes resource allocation, and strengthens resistance to climate change impacts. AI is not a luxury but a necessity for sustainable management as water scarcity and variability increase worldwide.

### Objectives

To tackle the challenges in water resources, advanced tools are required that can deliver accurate predictions, real-time monitoring, and data-driven support for decision-making. AI and RS are being used to track water availability, monitor surface water and groundwater, and forecast floods and droughts. Focus is placed on the integration of AI and remote sensing, as their combination reinforces data analysis and provides valuable support for adaptive water management strategies. By presenting practical case studies and recent applications, the chapter highlights the real-world value of these technologies, while also recognizing the challenges, limitations, and future directions that must be considered to guide further research and innovation in this evolving field.

## 2. REVIEW OF CURRENT PRACTICES IN WATER RESOURCES MANAGEMENT

Before the emergence of advanced technologies, water resources management primarily relied on conventional hydrological and statistical approaches. Traditional methods for water availability assessment and forecasting were mainly dependent on direct measurements, empirical models, and hydrological observations. All these methods laid the foundation for understanding water systems.

## Traditional Methods

### a) Ground-Based Hydrological Measurements

Traditional water availability assessment depends heavily on field-based observations. Rain gauges, streamflow monitoring stations, piezometers, and evaporation pans were commonly used to record precipitation, river discharge, groundwater levels, and evaporation rates (Ward & Trimble, 2003). These datasets form the backbone of hydrological analysis and provide localized, long-term records essential for water balance studies. River discharge, streamflow, and reservoir water volume were measured by observing physical parameters such as flow velocity, cross-sectional area, and water level. Techniques include float methods, current meters, weirs, and flumes to estimate discharge accurately at specific points. Structural devices like weirs and flumes are constructed to create controlled flow conditions, enabling flow to be measured based on water height over the structure using standard equations (Simonovic, 2012).

### b) Water Balance Approach

The water balance method is one of the earliest techniques used to estimate water availability. The water balance approach is based on the principle of mass conservation, where the difference between precipitation, evapotranspiration, surface runoff, and groundwater recharge provides an estimate of water availability in a basin. This method is relatively simple but requires reliable data inputs, which are often scarce in developing regions.

### c) Statistical and Time-Series Analysis

Statistical approaches such as regression analysis, trend analysis, and time-series forecasting models have traditionally been employed to predict streamflow, rainfall, or groundwater levels (Box George et al., 1976). Empirical models like the Rational Method, Cook's Method, and the Curve Number method estimate runoff volumes from rainfall data and watershed characteristics. Analysis depends on historical records to project future conditions. Statistical analysis often fails to capture the nonlinear dynamics of hydrological systems.

### d) Conceptual Hydrological Models

Several conceptual hydrological models have been widely used for runoff simulation and forecasting. All these models use equations to simulate the key physical processes of rainfall-runoff (Birkel & Barahona, 2019). To make the model more detailed, they require extensive calibration and large amounts of input data. All these models have served as the foundation for water resource assessment and have been effectively applied in many regions. However, they often require extensive field data, skilled operation, and can be limited in spatial and temporal resolution compared to newer technologies.

Traditional methods have greatly contributed to hydrology, yet they face important drawbacks. Building and maintaining ground-based monitoring networks is costly, and in ungauged basins with no historical flow records, their effectiveness is very limited. They also struggle to capture spatial variations in rainfall, soil, and land use over large areas. Modern technology offers more data-driven approaches, which are more flexible, scalable, and accurate ways to assess and forecast water availability.

## Role of Hydrological Modeling and Remote Sensing

Water resources management requires reliable tools to understand, simulate, and predict the behavior of hydrological systems. Hydrological modeling and remote sensing are the most used approaches in this regard. Hydrological models are mathematical representations of the water cycle that help in understanding how precipitation translates into runoff, infiltration, groundwater recharge, and evapotranspiration (Devia et al., 2015). HEC-HMS, SWAT, and MIKE SHE are crucial models for runoff simulation and flood forecasting as they predict river flows, peak discharges, and flood risks (Beven, 2012). Watershed or basin scales are simulated to estimate available surface and groundwater resources under different scenarios. Hydrological models can evaluate how deforestation, urbanization, or climate change affect runoff, groundwater recharge, and evapotranspiration (Singh & Woolhiser, 2002). Model outputs inform water allocation, reservoir operations, and drought management strategies.

Examples of widely used models include SWAT (Soil and Water Assessment Tool), HEC-HMS, MIKE SHE, and data-driven models that incorporate machine learning for enhanced accuracy. However, hydrological models rely heavily on ground-based data for calibration and validation.

Remote sensing provides spatially continuous and temporally frequent data on key hydrological variables, bridging the gap where field measurements are sparse. Satellites such as TRMM and GPM provide rainfall estimates, while MODIS and Landsat offer snow cover data essential for runoff forecasting (Kidd & Huffman, 2011). Thermal sensors and microwave satellites (e.g., SMOS, SMAP) provide large-scale measurements for agricultural water management. Missions like GRACE detect changes in terrestrial water storage, providing indirect insights into groundwater depletion (Rodell et al., 2009). Remote sensing also allows continuous monitoring of watershed changes, including urban expansion, deforestation, and agricultural practices (DeVantier & Feldman, 1993).

Individually, hydrological models and RS have limitations, but their integration produces more robust outcomes. RS data can be used to parameterize, calibrate, and validate hydrological models, improving accuracy in ungauged or data-scarce basins. Conversely, hydrological models help interpret RS data by linking satellite-derived parameters to physical processes.

## Limitations in Hydrological Modeling and Remote Sensing

Despite significant advances in hydrological modeling and remote sensing, several limitations still constrain their ability to deliver highly accurate and real-time predictions of water availability. These challenges arise from data quality, model uncertainties, technical gaps, and the inherent complexity of hydrological systems.

### a) Data Gaps and Measurement Uncertainties

Many regions, especially in developing countries, lack dense hydro meteorological networks for rainfall, streamflow, and groundwater. This makes calibration and validation of models difficult (Sivapalan et al., 2003). While remote sensing provides large-scale coverage, rainfall and soil moisture products may suffer from coarse spatial resolution, retrieval errors, or cloud contamination (Hou et al., 2014). Moreover, Hydrological models require accurate soil, land use, and climate inputs, which are often generalized or uncertain. This introduces errors in runoff and water balance simulations.

### b) Model Structural and Process Uncertainty

Models like HEC-HMS or SWAT use mathematical approximations of complex processes such as infiltration, evapotranspiration, and groundwater recharge. These simplifications can lead to biased outputs. Hydrological processes vary across scales, and models calibrated for small watersheds may not perform well at larger basin levels (Beven, 2012). Different sets of parameters may produce equally good model outputs, making it difficult to identify a "true" solution.

### c) Challenges in Real-Time Monitoring

Satellite-based products often come with time lags ranging from hours to days, reducing their usefulness for real-time flood or drought management. Many satellite-derived products are not cross-validated with local ground measurements, lowering confidence in real-time applications. Real-time monitoring requires robust data transmission, storage, and processing infrastructure, which is often unavailable in remote or resource-limited areas.

### d) Climate and Human-Induced Variability

The increasing frequency of flash floods, droughts, and intense rainfall events due to climate change poses a major challenge to prediction accuracy. Dam construction, irrigation withdrawals, and rapid urbanization have altered the natural hydrological cycle, which makes model predictions less precise if these interventions are not incorporated.

## 3. ARTIFICIAL INTELLIGENCE TECHNIQUES IN WATER RESOURCES

Unlike traditional methods, which focus on fixing formulas and assumptions, AI techniques can utilize exceptionally large data, capture nonlinear relationships, and adapt to dynamic environments. Tackling challenges due to climate variability, data scarcity, and growing human demands on water has become easier. AI also enables real-time forecasting for floods, droughts, and water demand.

*Figure 2. AI and its Domains*

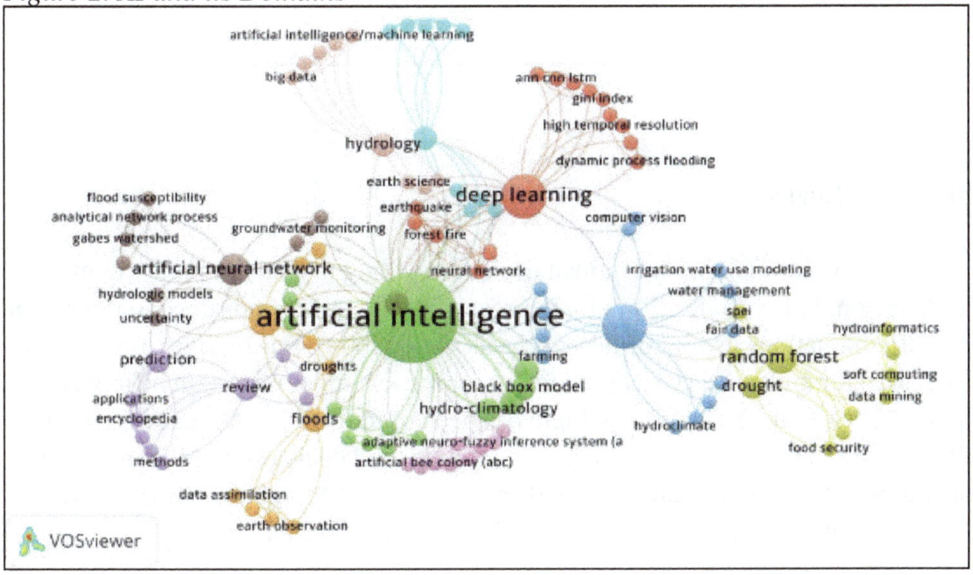

a) **Artificial Neural Networks**

Artificial Neural Networks (ANNs) are layers of interconnected nodes that process input information and generate outputs through hidden layers, enabling them to recognize patterns and capture complex relationships in data.(Maier et al., 2010). It is widely used in water resources due to its ability to capture complex and nonlinear relationships. ANNs are most widely used for rainfall–runoff modeling, streamflow forecasting, and groundwater level prediction. They are also used for water quality monitoring, flood forecasting, and drought assessment by analyzing large and often incomplete datasets.

*Figure 3. Layers of ANNs*

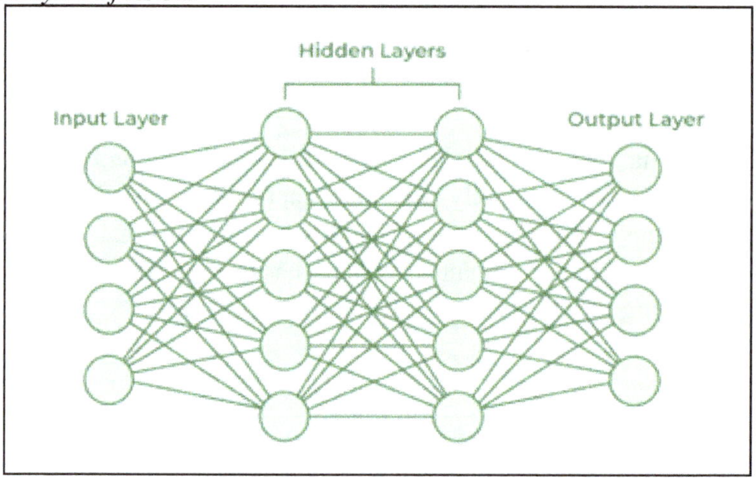


b) **Fuzzy Logic**

Fuzzy Logic (FL) mimics human decision-making to deal with uncertainty and imprecision. Fuzzy logic allows for degrees of truth, unlike traditional logic, which classifies situations as true or false. (Nayak et al., 2004). Fuzzy logic models have been used for multi-reservoir operation optimization under uncertain inflow conditions (Nayak et al., 2004). Fuzzy logic has been used for flood forecasting, drought risk assessment, groundwater potential mapping, and decision-making in reservoir operations. It enables researchers and decision-makers to combine expert knowledge with available data, providing more flexible and realistic solutions.

*Figure 4. Difference between Boolean logic and Fuzzy Logic*

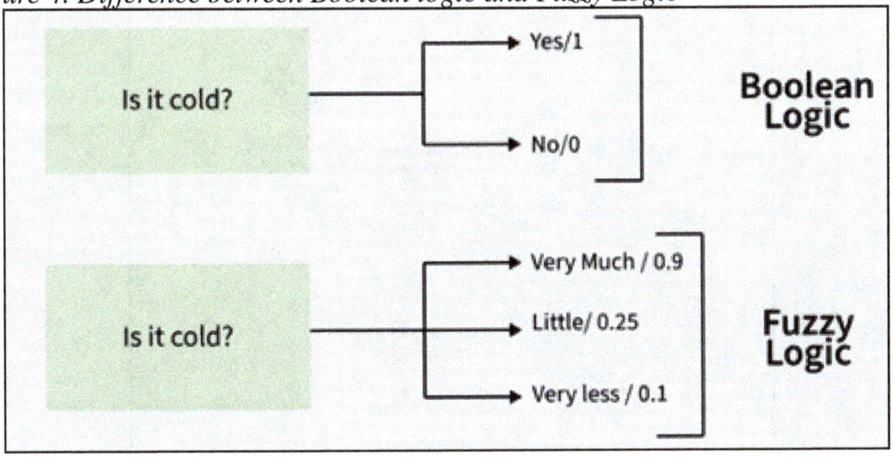


c) **Support Vector Machines**

Support Vector Machines (SVMs) work by finding an optimal boundary that separates different categories of data because they supervise learning algorithms. Streamflow forecasting, groundwater quality classification, and rainfall prediction are done by support vector machines (Behzad et al., 2010). SVMs have been applied to forecast monthly streamflow with high accuracy (Asefa et al., 2006). Their ability to perform well with limited datasets makes it unique. Moreover, SVMs are less prone to overfitting compared to some other AI techniques, ensuring reliable predictions in both training and testing phases. SVMs are becoming increasingly popular tools for researchers and practitioners due to their adaptability.

*Figure 5. Support Vector Machines finding an optimal boundary*

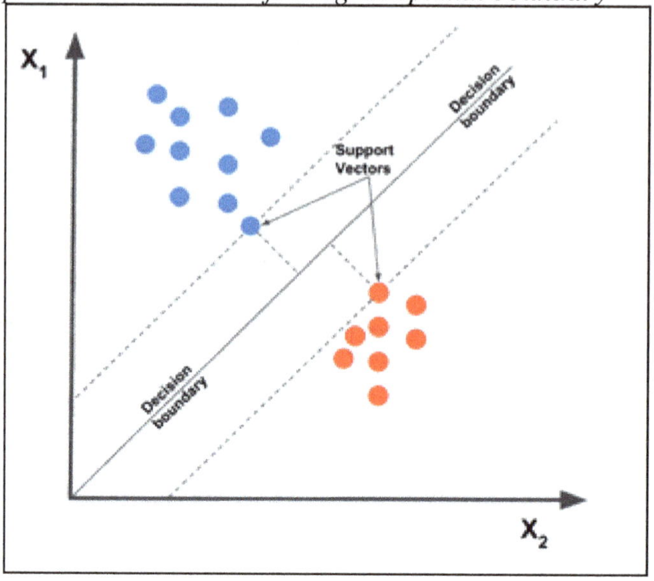


d) **Genetic Algorithms and Evolutionary Computation**

Genetic Algorithms are optimization techniques inspired by the process of natural selection. They use mechanisms such as selection, crossover, and mutation to evolve solutions to complex problems. More broadly, evolutionary computation refers to a family of algorithms that simulate biological evolution to search for optimal or near-optimal solutions (Nicklow et al., 2010). Genetic Algorithms are used for optimization of reservoir operations, groundwater pumping strategies, and calibration of hydrological models. GAs have been widely applied for optimal reservoir rule curve design to balance flood control and water supply (Oliveira & Loucks, 1997). The ability to search through a wide range of possible solutions and track the most suitable outcome even in the presence of uncertainty or a rough data set strengthens genetic algorithms and evolutionary computation. They are used in multi-objective optimization problems such as minimizing water losses while maximizing supply reliability.

*Figure 6. Genetic Algorithms working procedure*

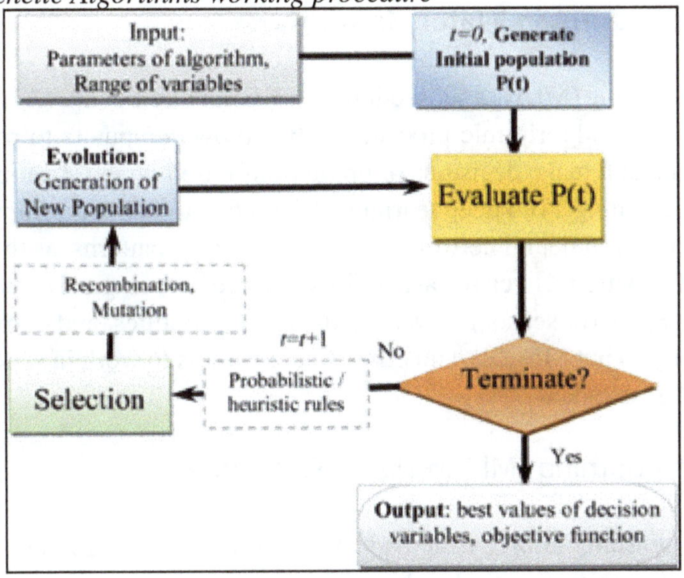


e) **Hybrid and Ensemble Approaches**

Hybrid approaches combine two or more AI methods, such as ANN + Fuzzy Logic, or SVM + Genetic Algorithms, to enhance their individual strengths. Ensemble approaches, on the other hand, integrate multiple models and average or weight predictions to improve accuracy and robustness. Hybrid ANN–SWAT models have been developed to improve runoff predictions by integrating physical and data-driven approaches (Abrahart & See, 2002). Similarly, assembly techniques involve running several models simultaneously and combining their outputs to provide more reliable outcomes.

All these approaches have been successfully applied in flood forecasting, rainfall run-off modelling, groundwater level prediction, and water quality assessment. The adaptability of ANNs, the rule-based reasoning of fuzzy logics, and the optimization power of GA's hybrid and ensemble methods provide the most robust and accurate outcomes, making them particularly valuable in decision-making in water source engineering.

## Machine Learning (ML), Deep Learning (DL), and AI Algorithms in Water Resources

Machine learning (ML) is a subsection of Artificial Intelligence (AI) and encompasses a variety of algorithmic procedures that allow computers to reliably learn patterns in data and make decisions or predictions based on those patterns without explicit programming. And deep learning (DL) refers to ML algorithms using one or more multilayered neural networks to model complex patterns, as the brain does to process information. Given the availability of large hydrological datasets through records of inexpensive sensors, advanced data from satellites, and climate models, the opportunity to put ML, DL, and other AI algorithms to work in water resources management is abundant.

### a) Machine Learning (ML) in Water Resources

Machine Learning allows systems to learn from data, improving their performance on tasks automatically. Machine Learning is made up of four components that involve: data preparation, feature selection, training a model, and evaluation to make predictions and/or classify data to predict outcomes or classify information. Machine learning, once identified, is a specific set of algorithms that allows us to learn from historical data to help the system predict or decide something, learned on its own without coding. The use of Machine Learning helps in streamflow and runoff forecasting, in water quality prediction, and in mapping of groundwater. It can be directly implemented to use pointers of information for an extensive amount of data. (Mosavi et al., 2018). ML works well with nonlinear systems and integrates multi-source data (e.g., rainfall, land use, soil).

### b) Deep Learning in Water Resources

Deep learning (DL) is a continuation of machine learning (ML), which employs deep neural networks with multiple "hidden" layers of interconnected neurons that can learn from one another. These layers extract features progressively from the input with both forward and backward propagation while learning to improve prediction accuracy through an iterative process. Deep learning approaches often employ Convolutional Neural Networks to analyze rainfall, while Recurrent Neural Networks are used to capture the temporal dynamics of flood events for spatiotemporal flood forecasting. (Sit et al., 2020).

## c) Other AI Algorithms in Water Resources

AI algorithms include Evolutionary Algorithms (EAs) and Hybrid AI Models. Evolutionary Algorithms include techniques like Genetic Algorithms (GAs) and Particle Swarm Optimization (PSO), which are widely used for multi-objective optimization in reservoir management. Hybrid AI models include the combination of physics-based hydrological models with AI techniques, e.g., SWAT + ANN, HEC-HMS + SVM. To take advantage of a data-driven model and a process-oriented model, these AI techniques are utilized. These approaches improve prediction accuracy by capturing the complex nonlinear interactions within hydrological systems that traditional models often fail to represent.

## AI for Rainfall-Runoff Modeling and Streamflow Forecasting

Rainfall-runoff modeling and streamflow forecasting are fundamental processes in the discipline of water resources engineering. Historically, these processes have been modeled using traditional, weathered, conceptual, or physically based hydrologic models such as HEC-HMS or SWAT. They offer a considerable degree of reliability; however, they do require significant calibration, high-quality input data, and do not inherently reflect the nonlinear and dynamic nature of the hydrology system. Artificial Intelligence (AI), particularly Machine Learning (ML) and Deep Learning (DL), represents a fresh opportunity for moisture forecasting to learn from past offences to understand and reveal implicit relationships in the rainfall runoff relationships.

Artificial Neural Networks (ANNs) are a typical choice for simulating rainfall runoff relationships. ANNs can capture rainfall inputs and map the output to simulated streamflow. ANNs have been shown to outperform traditional statistical models, particularly with respect to short-term flood forecasting (Dawson & Wilby, 2001). Support Vector Machines (SVM) achieve remarkable results for forecasting daily or monthly streamflow, particularly for basins with small datasets, where there is a risk of overfitting with neural networks (Asefa et al., 2006). Random Forests (RF) and Gradient Boosting capture the benefit of capturing the multi-variability and interactions often overlooked when building purely composite models and generate empirically based ensemble predictions, which can be powerful with minimal computational cost.

Recurrent Neural Networks (RNNs) are purpose-built for sequential data and can represent the temporal relationships in rainfall–runoff processes. Long short-term memory networks (LSTMs) are a type of RNN that are particularly well suited for hydrology because they can effectively model long-term dependencies in rainfall and streamflow time series (Kratzert et al., 2018). Benefits of Artificial Intelligence in Streamflow Forecasting. AI models can rapidly produce forecasts after training, which makes them suitable for real-time applications for flood early warning systems.

AI models rely on access to large, high-quality datasets for training. Inappropriate or limited datasets can limit performance. Many AI models, especially deep learning models, are opaque in terms of how predictions are generated, which can limit acceptability by decision-makers. AI models may perform well on training data, but performance can be poor in real-world forecasting when not validated.

## AI in Groundwater Level Prediction and Aquifer Recharge Analysis

Groundwater is perhaps one of the most important sources of freshwater that allows agriculture, domestic supply, and industry globally. However, groundwater resources and their management have become more complex with rising demand, over-consumption, and climate fluctuations. Assessing levels of groundwater and recharge rates is crucial for sustainable management and use of these resources, yet these assessments are difficult given the hidden nature of aquifers, sparse monitoring networks, and uncertainties about the recharge process.

Artificial Neural Networks are among the most widely used methods for forecasting groundwater levels. They work by linking available data such as rainfall, groundwater pumping, and historical water table records to generate predictions (Lee et al., 2019). In addition (Deka, 2014), Random Forests and other ensemble models provide further improvements by integrating multiple data sources such as climate variables, pumping records, and soil characteristics, thereby enhancing prediction accuracy. ANNs have been successfully implemented in India and China to forecast monthly groundwater levels, supporting irrigation planning and water allocation in agricultural regions. These practical implementations reveal that machine learning can bridge knowledge gaps and strengthen groundwater management strategies.

Recurrent Neural Networks (RNNs) and Long Short-Term Memory (LSTM) network models are particularly effective because aquifers typically respond to changes in rainfall and streamflow with a noticeable delay, which LSTMs are well-suited to capture. When working with spatial data, Convolutional Neural Networks are also useful through the integration of remote sensing products such as groundwater anomalies derived from GNACE satellites, land cover maps, soil moisture datasets, etc., with hydrological inputs to produce larger recharge patterns. If additional reliability is desired, hybrid options are also becoming popular, where the outputs from deep learning models augment physical models and deep learning models (e.g., MODFLOW and LSTM) (Pham et al., 2022). These hybrid frameworks not only enhance predictive accuracy but also provide deeper insights into the physical mechanisms governing aquifer recharge. Together, these deep learning methods enable more precise, data-driven strategies for managing groundwater sustainability in the face of increasing demand and climate variability.

While Artificial Intelligence has presented new possibilities for managing water resources, there are still notable challenges. Good datasets are often in short supply, specifically for groundwater or hydrology. This makes the models less accurate. Many advanced models also act as "black boxes" when predicting, giving little insight into the underlying processes. Often, a model trained in one area will not produce usable predictions in another area without some form of adjustment. In summary, there are significant challenges and opportunities to combine AI with robust field knowledge, increased data, and ready-to-interpret modeling processes, as a way to generate more reliable, effective outcomes.

### Use of AI in Flood and Drought Risk Assessment

Floods and droughts are two of the most devastating natural hazards affecting the availability of potable and irrigation water for people, agriculture, infrastructure, and livelihoods across the globe. Existing risk assessment practice that includes reliance on historical hydrological records, statistics, and physical models can be limited. Artificial Intelligence (AI) techniques, specifically Machine Learning (ML) and Deep Learning (DL), are now increasingly available tools to help researchers overcome challenges by informing and learning from diverse datasets to achieve new perspectives and improve risk forecasting (Mosavi et al., 2018).

AI has been used to improve rainfall-runoff modelling, forecast river discharge, and perform inundation mapping. New methods derived from DL, such as Convolutional Neural Networks, can learn spatial features from satellite imagery, allowing for near real-time flood mapping and monitoring inundation (Imran et al., 2023). extent; this is beneficial in regions where operational hydrological networks do not exist or are unavailable (Hao et al., 2018)

Flood and drought risk assessment enhances prediction accuracy while also strengthening disaster preparedness through real-time monitoring, early warning, and probabilistic risk assessments. AI enables water managers, policymakers, and local communities to carry out timely actions.

## 4. INTEGRATION OF AI WITH REMOTE SENSING AND GIS

The integration of Artificial Intelligence with Remote Sensing and Geographic Information Systems is increasing the possibilities of water resources management. Remote Sensing provides valuable information regarding land cover, vegetation coverage, and water bodies (Sabir et al., 2024). On the other hand, GIS offers a clear and accessible way to organize and analyze these data (Safdar et al., 2023). The combination of AI with RS and GIS provides a robust foundation for decision-making related to water availability, distribution, and risk management (Parmar, 2023).

Machine Learning and Deep Learning algorithms can be used to analyze high-resolution satellite images obtained from remote sensing for flood mapping and monitoring (Mosavi et al., 2018). The integration of AI, Remote Sensing (RS), and GIS can accurately forecast drought intensity and duration, while GIS aids in mapping drought-prone regions and identifying vulnerable communities, providing a comprehensive framework for data-driven water resource management.

## AI Models for Satellite Data Analysis

Satellite remote sensing provides large-scale images of the Earth's surface. The complexity of satellite data, which ranges from multispectral images to radar, provides distinct hurdles for conventional interpretation (Ma et al., 2015). AI models are used for evaluating satellite data. Random Forests and Support Vector Machines are frequently used to address land use and land cover concerns (Abdi, 2020). Furthermore, deep learning algorithms have proved extremely fruitful in evaluating high-resolution data for identifying water bodies (Maxwell et al., 2018).

AI models enhance data fusion capabilities by combining multiple satellite sources for more reliable outputs. This can improve the detection of flooding and near-real-time flood mapping. Hybrid and ensemble AI models for satellite data analysis minimize uncertainty and improve prediction robustness. Ensembles of CNNs and Random Forests can improve drought monitoring by combining vegetation indices with soil moisture data from satellites. AI integration with satellite data aids in water body mapping, drought forecasting, groundwater monitoring, and flood risk assessment. As satellite missions expand, AI-driven analysis will continue to play a crucial role in sustainable water resource management.

## Smart Hydrological Modeling Using AI + RS Data Fusion

By combining Artificial Intelligence with Remote Sensing data, intelligent hydrological modeling is becoming an increasingly effective approach to manage and understand complicated hydrological systems (Kharat, Darade, et al. 2025). Ground observations can be limited in hydrological models because of errors from spatial variability and gaps in actual data. With the integration of AI methods and remotely sensed data, the opportunity to obtain real-time, higher-resolution values of rainfall, effective rainfall, soil moisture, land use, evapotranspiration, and surface waters has developed. RS provides spatial coverage and continuous monitoring, while AI improves predictive power, pattern acknowledgement, and data integration (See & Abrahart, 2001). AI-RS fusion could support real-time decision-making.

In short, the combination of AI and remote sensing in hydrological modeling not only overcomes the limitations of conventional approaches but also provides a smarter, more holistic framework to manage water resources.

## 5. CASE STUDIES

Watersheds are complex systems influenced by rainfall, land use, soil, and topography, making traditional monitoring approaches time-consuming and inaccurate. AI improves GIS-based watershed studies by analyzing massive datasets, revealing hidden patterns, and boosting prediction skills. Several examples demonstrate the advantages of this connection.

### 1. Rainfall–Runoff Modeling in the Upper Indus Basin, Pakistan

AI models are being utilized in GIS frameworks to improve rainfall-runoff simulations, particularly in data-scarce regions like the Upper Indus Basin. The upper Indus basin is the most critical source of water in South Asia, which supplies water to millions of people. It contributes to the flow of the Indus River, which sustains agriculture, hydropower generation, domestic consumption, and industrial usage. The Indus basin is characterized by complex hydrological processes driven by a combination of snowmelt, glacier melt, and monsoon rainfall. Accurate rainfall runoff in the morning is essential for effective flood forecasting, reservoir operation, irrigation planning, and long-term water source management.

Traditionally, hydrologists have relied on conceptual models such as HEC-HMS to stimulate rainfall runoff modelling. HEC-HMS has been widely applied in the Indus Basin for flood forecasting and reservoir management, but the application of such models in the UIB is constrained by limited data availability, parameter uncertainty, and calibration challenges. These issues often reduce the reliability of predictions.

To overcome these limitations, researchers have increasingly explored the potential of artificial intelligence models. AI techniques, particularly artificial neural networks, are capable of learning complex nonlinear relationships between inputs and outputs without the need for extensive physical parameterization.

(Rauf & Ghumman, 2018) Studied both the HEC-HMS conceptual model and an ANN model to predict streamflow using daily precipitation, temperature, and streamflow data from the Upper Indus River. HEC HMS provided physical interpretable outcomes while ANN's model captured nonlinear relationships more effectively and delivered slightly improved predictive performance in some scenarios. Both approaches provided reliable predictions, highlighting the potential of combining traditional hydrological models with AI techniques, despite the need for careful validation of their accuracy.

The case study demonstrated that both the HEC-HMS and the Artificial Neural Network (ANN) models produced reliable streamflow predictions, but the ANN showed superior performance. The HEC-HMS model achieved a Nash–Sutcliffe Efficiency of 0.74 and an $R^2$ value of 0.78, while the ANN reached an NSE of 0.82 and an $R^2$ of 0.85. The higher efficiency and correlation in the ANN highlight its ability to better capture the nonlinear and dynamic nature of rainfall–runoff processes, particularly during peak flow events where traditional conceptual models often struggle. However, despite this improved accuracy, the authors pointed out that ANN models lack physical interpretability, which limits their direct use for understanding hydrological processes.

The case study highlights that **AI is not necessarily a replacement for traditional hydrological models but rather a powerful complement** in data-limited regions like the UIB. While models remain crucial for understanding hydrological processes and ensuring transferability across scenarios, AI offers enhanced predictive capacity, especially under uncertain and changing conditions.

*Figure 7. ANN architecture for Rainfall–Runoff Modeling in the Upper Indus Basin*

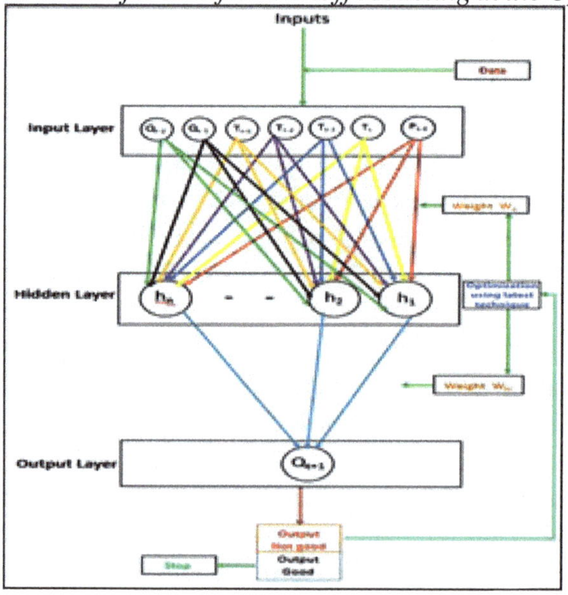

## 6. CHALLENGES AND LIMITATIONS

AI's global widespread use is hindered by data and model transferability issues, particularly in underdeveloped nations, where accurate hydrological data has limited spatial and temporal ranges (Mosavi et al., 2018). Computational cost is a significant barrier in deep learning, as it requires large datasets and processing capabilities, leading to overfitting and incorrect predictions (Sit et al., 2020). AI adoption in water management faces challenges due to a lack of interdisciplinary teams, talent shortages, and inadequate training opportunities. Addressing these issues will involve increased open data sharing, computational capabilities investments, and capacity-building efforts among hydrologists, computer scientists, and policymakers.

## 7. FUTURE DIRECTIONS

Artificial Intelligence (AI) is expected to play a significant role in water resources management, particularly in the use of cloud-based platforms and IoT sensors. This integration will enable faster analytics, decision-making, and up-to-date information for hydrologists and policymakers, enhancing disaster preparedness and disaster risk management (Seel et al., 2025). X-Aid (XAI) has been used in recent research to assess deep learning models trained on hydrological data, enhancing our understanding of processes and functions in watersheds. This approach increases trust and ensures AI suggestions align with physical hydrological concepts and policy frameworks (Arrieta et al., 2020).

Furthermore, expanding the implications for policymakers and addressing data constraints, particularly in developing countries, would significantly enhance the study's impact and applicability. Artificial Intelligence in water resource engineering is not only a technical advancement but also a decision-support tool for governments and planning agencies. By integrating AI into early warning systems, disaster preparedness can be improved through more accurate flood and drought forecasts, thereby reducing human and economic losses. AI-based groundwater prediction can inform extraction policies, helping to prevent overexploitation in agricultural regions. Similarly, reservoir operation models powered by AI can guide policymakers in balancing hydropower generation, irrigation, and environmental flow requirements under uncertain climatic conditions.

AI models are crucial for meeting water demands in developing nations, offering a global solution to sustainability. These flexible models can work in various hydroclimatic scenarios, leading to a more intelligent and inclusive approach to addressing global water security issues (Sit et al., 2020). However, a major challenge in developing countries lies in data availability and quality. Hydrological and meteorological records are often sparse, inconsistent, or missing due to limited monitoring infrastructure. This constrains the training and validation of AI models, which require large and reliable datasets. Policymakers therefore need to prioritize investments in monitoring networks, satellite data integration, and open-data platforms. Together, these directions enhance the reliability, explainability, and accessibility of AI-driven water resources management globally, particularly benefiting data-scarce and resource-constrained environments.

## 8. CONCLUSION

Artificial Intelligence (AI), Remote Sensing (RS), and Geographic Information Systems (GIS) have a transformative impact on water resource management. AI-based technologies, which include machine learning, deep learning, and hybrid approaches, have shown that they can better depict complicated, nonlinear hydrological systems than conventional methodologies. AI also improves operational efficiency by enabling predictive maintenance, early leak detection, and improved infrastructure management, all of which result in cost savings and environmental benefits.

Cloud platforms and IoT sensors will increase the availability of real-time data collection and forecasting, while Explainable AI will be employed to ensure that model results are trustworthy and scientifically interpretable. Furthermore, it will be critical to construct scalable and transferable AI models to assist developing countries where a lack of data is an urgent problem. AI may be utilized to establish sustainable and equitable methods for water management through technical innovation and interdisciplinary collaboration. To ensure resilient and sustainable water management benefits communities in many parts of the world, governments and organizations must prioritize open data sharing and integrated smart water systems. The world is witnessing a strong array of AI-based tools to facilitate sustainability, resilience, and effective operations within water resource management.

## REFERENCES


Abdi, A. M. (2020). Land cover and land use classification performance of machine learning algorithms in a boreal landscape using Sentinel-2 data. *GIScience & Remote Sensing*, *57*(1), 1–20. 10.1080/15481603.2019.1650447. DOI: 10.1080/15481603.2019.1650447

Abrahart, R. J., & See, L. (2002). Multi-model data fusion for river flow forecasting: An evaluation of six alternative methods based on two contrasting catchments. *Hydrology and Earth System Sciences*, *6*(4), 655–670. 10.5194/hess-6-655-2002. DOI: 10.5194/hess-6-655-2002

Arrieta, A. B., Díaz-Rodríguez, N., Del Ser, J., Bennetot, A., Tabik, S., Barbado, A., García, S., Gil-López, S., Molina, D., & Benjamins, R. (2020). Explainable Artificial Intelligence (XAI): Concepts, taxonomies, opportunities and challenges toward responsible AI. *Information Fusion*, *58*, 82–115. 10.1016/j.inffus.2019.12.012. DOI: 10.1016/j.inffus.2019.12.012

Asefa, T., Kemblowski, M., McKee, M., & Khalil, A. (2006). Multi-time scale stream flow predictions: The support vector machines approach. *Journal of Hydrology (Amsterdam)*, *318*(1-4), 7–16. 10.1016/j.jhydrol.2005.06.001. DOI: 10.1016/j.jhydrol.2005.06.001

Behzad, M., Asghari, K., & Coppola, E. A.Jr. (2010). Comparative study of SVMs and ANNs in aquifer water level prediction. *Journal of Computing in Civil Engineering*, *24*(5), 408–413. 10.1061/(ASCE)CP.1943-5487.0000043. DOI: 10.1061/(ASCE)CP.1943-5487.0000043

Beven, K. J. (2012). *Rainfall-runoff modelling: the primer*. John Wiley & Sons. DOI: 10.1002/9781119951001

Birkel, C., & Barahona, A. C. (2019). Rainfall-runoff modeling: a brief overview. *Reference module in earth systems and environmental sciences*. 10.1016/B978-0-12-409548-9.11595-7

Boudreau, K., Robinson, M., & Farooqi, Z. (2022). IPCC Sixth assessment report: climate change 2021: the physical science basis summary for policymakers. *Canadian Journal of Emergency Management*, *2*(1). Advance online publication. 10.25071/6sw6za31. DOI: 10.25071/6sw6za31

Box George, E., Jenkins Gwilym, M., Reinsel Gregory, C., & Ljung Greta, M. (1976). *Time series analysis: forecasting and control*. Holden Bay., 10.2307/3008255

Dawson, C., & Wilby, R. (2001). Hydrological modelling using artificial neural networks. *Progress in Physical Geography*, *25*(1), 80–108. 10.1177/030913330102500104. DOI: 10.1177/030913330102500104

Deka, P. C. (2014). Support vector machine applications in the field of hydrology: A review. *Applied Soft Computing*, *19*, 372–386. 10.1177/030913330102500104. DOI: 10.1016/j.asoc.2014.02.002

DeVantier, B. A., & Feldman, A. D. (1993). Review of GIS applications in hydrologic modeling. *Journal of Water Resources Planning and Management*, *119*(2), 246–261. 10.1061/(ASCE)0733-9496(1993)119:2(246). DOI: 10.1061/(ASCE)0733-9496(1993)119:2(246)

Devia, G. K., Ganasri, B. P., & Dwarakish, G. S. (2015). A review on hydrological models. *Aquatic Procedia*, *4*, 1001–1007. 10.1016/j.aqpro.2015.02.126. DOI: 10.1016/j.aqpro.2015.02.126

Hao, Z., Singh, V. P., & Xia, Y. (2018). Seasonal drought prediction: Advances, challenges, and future prospects. *Reviews of Geophysics*, *56*(1), 108–141. 10.1002/2016RG000549. DOI: 10.1002/2016RG000549

Hirabayashi, Y., Mahendran, R., Koirala, S., Konoshima, L., Yamazaki, D., Watanabe, S., Kim, H., & Kanae, S. (2013). Global flood risk under climate change. *Nature Climate Change*, *3*(9), 816–821. DOI: 10.1038/nclimate1911

Hou, A. Y., Kakar, R. K., Neeck, S., Azarbarzin, A. A., Kummerow, C. D., Kojima, M., Oki, R., Nakamura, K., & Iguchi, T. (2014). The global precipitation measurement mission. *Bulletin of the American Meteorological Society*, *95*(5), 701–722. 10.1175/BAMS-D-13-00164.1. DOI: 10.1175/BAMS-D-13-00164.1

Immerzeel, W. W., Van Beek, L. P., & Bierkens, M. F. (2010). Climate change will affect the Asian water towers. *science, 328*(5984), 1382-1385. 10.1126/science.1183188

Imran, M., Majeed, M. D., Zaman, M., Shahid, M. A., Zhang, D., Zahra, S. M., Sabir, R. M., Safdar, M., & Maqbool, Z. (2023). Artificial neural networks and regression modeling for water resources management in the upper Indus Basin. *Environmental Sciences Proceedings, 25*(1), 53. 10.3390/ECWS-7-14199

Kidd, C., & Huffman, G. (2011). Global precipitation measurement. *Meteorological Applications*, *18*(3), 334–353. 10.1002/met.284. DOI: 10.1002/met.284

Kratzert, F., Klotz, D., Brenner, C., Schulz, K., & Herrnegger, M. (2018). Rainfall–runoff modelling using long short-term memory (LSTM) networks. *Hydrology and Earth System Sciences*, *22*(11), 6005–6022. 10.5194/hess-22-6005-2018. DOI: 10.5194/hess-22-6005-2018

Kundzewicz, Z. W., Mata, L., Arnell, N. W., Döll, P., Jimenez, B., Miller, K., Oki, T., Şen, Z., & Shiklomanov, I. (2008). The implications of projected climate change for freshwater resources and their management. *Hydrological Sciences Journal*, *53*(1), 3–10. 10.1623/hysj.53.1.3. DOI: 10.1623/hysj.53.1.3

Lee, S., Lee, K.-K., & Yoon, H. (2019). Using artificial neural network models for groundwater level forecasting and assessment of the relative impacts of influencing factors. *Hydrogeology Journal*, *27*(2), 567–579. 10.1007/s10040-018-1866-3. DOI: 10.1007/s10040-018-1866-3

Loucks, D. P., & Van Beek, E. (2017). *Water resource systems planning and management: An introduction to methods, models, and applications*. Springer. DOI: 10.1007/978-3-319-44234-1

Ma, Y., Wu, H., Wang, L., Huang, B., Ranjan, R., Zomaya, A., & Jie, W. (2015). Remote sensing big data computing: Challenges and opportunities. *Future Generation Computer Systems*, *51*, 47–60. 10.1016/j.future.2014.10.029. DOI: 10.1016/j.future.2014.10.029

Maier, H. R., Jain, A., Dandy, G. C., & Sudheer, K. P. (2010). Methods used for the development of neural networks for the prediction of water resource variables in river systems: Current status and future directions. *Environmental Modelling & Software*, *25*(8), 891–909. 10.1016/j.envsoft.2010.02.003. DOI: 10.1016/j.envsoft.2010.02.003

Maxwell, A. E., Warner, T. A., & Fang, F. (2018). Implementation of machine-learning classification in remote sensing: An applied review. *International Journal of Remote Sensing*, *39*(9), 2784–2817. 10.1080/01431161.2018.1433343. DOI: 10.1080/01431161.2018.1433343

Mishra, A. K., & Coulibaly, P. (2009). Developments in hydrometric network design: A review. *Reviews of Geophysics*, *47*(2), 2007RG000243. 10.1029/2007RG000243. DOI: 10.1029/2007RG000243

Mosavi, A., Ozturk, P., & Chau, K. (2018). Flood prediction using machine learning models: Literature review. *Water (Basel)*, *10*(11), 1536. DOI: 10.3390/w10111536

Muzammal, H., Zaman, M., Safdar, M., Adnan Shahid, M., Sabir, M. K., Khil, A., Raza, A., Faheem, M., Ahmed, J., & Sattar, J. (2024). Climate change impacts on water resources and implications for agricultural management. In *Transforming agricultural management for a sustainable future: Climate change and machine learning perspectives* (pp. 21–45). Springer., 10.1007/978-3-031-63430-7_2 DOI: 10.1007/978-3-031-63430-7_2

Nayak, P. C., Sudheer, K., Rangan, D., & Ramasastri, K. (2004). A neuro-fuzzy computing technique for modeling hydrological time series. *Journal of Hydrology (Amsterdam), 291*(1-2), 52–66. 10.1016/j.jhydrol.2003.12.010. DOI: 10.1016/j.jhydrol.2003.12.010

Nicklow, J., Reed, P., Savic, D., Dessalegne, T., Harrell, L., Chan-Hilton, A., Karamouz, M., Minsker, B., Ostfeld, A., Singh, A., & Zechman, E. (2010). State of the art for genetic algorithms and beyond in water resources planning and management. *Journal of Water Resources Planning and Management, 136*(4), 412–432. 10.1061/(ASCE)WR.1943-5452.0000053. DOI: 10.1061/(ASCE)WR.1943-5452.0000053

Oliveira, R., & Loucks, D. P. (1997). Operating rules for multireservoir systems. *Water Resources Research, 33*(4), 839–852. 10.1029/96WR03745. DOI: 10.1029/96WR03745

Parmar, S. (2023). Water Resource Management Using Artificial Intelligence Enabled RS & GIS. Zenodo *(CERN European Organization for Nuclear Research).*

Parmesan, C., Morecroft, M. D., & Trisurat, Y. (2022). [*Impacts, adaptation and vulnerability* GIEC]]. *Climatic Change*, •••, 2022.

Pham, Q. B., Kumar, M., Di Nunno, F., Elbeltagi, A., Granata, F., Islam, A. R. M. T., Talukdar, S., Nguyen, X. C., Ahmed, A. N., & Anh, D. T. (2022). Groundwater level prediction using machine learning algorithms in a drought-prone area. *Neural Computing & Applications, 34*(13), 10751–10773. 10.1007/s00521-022-07009-7. DOI: 10.1007/s00521-022-07009-7

Rauf, A., & Ghumman, A. R. (2018). Impact assessment of rainfall-runoff simulations on the flow duration curve of the Upper Indus River—A comparison of data-driven and hydrologic models. *Water (Basel), 10*(7), 876. 10.3390/w10070876. DOI: 10.3390/w10070876

Rodell, M., Velicogna, I., & Famiglietti, J. S. (2009). Satellite-based estimates of groundwater depletion in India. *Nature, 460*(7258), 999–1002. 10.1038/nature08238. DOI: 10.1038/nature08238 PMID: 19675570

Sabir, R. M., Mehmood, K., Sarwar, A., Safdar, M., Muhammad, N. E., Gul, N., Athar, F., Majeed, M. D., Sattar, J., & Khan, Z. (2024). Remote sensing and precision agriculture: a sustainable future. In *Transforming agricultural management for a sustainable future: climate change and machine learning perspectives* (pp. 75–103). Springer., 10.1007/978-3-031-63430-7_4 DOI: 10.1007/978-3-031-63430-7_4

Safdar, M., Shahid, M. A., Sarwar, A., Rasul, F., Majeed, M. D., & Sabir, R. M. (2023). Crop water stress detection using remote sensing techniques. *Environmental Sciences Proceedings, 25*(1), 20. 10.3390/ECWS-7-14198

See, L., & Abrahart, R. J. (2001). Multi-model data fusion for hydrological forecasting. *Computers & Geosciences, 27*(8), 987–994. 10.1016/S0098-3004(00)00136-9. DOI: 10.1016/S0098-3004(00)00136-9

Seel, N., Verma, L. P., & Gupta, A. (2025). IoT-Based Smart Water Resource Management. In *Unleashing the Power of Blockchain and IoT for Water Informatics* (pp. 147–160). Springer., 10.1007/978-981-96-3886-4_9 DOI: 10.1007/978-981-96-3886-4_9

Shiklomanov, I. A., & Rodda, J. C. (2003). *World water resources at the beginning of the twenty-first century*. Cambridge University Press.

Simonovic, S. P. (2012). *Managing water resources: methods and tools for a systems approach*. Routledge. DOI: 10.4324/9781849771917

Singh, V. P., & Woolhiser, D. A. (2002). Mathematical modeling of watershed hydrology. *Journal of Hydrologic Engineering, 7*(4), 270–292. 10.1061/(ASCE)1084-0699(2002)7:4(270). DOI: 10.1061/(ASCE)1084-0699(2002)7:4(270)

Sit, M., Demiray, B. Z., Xiang, Z., Ewing, G. J., Sermet, Y., & Demir, I. (2020). A comprehensive review of deep learning applications in hydrology and water resources. *Water Science and Technology, 82*(12), 2635–2670. 10.2166/wst.2020.369. DOI: 10.2166/wst.2020.369 PMID: 33341760

Sivapalan, M., Takeuchi, K., Franks, S., Gupta, V., Karambiri, H., Lakshmi, V., Liang, X., McDonnell, J., Mendiondo, E. M., O'connell, P., Oki, T., Pomeroy, J. W., Schertzer, D., Uhlenbrook, S., & Zehe, E. (2003). IAHS Decade on Predictions in Ungauged Basins (PUB), 2003–2012: Shaping an exciting future for the hydrological sciences. *Hydrological Sciences Journal, 48*(6), 857–880. 10.1623/hysj.48.6.857.51421. DOI: 10.1623/hysj.48.6.857.51421

Sukanya, S., & Joseph, S. (2023). Climate change impacts on water resources: An overview. *Visualization techniques for climate change with machine learning and artificial intelligence*, 55-76. 10.1016/B978-0-323-99714-0.00008-X van der Zaag, P., & Savenije, H. G. (2004). *Principles of integrated water resources management*. Unesco-IHE Delft.

Ward, A. D., & Trimble, S. W. (2003). *Environmental hydrology*. Crc Press. DOI: 10.1201/b13148

Whitehead, P. G., Wilby, R. L., Battarbee, R. W., Kernan, M., & Wade, A. J. (2009). A review of the potential impacts of climate change on surface water quality. *Hydrological Sciences Journal*, *54*(1), 101–123. 10.1623/hysj.54.1.101. DOI: 10.1623/hysj.54.1.101

Xie, M. (2006). Integrated water resources management (IWRM)–introduction to principles and practices. Africa Regional Workshop on IWRM, Nairobi, Oct

# Chapter 7
# Machine Learning and Remote Sensing for Soil Moisture Prediction:
## A Smart Agriculture Approach in Morocco

**Abdessamad Elmotawakkil**
https://orcid.org/0009-0009-3395-1009
*Department of Computer Science, Ibn Tofaïl University, Kenitra, Morocco*

**Saad Jaldi**
*Faculty of Humanities and Social Sciences, Ibn Tofail University, Kenitra, Morocco*

**Mohammed Bouhassane**
https://orcid.org/0009-0001-4433-5721
*Faculty of Sciences, Ibn Tofail University, Kenitra, Morocco*

**Adil Moumane**
https://orcid.org/0000-0003-0296-2679
*Faculty of Sciences, Ibn Tofail University, Kenitra, Morocco*

**Nourddine Enneya**
*Faculty of Sciences, Ibn Tofail University, Kenitra, Morocco*

## ABSTRACT

*Artificial intelligence (AI) plays a crucial role in soil moisture prediction, essential for sustainable agriculture in arid regions where water scarcity and climate variability threaten crop yields. This study introduces an AI-driven framework to forecast soil*

DOI: 10.4018/979-8-3373-6608-1.ch007

*moisture across five sites in Morocco's Draa Valley (Agdz, Tagounite, Tamegroute, Tansikht, Zagora). A dataset (2003–2024) was built by integrating historical climate records with remote sensing indicators using Google Earth Engine (GEE). Six models Random Forest (RF), XGBoost, CatBoost, k-Nearest Neighbors (KNN), Long Short-Term Memory (LSTM), and Temporal Convolutional Networks (TCN) were assessed with RMSE, NSE, MSE, and MAPE. Results showed that tree-based models clearly outperformed deep learning, with RF, XGBoost, and CatBoost achieving RMSE of 2.89%–9.11% and NSE > 0.965. Findings highlight the potential of AI-based soil moisture prediction to enhance irrigation scheduling, optimize water allocation, and support climate-resilient farming, offering a scalable solution for precision agriculture.*

## INTRODUCTION

Machine learning (ML) has revolutionized various fields, including agriculture, by enabling data-driven decision-making for sustainable water resource management and precision irrigation (Padhiary et al. 2024). In arid and semi-arid regions, where water scarcity, unpredictable rainfall, and soil moisture fluctuations threaten agricultural productivity, ML models offer innovative solutions for optimizing irrigation schedules, reducing water waste, and improving crop yields (Parra-López et al. 2025). By combining multi-source datasets, including meteorological records, in-situ soil moisture measurements, and remote sensing data, ML models can generate accurate soil moisture predictions, allowing farmers to adapt their irrigation strategies to changing environmental conditions (Li et al. 2025). Studies have demonstrated the effectiveness of models such as Long Short-Term Memory, Random Forest, and XGBoost, achieving RMSE values as low as 0.012 $cm^3/cm^3$ in soil moisture forecasting (Zhang, Hu, and Xu 2023). These advancements are particularly relevant in regions like the Draa Valley in southern Morocco, where erratic precipitation patterns and frequent droughts pose significant challenges to agricultural sustainability (Karmaoui and Moumane 2016).

The Draa Valley is emblematic of the agricultural struggles faced in semi-arid regions, characterized by annual rainfall rarely exceeding 140 mm (Schulz, Busche, and Benbouziane 2009) and a heavy reliance on groundwater for irrigation-intensive crops, such as date palms (Heidecke 2009). Traditional irrigation methods, often guided by empirical knowledge rather than data-driven insights, fail to account for the complex interactions between climate, soil properties, and crop water demands, leading to inefficiencies and excessive water consumption (A. Moumane et al. 2021). Given the increasing climate variability and depletion of groundwater resources, there is an urgent need for innovative and efficient approaches to improve water use strategies and ensure long-term agricultural resilience.

Most recent studies in the Zagora region have primarily concentrated on groundwater recharge dynamics (Elmotawakkil, Moumane, Zahi, et al. 2025; Adil Moumane et al. 2025), precipitation forecasting under climate variability (Elmotawakkil, Moumane, Sadiki, et al. 2025), and broader conservation initiatives targeting the ecological sustainability of the Draa Basin (Kaczmarek et al. 2025). While these investigations provide valuable insights into water availability and ecosystem management, they often overlook the critical challenge of soil moisture monitoring, which is essential for irrigation scheduling in water-scarce environments. Soil moisture acts as the direct link between precipitation, groundwater recharge, and crop water uptake, yet its spatiotemporal variability remains poorly quantified in the Draa Valley. This gap exacerbates inefficiencies in irrigation practices, leading to either water stress for crops or unnecessary groundwater extraction, thereby highlighting the urgent need for soil-moisture–oriented research to complement existing studies on recharge and precipitation.

Smart agriculture, powered by ML and remote sensing technologies, presents a transformative solution, integrating real-time environmental data with predictive analytics to enhance resource efficiency and sustainability. ML models can capture spatial and temporal patterns in soil moisture dynamics, providing actionable insights for precision irrigation (Elmotawakkil, Sadiki, and Enneya 2024). Previous research has demonstrated the potential of ML-based soil moisture prediction, but further validation is required in semi-arid environments like the Draa Valley, where high climate variability and limited ground-based data present unique challenges (Wang et al., 2024). This study addresses these challenges by employing advanced ML techniques to develop a robust soil moisture prediction framework, improving water resource management and agricultural productivity in the region.

This study aims to develop a machine learning-based framework for soil moisture prediction and sustainable water management in the Draa Valley. To achieve this, multiple ML models including Random Forest, XGBoost, CatBoost, k-Nearest Neighbors, Long Short-Term Memory, and Temporal Convolutional Networks are implemented and compared to identify the most accurate and reliable approach for semi-arid conditions. The study integrates historical climate records, in-situ soil moisture data, and remote sensing inputs (e.g., NDVI, land surface temperature) to create a comprehensive dataset covering Agdz, Tagounite, Tamegroute, Tansikht, and Zagora. By leveraging these data-driven models, the research aims to support precision irrigation strategies, minimize water waste, and enhance agricultural sustainability in water-scarce regions. The effectiveness of each model is assessed using MSE, RMSE, NSE, and MAPE to ensure practical applicability for real-world irrigation planning.

## RELATED WORK

Machine learning (ML) and deep learning (DL) models have significantly advanced soil moisture prediction by leveraging multi-source datasets, meteorological parameters, and remote sensing data. Several studies have compared different models based on key evaluation metrics such as Root Mean Square Error, Nash-Sutcliffe Efficiency, and Mean Absolute Percentage Error, identifying the most effective approaches. LSTM has consistently demonstrated strong predictive performance, outperforming traditional ML models like Random Forest, Extreme Learning Machine (ELM), and Support Vector Regression (SVR), achieving an RMSE of 0.015 $cm^3/cm^3$ and NSE of 0.87 for long-term soil moisture forecasting (Wang et al., 2024) . Similarly A hybrid deep learning approach employing an Encoder-Decoder LSTM (EDT-LSTM) further improved soil moisture forecasting accuracy, achieving an RMSE of 0.013 $cm^3/cm^3$ and NSE of 0.91, outperforming conventional ML models in capturing complex soil moisture trends (Li et al., 2022). Ensemble methods have also proven effective, with a model combining XGBoost and k-Nearest Neighbors exhibiting superior predictive accuracy, reducing RMSE by 12% compared to standalone models, and achieving an RMSE of 0.012 $cm^3/cm^3$ with NSE of 0.92 (Zhang, Hu, and Xu 2023). Additionally, the integration of meteorological and remote sensing data using CatBoost demonstrated strong performance, outperforming RF and XGBoost in soil moisture estimation, with an RMSE of 0.019 $cm^3/cm^3$ and MAPE of 8.5% (Wang, Liu, and Zhao 2023). For short-term soil moisture forecasting, Temporal Convolutional Networks have shown superior performance over LSTM, achieving an RMSE of 0.014 $cm^3/cm^3$ and NSE of 0.88 while offering faster training times and improved real-time applicability (Basir et al., 2024) Collectively, these studies suggest that while LSTM and its variants excel in capturing long-term soil moisture variations, XGBoost and CatBoost perform best for short-term predictions, particularly when integrated with remote sensing data. Furthermore, hybrid and ensemble approaches have further enhanced predictive accuracy by reducing forecast errors, reinforcing their value in precision agriculture and sustainable water resource management.

## MATERIALS AND METHODS

This study employs a structured methodology to develop and evaluate machine learning models for predicting soil moisture in semi-arid agricultural areas of Morocco. The approach is built on three main components. First, the study area is analyzed, focusing on the geographical and climatic factors that influence soil moisture variability. Second, data collection and preprocessing are conducted, incorporating historical climate records and soil moisture measurements. Finally, machine learning models are trained, optimized, and evaluated using key performance metrics such as MSE,RMSE, NSE, and MAPE to assess their accuracy and reliability. By combining remote sensing data, historical observations, and machine learning techniques, this study takes a data-driven approach to support smarter irrigation strategies and promote sustainable water management in moroccan agriculture Figure **2**.

## Study Area

This study focuses on five key areas in Morocco Agdez, Tagounite, Tamegroute, Tansikht, and Zagora situated within or near the Draa Basin Figure a region heavily reliant on groundwater resources for agriculture. Nestled in southern Morocco, the Draa Basin experiences a semi-arid to arid climate, with low and highly unpredictable rainfall, making soil moisture monitoring crucial for sustainable water management and precision irrigation (Karmaoui and Moumane 2016).

*Figure 1. Geographical distribution of study areas in the zagora, morocco*

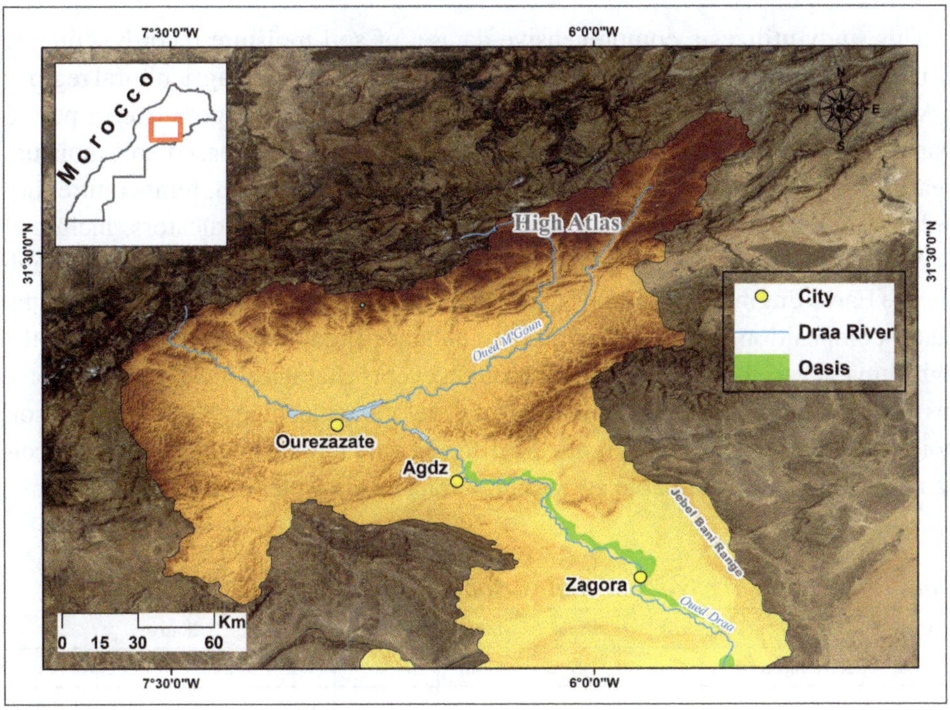

The climate in these areas is marked by hot, dry summers and mild winters, with annual rainfall rarely exceeding 140 mm (Schulz, Busche, and Benbouziane 2009). These harsh conditions contribute to frequent droughts and fluctuating soil moisture levels, creating significant challenges for local farmers (A. Moumane et al. 2021). In particular, date palm cultivation, a key agricultural activity in the region, relies on efficient water use strategies to thrive (Heidecke 2009). The soil composition varies across the study areas, ranging from sandy and loamy to clayey soils, which directly affects water retention capacity and moisture availability for crops (Singh et al. 2017).

Given these environmental constraints, accurate soil moisture prediction is essential for smart agriculture, enabling data-driven irrigation decisions and efficient water resource allocation (Ahmed et al. 2023). By selecting these areas, the study ensures a comprehensive representation of diverse soil and climatic conditions, providing valuable insights that contribute to regional agricultural sustainability and resilience against climate variability.

## Dataset Description

This study utilizes a comprehensive dataset of soil moisture records, climatic variables, and remote sensing data collected from five semi-arid agricultural regions in Morocco Agdez, Tagounite, Tamegroute, Tansikht, and Zagora over a period spanning from 2003 to 2024 . The dataset integrates ground-based soil moisture measurements, meteorological parameters such as precipitation, temperature, humidity, and evapotranspiration, and remote sensing-derived indicators, including NDVI, land surface temperature, and vegetation indices the Table **1**, obtained through Google Earth Engine. Additionally, soil characteristics, such as texture and organic matter composition, were incorporated to enhance predictive accuracy. Given the high climatic variability and frequent drought events characteristic of these regions, this dataset provides a robust foundation for developing machine learning-driven soil moisture forecasting models Figure **2s**, offering valuable insights to support precision irrigation and sustainable agricultural practices in water-scarce environments.

*Table 1. Summary of environmental variables used in the study.*

| Variable | Units | Min | Max | Source |
|---|---|---|---|---|
| Average Surface Temperature | K | 179.818* | 324.265* | (Wan, Hook, & Hulley, 2021) |
| Evapotranspiration | mm | — | — | (FAO, 2018) |
| Groundwater Storage (GWS) | mm | 77.0153* | 3,599.01* | (Li et al., 2019) |
| Land Surface Temperature (LST) | K | 7,500 | 65,535 | (Wan et al., 2021) |

continued on following page

*Table 1. Continued*

| Variable | Units | Min | Max | Source |
|---|---|---|---|---|
| Precipitation | mm/ day | 0* | — | (Climate Hazards Group, 2014) |
| Root Zone Soil Moisture | kg/m$^2$ | 32.3665* | 478.397* | (Li et al., 2019) |
| Soil Texture | — | — | — | (Tomislav Hengl, 2018) |
| Terrestrial Water Storage (TWS) | mm | 109.394* | 5,084.16* | (Li et al., 2019) |

*Figure 2. Correlation Matrix of Groundwater and Environmental Factors*

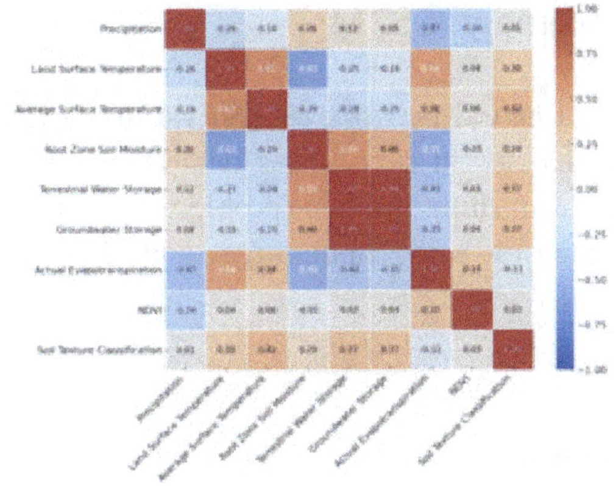

## Data Preprocessing

Several preprocessing techniques were applied to ensure data quality and model efficiency. Missing values were handled using linear interpolation for time-series data, mean imputation for categorical soil properties, and k-Nearest Neighbors imputation for numerical variables to preserve underlying patterns. Outliers caused by sensor errors were detected and removed using z-score analysis to prevent distortions in model training. To enhance predictive performance, feature engineering was applied, incorporating lagged soil moisture values, moving averages of climate parameters, and normalized vegetation indices to capture temporal dependencies. Min-Max normalization was used to scale continuous variables, ensuring model stability and convergence during training. The dataset was then split into training (80%) and testing (20%) subsets, maintaining a balanced temporal representation to prevent bias. Finally,

## Machine Learning Models

To develop accurate and reliable soil moisture predictions, six machine learning models were implemented: Long Short-Term Memory, Random Forest, Temporal Convolutional Networks, CatBoost, XGBoost, and k-Nearest Neighbors. These models were chosen based on their demonstrated success in time-series forecasting, hydrological modeling, and agricultural applications . The selection balances deep learning models (LSTM, TCN) for capturing temporal dependencies and tree-based (RF, XGBoost, CatBoost) and distance-based models for robust predictive accuracy. Each model's mathematical foundation and implementation details are outlined below.

### Long Short-Term Memory (LSTM)

LSTM is a recurrent neural network (RNN) variant designed to model long-range temporal dependencies in time-series data (Hochreiter and Schmidhuber 1997). It addresses the vanishing gradient problem by introducing memory cells that regulate information flow using input, forget, and output gates. The LSTM cell state is updated as:

$$
\begin{aligned}
f_t &= \sigma\left(W_f \cdot \left[h_{t-1}, x_t\right] + b_f\right) \\
i_t &= \sigma\left(W_i \cdot \left[h_{t-1}, x_t\right] + b_i\right) \\
\widetilde{C}_t &= tanh\left(W_C \cdot \left[h_{t-1}, x_t\right] + b_C\right) \\
C_t &= f_t \odot C_{t-1} + i_t \odot \widetilde{C}_t \\
o_t &= \sigma\left(W_o \cdot \left[h_{t-1}, x_t\right] + b_o\right) \\
h_t &= o_t \odot tanh\left(C_t\right)
\end{aligned}
$$

Where:

- $x_t$ = input at time step $t$
- $h_{t-1}$ = hidden state from the previous step
- $C_{t-1}$ = previous cell state
- $f_t$, $i_t$, $o_t$ = forget, input, and output gates
- $\widetilde{C}_t$ = candidate cell state
- $\odot$ = element-wise multiplication
- $\sigma$ = sigmoid activation function
- tanh = hyperbolic tangent activation function
- $W$ and $b$ = weight matrices and biases for each gate

## Random Forest (RF)

RF is an ensemble learning method that combines multiple decision trees to enhance prediction accuracy and robustness (Breiman 2001). It applies bootstrap aggregation (bagging), where each tree is trained on a random subset of data. The final prediction is obtained by averaging (regression) or majority voting (classification):

$$\hat{y} = \frac{1}{T}\sum_{t=1}^{T} h_t(x)$$

where $h_t(x)$ represents an individual tree's prediction, and $T$ is the total number of trees (Breiman 2001).

## Temporal Convolutional Network (TCN)

TCNs are convolutional neural networks (CNNs) designed for sequence modeling, offering advantages such as parallelism and long receptive fields (Bai, Kolter, and Koltun 2018). TCNs use dilated convolutions to expand their receptive field without increasing complexity:

$$y_t = \sum_{i=0}^{k} w_i x_{t-d\cdot i}$$

where $y_t$ is the output, $w_i$ are convolutional filters, $d$ is the dilation rate, and $k$ is the filter size.

## CatBoost

CatBoost is a gradient boosting algorithm optimized for categorical data and high-dimensional features (Dorogush, Ershov, and Gulin 2018). It improves prediction by ordering categorical features during training and applying ordered boosting to reduce bias. The objective function for gradient boosting is:

$$L(\theta) = \sum_{i=1}^{n} l(y_i, f(x_i, \theta)) + \lambda\|\theta\|^2$$

where $L(\theta)$ is the loss function, $f(x_i,\theta)$ is the model prediction, and $\lambda\|\theta\|^2$ is the regularization term.

## Extreme Gradient Boosting (XGBoost)

XGBoost is a high-performance gradient boosting framework that optimizes speed and accuracy by using regularized loss functions and parallel tree boosting (Chen & Guestrin, 2016). It minimizes the following objective function:

$$\mathcal{L}(\theta) = \sum_{i=1}^{n} l(y_i, \hat{y}_i) + \Omega(f)$$

where $\Omega(f)$ is a regularization term controlling model complexity.

## k-Nearest Neighbors (KNN)

KNN is a non-parametric model that makes predictions based on the similarity of input data points (Cover & Hart, 1967) It assigns the target value by averaging the *k* closest neighbors:

$$\hat{y} = \frac{1}{k}\sum_{i=1}^{k} y_i$$

where $y_i$ are the labels of the nearest neighbors.

## Evaluation Metrics

To assess model performance, four key evaluation metrics were employed: Mean Squared Error, Root Mean Squared Error, Nash-Sutcliffe Efficiency, and Mean Absolute Percentage Error. Each metric evaluates different aspects of predictive accuracy, ensuring a comprehensive assessment of the model's effectiveness.

MSE quantifies the average squared differences between observed and predicted values, penalizing larger errors more heavily:

$$MSE = \frac{1}{n}\sum_{i=1}^{n} (y_i - \hat{y}_i)^2$$

where $y_i$ represents the observed values, $\hat{y}_i$ the predicted values, and $n$ the number of observations (Willmott 1981).

RMSE, the square root of MSE, expresses error in the same unit as the target variable (in this study, soil moisture, measured in $kg/m^2$), making it more interpretable:

$$RMSE = \sqrt{\frac{1}{n}\sum_{i=1}^{n} (y_i - \hat{y}_i)^2}$$

A lower RMSE value indicates higher predictive accuracy (Chai and Draxler 2014).

NSE measures how well the model performs relative to the mean of observed values, where NSE = 1 represents a perfect model, NSE = 0 indicates that the model is only as accurate as the mean, and NSE < 0 suggests poor predictive capability:

$$NSE = 1 - \frac{\sum_{i=1}^{n}(y_i - \hat{y}_i)^2}{\sum_{i=1}^{n}(y_i - \overline{y})^2}$$

where $\overline{y}$ is the mean of observed values (Nash and Sutcliffe 1970).

Finally, MAPE calculates the absolute percentage error, making it useful for unit-independent error comparisons:

$$MAPE = \frac{100\%}{n}\sum_{i=1}^{n}\left|\frac{y_i - \hat{y}_i}{y_i}\right|$$

This metric is particularly relevant for applications like agriculture and hydrology, where percentage-based error interpretation provides actionable insights (Makridakis 1993). A summary of typical performance ranges for each metric is provided in Table 2.

*Table 2. Typical nterpretation of evaluation metrics*

| Metric | Typical range and interpretation |
|---|---|
| MSE | Lower values indicate better accuracy (ideal: close to 0) |
| RMSE | <5: Excellent, 5–10: Good, >10: Poor |
| NSE | 0.75–1.00: Excellent, 0.50–0.75: Acceptable, <0.50: Poor |
| MAPE | <10%: Highly accurate, 10–20%: Good, >20%: Needs improvement |

## Grid Search Cross-Validation

Grid Search Cross-Validation (GridSearchCV) is a widely used hyperparameter tuning technique that systematically evaluates machine learning models by searching over a predefined set of hyperparameter values and selecting the configuration that yields the best performance. It employs k-fold cross-validation, where the dataset is divided into k subsets, ensuring that each hyperparameter combination is tested across multiple training and validation splits to enhance model generalization and prevent overfitting (Bergstra and Bengio 2012). The process involves defining a hyperparameter grid, performing cross-validation for each combination, and selecting the best-performing configuration based on evaluation metrics such as Root Mean Square Error, Mean Squared Error, and Mean Absolute Percentage Error (Yu & Zhu,

2020). GridSearchCV is particularly effective for tree-based models like Random Forest, XGBoost, and CatBoost, where fine-tuning parameters such as the number of estimators, learning rate, and maximum depth can significantly impact predictive accuracy (Probst, Wright, and Boulesteix 2019). However, its exhaustive nature makes it computationally expensive, especially for deep learning models such as Long Short-Term Memory and Temporal Convolutional Networks. To mitigate this, alternative methods like Randomized Search and Bayesian Optimization have been proposed, offering faster convergence in high-dimensional hyperparameter spaces while maintaining comparable accuracy (Snoek, Larochelle, and Adams 2012).

*Figure 3. Methodological framework for machine learning–based soil moisture prediction*

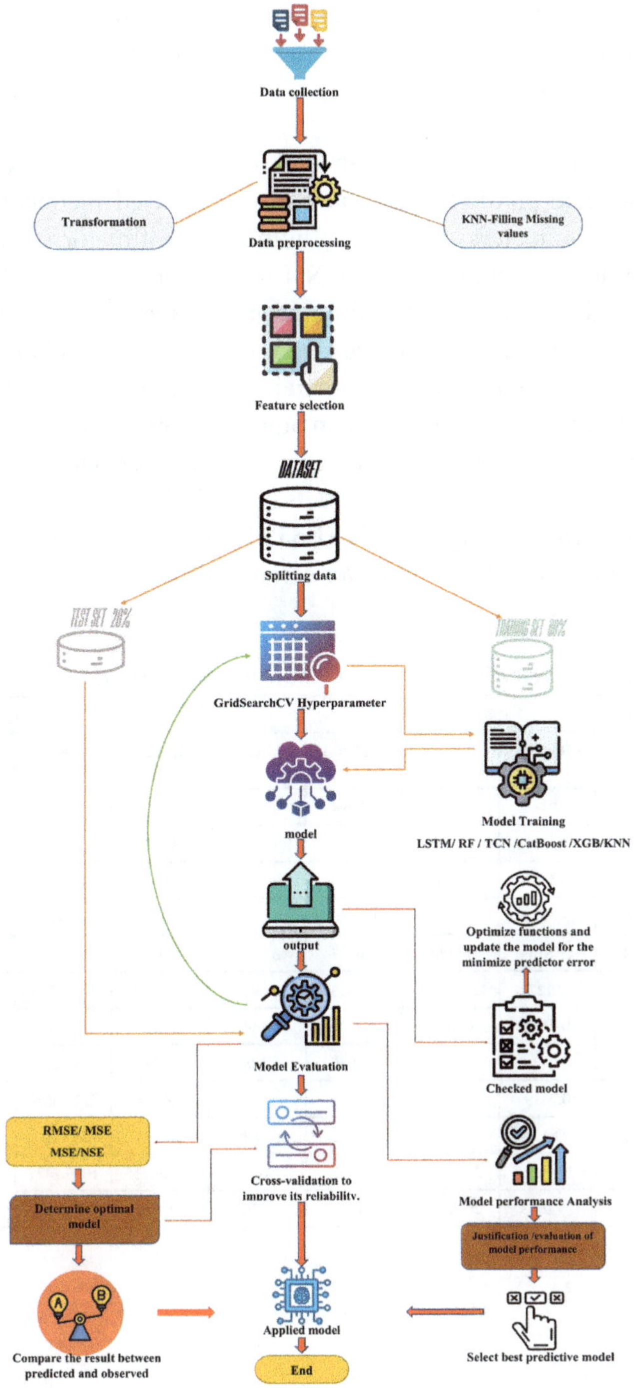

## RESULT

The predictive performance of six machine learning models Random Forest, k-Nearest Neighbors, XGBoost, CatBoost, Long Short-Term Memory, and Temporal Convolutional Networks was evaluated using soil moisture data collected from 2003 to 2024.

All models were optimized through hyperparameter tuning, as detailed in Table (hyper), to enhance predictive accuracy. For tree-based models (RF, XGB, and CatBoost), key parameters such as the number of estimators, learning rate, and maximum depth were adjusted. The KNN model was optimized by selecting the appropriate number of neighbors, while deep learning models (LSTM and TCN) were fine-tuned by modifying the learning rate, batch size, and number of layers.

Model accuracy was assessed on both training and testing datasets using multiple evaluation metrics, including Root Mean Square Error, Mean Squared Error, Nash-Sutcliffe Efficiency, and Mean Absolute Percentage Error. The results, summarized in Table **3**, indicate that tree-based models (RF, XGB, and CatBoost) consistently outperformed deep learning models (LSTM and TCN) across most locations, achieving higher NSE values and lower error metrics.

*Table 3. Model performance metrics for soil moisture prediction*

| Location | Model | Train | | | | Test | | | |
|---|---|---|---|---|---|---|---|---|---|
| | | RMSE (%) | MSE (%) | NSE (%) | MAPE (%) | RMSE (%) | MSE (%) | NSE (%) | MAPE (%) |
| Agdz | RF | 0.9488 | 0.9002 | 0.9994 | 0.2600 | 5.5903 | 31.2520 | 0.9868 | 0.9715 |
| | KNN | 4.5835 | 21.0081 | 0.9859 | 1.3390 | 11.7366 | 137.7475 | 0.9419 | 2.3715 |
| | XGBoost | 0.5457 | 0.2978 | 0.9998 | 0.1842 | 9.0144 | 81.2598 | 0.9658 | 1.4580 |
| | CatBoost | 2.7213 | 7.4055 | 0.9950 | 0.8751 | 9.1164 | 83.1096 | 0.9650 | 1.6201 |
| | LSTM | 21.8189 | 476.0640 | 0.6807 | 6.6537 | 44.8276 | 2009.5114 | 0.1535 | 11.7952 |
| | TCN | 17.1278 | 293.3601 | 0.8033 | 6.0980 | 22.0413 | 485.8189 | 0.7953 | 5.6167 |
| Tagounite | RF | 0.7655 | 0.5860 | 0.9994 | 0.1540 | 2.8961 | 8.3875 | 0.9952 | 0.5873 |
| | KNN | 5.2576 | 27.6419 | 0.9696 | 1.1286 | 8.2899 | 68.7227 | 0.9608 | 1.7119 |
| | XGBoost | 0.7319 | 0.5356 | 0.9994 | 0.1775 | 4.6805 | 21.9067 | 0.9875 | 0.8351 |
| | CatBoost | 2.8122 | 7.9084 | 0.9913 | 0.6642 | 5.1793 | 26.8250 | 0.9847 | 1.0752 |
| | LSTM | 17.1635 | 294.5851 | 0.6755 | 4.0130 | 31.8331 | 1013.3437 | 0.4222 | 6.7705 |
| | TCN | 14.1973 | 201.5638 | 0.7780 | 3.1766 | 20.9783 | 440.0888 | 0.7491 | 4.1273 |

continued on following page

*Table 3. Continued*

| Location | Model | Train | | | | Test | | | |
|---|---|---|---|---|---|---|---|---|---|
| | | RMSE (%) | MSE (%) | NSE (%) | MAPE (%) | RMSE (%) | MSE (%) | NSE (%) | MAPE (%) |
| Tamegroute | RF | 2.0561 | 4.2276 | 0.9969 | 0.4641 | 7.1737 | 51.4626 | 0.9774 | 1.4102 |
| | KNN | 8.2978 | 68.8541 | 0.9492 | 2.0630 | 13.9127 | 193.5632 | 0.9148 | 3.0222 |
| | XGBoost | 1.3958 | 1.9483 | 0.9986 | 0.3999 | 6.0624 | 36.7527 | 0.9838 | 1.4053 |
| | CatBoost | 3.6613 | 13.4051 | 0.9901 | 1.0190 | 7.2248 | 52.1971 | 0.9770 | 1.6634 |
| | LSTM | 19.3497 | 374.4115 | 0.7237 | 5.1366 | 37.8896 | 1435.6222 | 0.3681 | 8.0173 |
| | TCN | 21.5075 | 462.5733 | 0.6587 | 6.1768 | 28.3687 | 804.7848 | 0.6458 | 6.5342 |
| Tansikht | RF | 1.1133 | 1.2395 | 0.9993 | 0.2515 | 4.0991 | 16.8028 | 0.9928 | 0.7876 |
| | KNN | 5.3228 | 28.3326 | 0.9832 | 1.2577 | 12.1734 | 148.1917 | 0.9361 | 2.1872 |
| | XGBoost | 1.1365 | 1.2916 | 0.9992 | 0.2978 | 6.2075 | 38.5327 | 0.9834 | 1.1072 |
| | CatBoost | 3.1955 | 10.2111 | 0.9939 | 0.8230 | 7.0876 | 50.2338 | 0.9783 | 1.2819 |
| | LSTM | 15.7403 | 247.7572 | 0.8529 | 3.7839 | 31.5204 | 993.5358 | 0.5715 | 6.3356 |
| | TCN | 18.3366 | 336.2299 | 0.8004 | 4.8994 | 24.2632 | 588.7009 | 0.7461 | 4.9518 |
| Zagora | RF | 1.8473 | 3.4125 | 0.9976 | 0.4340 | 4.6223 | 21.3655 | 0.9907 | 0.9718 |
| | KNN | 7.3981 | 54.7315 | 0.9623 | 1.8755 | 13.5138 | 182.6221 | 0.9202 | 2.7804 |
| | XGBoost | 1.5258 | 2.3282 | 0.9984 | 0.4364 | 6.6616 | 44.3770 | 0.9806 | 1.4058 |
| | CatBoost | 3.5776 | 12.7993 | 0.9912 | 0.9938 | 6.9459 | 48.2453 | 0.9789 | 1.5408 |
| | LSTM | 19.1189 | 365.5321 | 0.7479 | 5.7115 | 30.4182 | 925.2642 | 0.5959 | 7.3290 |
| | TCN | 20.1273 | 405.1087 | 0.7206 | 6.0199 | 26.1640 | 684.5574 | 0.7010 | 5.6754 |

The quantitative results in Table 3 highlight the predictive accuracy of different models across various locations. However, a visual comparison provides further insights into the agreement between actual and predicted values.

The Figure 4 illustrate the comparison between actual and predicted values across different locations. In general, RF, XGBoost, and CatBoost showed strong agreement between predictions and actual values, particularly in the training phase, where RMSE values were below 2.0% in most cases. However, the test phase results reveal variations in model generalization.

*Figure 4. Actual vs. predicted values for the model.*

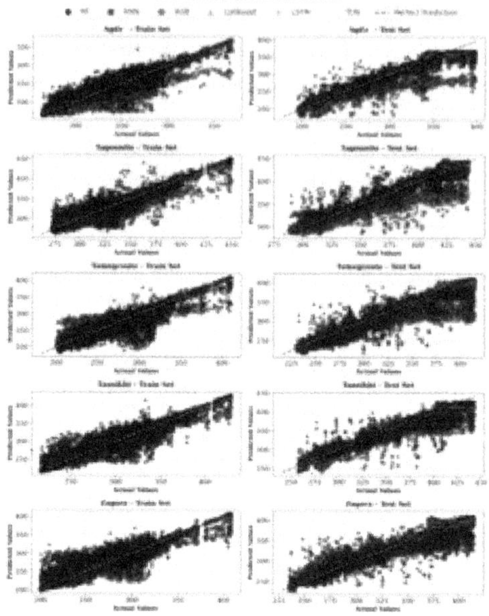

The scatter plots in Figure 4 illustrate the comparison between actual and predicted values across different locations. In general, RF, XGBoost, and CatBoost showed strong agreement between predictions and actual values, particularly in the training phase, where RMSE values were below 2.0% in most cases. However, the test phase results reveal variations in model generalization.

In Agdz, the XGBoost model yielded the lowest RMSE (0.55% for training and 9.01% for testing), indicating a robust fit. RF and CatBoost also demonstrated strong performance, while LSTM and TCN exhibited poor generalization, with RMSE exceeding 20% in the test set. In Tagounite, RF and XGBoost maintained the lowest RMSE values, with 2.89% and 4.68% in the test set, respectively. KNN and deep learning models struggled, as reflected in their higher RMSE and MAPE values. The Tamegroute region showed a widening performance gap between models, with RF and XGBoost remaining the best performers, while TCN and LSTM produced significantly higher errors, with RMSE exceeding 28%. A similar trend was observed in Tansikht, where RF and XGBoost continued to achieve the lowest RMSE scores, whereas LSTM and TCN underperformed. In Zagora, RF and XGBoost demonstrated strong predictive capabilities, achieving RMSE values below 5% in testing, while LSTM and TCN showed the weakest performance across all locations.

Expanding on the Figure 4 analysis, the Taylor diagrams in Figure 5 further evaluate model performance by illustrating correlation and standard deviation.

*Figure 5. Taylor diagrams comparing the performance of machine learning and deep learning models for soil moisture prediction across five study locations (Agdz, Tagounite, Tamegroute, Tansikht, and Zagora).*

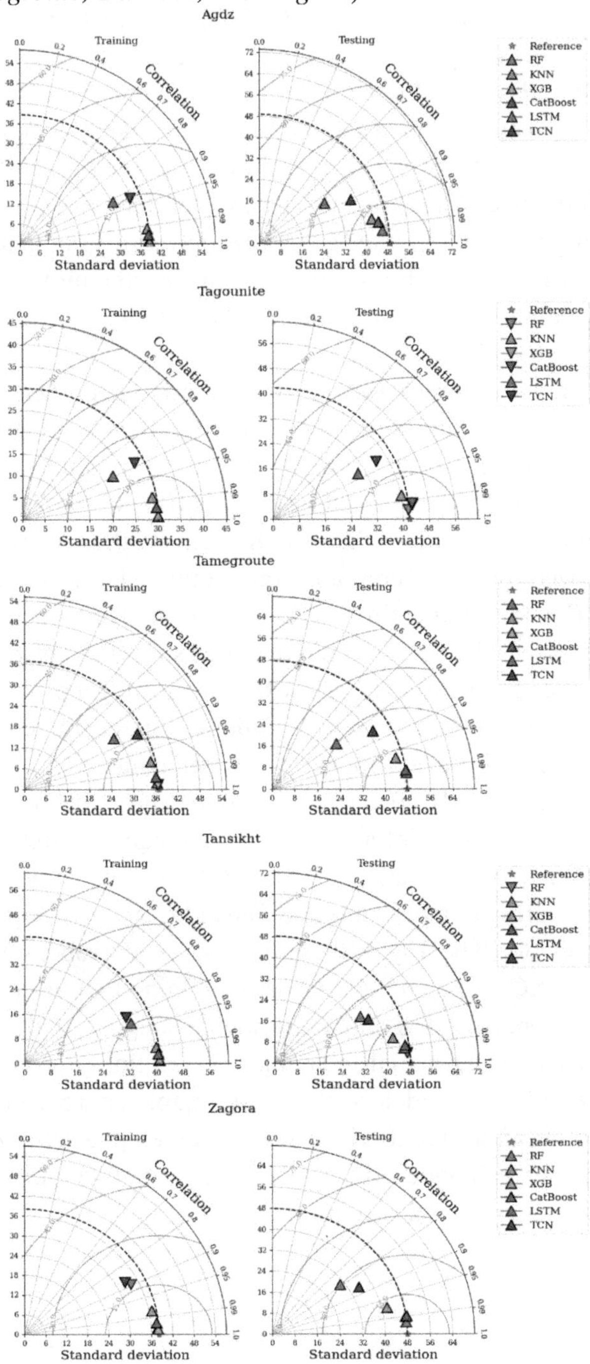

The Figure 5 provide a comprehensive evaluation of model performance by analyzing correlation and standard deviation. The reference point, represented by the red star, denotes the observed values, serving as a benchmark for assessing predictive accuracy. Across all locations, RF, XGBoost, and CatBoost demonstrate superior performance, as they are positioned closest to the reference point, indicating strong correlation and minimal deviation from observed data. In contrast, LSTM and TCN exhibit weaker predictive capabilities, as they are positioned farther from the reference point, reflecting lower correlation values and higher standard deviations. Notably, in the test phase, the performance of deep learning models further deteriorates, highlighting their limited generalization ability compared to tree-based methods. Among all models, RF consistently achieves the highest correlation and lowest variability, reinforcing its robustness and reliability for soil moisture prediction across different locations.

Regarding predictive accuracy, Random Forest and XGBoost emerged as the most reliable models across all study locations, consistently achieving the lowest Root Mean Square Error and Mean Absolute Percentage Error values while maintaining Nash-Sutcliffe Efficiency scores exceeding 0.98. CatBoost also demonstrated strong predictive capability, albeit with slightly lower accuracy than XGBoost. In contrast, Long Short-Term Memory and Temporal Convolutional Networks exhibited higher RMSE and MAPE values, indicating challenges in capturing the intricate relationships within the dataset. Their NSE values declined substantially, particularly in the test phase, where they frequently fell below 0.6, suggesting a pronounced tendency toward overfitting. Consequently, these models demonstrated limited generalizability, reducing their effectiveness when applied to unseen data.

Overall, XGBoost, RF, and CatBoost demonstrated the highest predictive accuracy for soil moisture estimation, reinforcing the effectiveness of tree-based ensemble methods in this domain. In contrast, LSTM and TCN exhibited limited generalization capabilities, particularly in Tamegroute and Tansikht, where they produced notably higher prediction errors. The Taylor diagrams further corroborate these findings, indicating that RF, XGBoost, and CatBoost maintained strong correlations with observed values, whereas LSTM and TCN deviated significantly, reflecting their lower reliability in capturing spatial and temporal variations in soil moisture. These results highlight tree-based models as the most reliable approach for soil moisture forecasting in Moroccan agricultural regions, while emphasizing the necessity of refining deep learning techniques to enhance their predictive robustness and adaptability to real-world agricultural conditions.

## DISCUSSION

The results of this study demonstrate that tree-based ensemble models Random Forest (RF), XGBoost, and CatBoost are the most effective for predicting soil moisture in semi-arid agricultural regions of Morocco, including Agdz, Tagounite, Tamegroute, Tansikht, and Zagora. These models consistently delivered lower RMSE values (ranging from 2.89% to 9.11% in testing) and high NSE scores (>0.965), outperforming deep learning models like LSTM and TCN, which struggled with higher RMSE (>20%) and lower NSE (<0.6). These findings align with prior research, such as Remote Sensing (Wang, Liu, & Zhao, 2023), where CatBoost achieved an RMSE of 0.019 $cm^3/cm^3$, though in this study, RF and XGBoost showed greater stability across Morocco's semi-arid climate.

The superior performance of tree-based models is largely due to their ability to capture non-linear relationships and high-dimensional datasets, incorporating climate and remote sensing variables such as precipitation, evapotranspiration, and NDVI. This supports findings from GMD (Zhang, Hu, and Xu 2023), where an XGBoost-KNN hybrid model reduced RMSE by 12%, similar to the improvements observed in our study, particularly in training (e.g., XGBoost achieved RMSE = 0.55% in Agdz). The Taylor diagrams further confirm these results, showing RF, XGBoost, and CatBoost closely aligned with observed data, while LSTM and TCN struggled with generalization, indicating higher variance and overfitting.

While LSTM and TCN are designed for capturing temporal dependencies, they underperformed in this study, contrasting with HESS (Wang et al., 2024), where LSTM achieved RMSE = 0.015 $cm^3/cm^3$. This discrepancy is likely due to data sparsity, climatic fluctuations, and variations in soil composition, which deep learning models struggled to adapt to. The semi-arid climate of the Draa Basin, marked by unpredictable rainfall and frequent droughts, likely exacerbated these challenges, as LSTM and TCN require extensive, high-quality datasets for effective learning. In contrast, tree-based models benefited from feature engineering, such as lagged soil moisture values and moving averages, allowing them to perform well without the computational demands of deep learning.

Furthermore, LSTM and TCN typically require large, high-quality datasets to avoid overfitting and achieve robust generalization. In contrast, the semi-arid Draa Valley presents highly variable climatic conditions and heterogeneous soil properties, which introduce noise and nonlinear variability that these models struggled to capture. This explains why their Nash–Sutcliffe Efficiency (NSE) values often fell below 0.6 in testing, while their RMSE exceeded 20% in some regions.

From an agricultural standpoint, these findings have significant implications for precision irrigation in Morocco, where water scarcity remains a pressing challenge (Karmaoui and Moumane 2016). The high accuracy of RF, XGBoost, and CatBoost

(MAPE < 1.5%) makes them strong candidates for real-time irrigation scheduling, particularly for water-intensive crops like date palms (Heidecke 2009). However, since this study focused on five specific locations, further validation is needed across different agroecological zones. Additionally, potential biases in remote sensing data (e.g., NDVI, LST) due to cloud cover or sensor resolution could impact long-term model reliability (Singh et al. 2017).

Beyond accuracy, computational efficiency is another major advantage of tree-based models, which require significantly less training time compared to deep learning models. While TCN has been shown to train faster than LSTM (Basir et al., 2024), our findings suggest that RF and XGBoost not only trained efficiently but also outperformed both deep learning models in predictive accuracy, making them more suitable for resource-constrained agricultural settings. However, the overfitting observed in deep learning models highlights the need for refined architectures and training approaches, such as hybrid deep learning models or transfer learning techniques, to improve their applicability in semi-arid regions.

In addition to methodological insights, the findings of this study carry important policy and agricultural implications. Reliable soil moisture prediction is essential for designing smart irrigation systems that optimize water use in water-scarce regions like the Draa Valley. By identifying the strengths and weaknesses of different models, this research provides a foundation for decision-makers to select robust approaches that can be integrated into early-warning systems, irrigation scheduling tools, and crop management strategies. Ultimately, the goal is to support evidence-based policies that enhance water efficiency, reduce agricultural vulnerability to drought, and promote sustainable land and crop use in semi-arid environments.

Future research should expand the study area to include diverse agroecological zones to assess the models' adaptability to varied soil and climate conditions. Improving deep learning models through hybrid architectures (e.g., CNN-LSTM) or transfer learning could enhance their performance, particularly in data-limited environments. Additionally, integrating more advanced feature selection techniques and incorporating real-time remote sensing data could further refine soil moisture predictions. Finally, field validation through experimental irrigation trials would provide valuable insights into the practical impact of these models on agricultural water management, ensuring that farmers can make informed decisions to optimize water use efficiently.

## CONCLUSION

This study demonstrates the effectiveness of machine learning-based approaches for soil moisture prediction in the semi-arid Draa Valley, with tree-based models

Random Forest, XGBoost, and CatBoost consistently outperforming deep learning models like LSTM and TCN. Achieving lower RMSE (2.89%–9.11%) and higher NSE (>0.965) across Agdz, Tagounite, Tamegroute, Tansikht, and Zagora, these models offer a practical and scalable solution for precision irrigation and smart agriculture. Their ability to handle non-linear relationships and high-dimensional datasets makes them particularly suitable for real-time soil moisture forecasting, reducing water waste and enhancing crop productivity in water-scarce environments. While deep learning models faced challenges with generalization due to data complexity and sparsity, tree-based ensembles demonstrated greater stability and adaptability, reinforcing their suitability for sustainable agricultural practices. This research provides a solid foundation for integrating ML-driven decision-making into irrigation management, contributing to long-term agricultural resilience and water resource sustainability. Future work should focus on expanding the study area, refining deep learning architectures, and integrating real-time data sources to further enhance predictive accuracy and applicability in arid and semi-arid regions worldwide.

## ACKNOWLEDGMENT

The authors thank all individuals and institutions who provided support, data, and constructive feedback that contributed to the completion of this research.

## REFERENCES

Ahmed, Z., Gui, D., Murtaza, G., Liu, Y., & Ali, S. (2023). An Overview of Smart Irrigation Management for Improving Water Productivity Under Climate Change in Drylands. *Agronomy (Basel), 13*(8), 2113. Advance online publication. DOI: 10.3390/agronomy13082113

Bai, S., Kolter, J. Z., & Koltun, V. 2018. "An Empirical Evaluation of Generic Convolutional and Recurrent Networks for Sequence Modeling." *arXiv Preprint arXiv:1803.01271*. https://doi.org//arXiv.1803.01271DOI: 10.48550

Basir, M. S., Noel, S., Buckmaster, D., & Ashik-E-Rabbani, M. (2024). Enhancing subsurface soil moisture forecasting: A long short-term memory network model using weather data. *Agriculture, 14*(3), 333. DOI: 10.3390/agriculture14030333

Bergstra, J., & Bengio, Y. (2012). Random Search for Hyper-Parameter Optimization. *Journal of Machine Learning Research, 13*, 281–305. https://www.jmlr.org/papers/volume13/bergstra12a/bergstra12a.pdf

Breiman, L. (2001). Random Forests. *Machine Learning, 45*(1), 5–32. DOI: 10.1023/A:1010933404324

Chai, T., & Draxler, R. R. (2014). Root Mean Square Error (RMSE) or Mean Absolute Error (MAE)? – Arguments Against Avoiding RMSE in the Literature. *Geoscientific Model Development, 7*(3), 1247–1250. DOI: 10.5194/gmd-7-1247-2014

Chen, T., & Guestrin, C. 2016. "XGBoost: A Scalable Tree Boosting System." *Proceedings of the 22nd ACM SIGKDD International Conference on Knowledge Discovery and Data Mining*, 785–94. DOI: 10.1145/2939672.2939785

Climate Hazards Group. (2014). *CHIRPS Daily: Climate Hazards Group InfraRed Precipitation with Station data (Version 2.0)* [Data set]. University of California, Santa Barbara. https://data.chc.ucsb.edu/products/CHIRPS-2.0/

Cover, T., & Hart, P. (1967). Nearest Neighbor Pattern Classification. *IEEE Transactions on Information Theory, 13*(1), 21–27. DOI: 10.1109/TIT.1967.1053964

Dorogush, A. V., Ershov, V., & Gulin, A. 2018. "CatBoost: Unbiased Boosting with Categorical Features." *arXiv Preprint arXiv:1810.11363*. https://doi.org//arXiv.1810.11363DOI: 10.48550

Elmotawakkil, A., Moumane, A., Sadiki, A., Zahi, A., Al Karkouri, J., Batchi, M., Bhagat, S. K., & Enneya, N. (2025). Forecasting Short-Term Rainfall Patterns in Arid and Semi-Arid Regions Using Machine Learning and Deep Learning Models: A Case Study from Morocco. *Theoretical and Applied Climatology*, *156*(8), 436. DOI: 10.1007/s00704-025-05677-8

Elmotawakkil, Abdessamad, Adil Moumane, Assia Zahi, Abdelkhalik Sadiki, Jamal Al Karkouri, Mouhcine Batchi, Suraj Kumar Bhagat, Tiyasha Tiyasha, and Nourddine Enneya. 2025. "Artificial Intelligence for Groundwater Recharge Prediction in an Arid Region: Application of Tabular Deep Learning Models in the Feija Basin, Morocco." *Frontiers in Remote Sensing* Volume 6 - 2025. .DOI: 10.3389/frsen.2025.1622360

Elmotawakkil, A., Sadiki, A., & Enneya, N. (2024). Predicting Groundwater Level Based on Remote Sensing and Machine Learning: A Case Study in the Rabat-Kénitra Region. *Journal of Hydroinformatics*, *26*(10), 2639–2667. DOI: 10.2166/hydro.2024.494

FAO. 2018. *WaPOR Database Methodology: Level 1. Remote Sensing for Water Productivity Technical Report: Methodology Series*. Rome: FAO. https://openknowledge.fao.org/server/api/core/bitstreams/eb122ea8-99a6-4719-97ff-cadd09a5eced/content

Heidecke, C. (2009). *Economic analysis of water use and management in the Middle Drâa Valley in Morocco*. Peter Lang International Academic Publishers. https://books.google.co.ma/books?id=h0l50AEACAAJ

Hengl, T. (2018). Soil texture classes (USDA system) for 6 soil depths (0, 10, 30, 60, 100 and 200 cm) at 250 m (v0.2) [Data set]. Zenodo. DOI: 10.5281/zenodo.2525817

Hochreiter, S., & Schmidhuber, J. (1997). Long Short-Term Memory. *Neural Computation*, *9*(8), 1735–1780. DOI: 10.1162/neco.1997.9.8.1735 PMID: 9377276

Kaczmarek, N., Mahjoubi, I., Naimi, M., Bossenbroek, L., Moumane, A., Znari, M., Luis, M. S.-N. S., Frör, O., & Berger, E. (2025). Nature Conservation in the Draa Basin (Morocco): History, Present Situation, and Future Challenges. *Journal for Nature Conservation*, *88*, 127038. DOI: 10.1016/j.jnc.2025.127038

Karmaoui, A., & Moumane, A. (2016). Changes in the Environmental Vulnerability of Oasean System (Desert Oasis), Pilot Study in Middle Draa Valley, Morocco. *Expert Opinion on Environmental Biology, 5*(3). Advance online publication. DOI: 10.4172/2325-9655.1000135

Li, B., Rodell, M., Kumar, S., Beaudoing, H., Getirana, A., Zaitchik, B. F., de Goncalves, L. G., Cossetin, C., Bhanja, S., Mukherjee, A., Tian, S., Tangdamrongsub, N., Long, D., Nanteza, J., Lee, J., Policelli, F., Goni, I. B., Daira, D., Bila, M., & Bettadpur, S. (2019). Global GRACE data assimilation for groundwater and drought monitoring: Advances and challenges. *Water Resources Research, 55*(9), 7564–7586. DOI: 10.1029/2018WR024618

Li, Q., Li, Z., Shangguan, W., Wang, X., Li, L., & Yu, F. (2022). Improving soil moisture prediction using a novel encoder–decoder model with residual learning. *Computers and Electronics in Agriculture, 195*, 106816. DOI: 10.1016/j.compag.2022.106816

Li, S., Zhu, P., Song, N., Li, C., & Wang, J. (2025). Regional Soil Moisture Estimation Leveraging Multi-Source Data Fusion and Automated Machine Learning. *Remote Sensing (Basel), 17*(5), 837. Advance online publication. DOI: 10.3390/rs17050837

Makridakis, S. (1993). Accuracy Measures: Theoretical and Practical Concerns. *International Journal of Forecasting, 9*(4), 527–529. DOI: 10.1016/0169-2070(93)90079-3

Moumane, A., El Ghazali, F. E., Al Karkouri, J., Delorme, J., Batchi, M., Chafiki, D., & Karmaoui, A. (2021). Monitoring Spatiotemporal Variation of Groundwater Level and Salinity Under Land Use Change Using Integrated Field Measurements, GIS, Geostatistical, and Remote-Sensing Approach: Case Study of the Feija Aquifer, Middle Draa Watershed, Moroccan Sahara. *Environmental Monitoring and Assessment, 193*(769), 769. Advance online publication. DOI: 10.1007/s10661-021-09581-2 PMID: 34735624

Moumane, A., Elmotawakkil, A., Hasan, M. M., Kranjčić, N., Batchi, M., & Karkouri, J. A. & M. Youssef, Y. (2025). Integrating GIS, Remote Sensing, and Machine Learning to Optimize Sustainable Groundwater Recharge in Arid Mediterranean Landscapes: A Case Study from the Middle Draa Valley, Morocco. *Water (Basel), 17*(15), 2336.

Nash, J. E., & Sutcliffe, J. V. (1970). River Flow Forecasting Through Conceptual Models: Part i - a Discussion of Principles. *Journal of Hydrology (Amsterdam), 10*(3), 282–290. DOI: 10.1016/0022-1694(70)90255-6

Padhiary, M., Saha, D., Kumar, R., Sethi, L. N., & Kumar, A. (2024). Enhancing Precision Agriculture: A Comprehensive Review of Machine Learning and AI Vision Applications in All-Terrain Vehicle for Farm Automation. *Smart Agricultural Technology*, *8*, 100483. DOI: 10.1016/j.atech.2024.100483

Parra-López, C., Ben Abdallah, S., Garcia-Garcia, G., Hassoun, A., Trollman, H., Jagtap, S., Gupta, S., Aït-Kaddour, A., Makmuang, S., & Carmona-Torres, C. (2025). Digital Technologies for Water Use and Management in Agriculture: Recent Applications and Future Outlook. *Agricultural Water Management*, *309*, 109347. DOI: 10.1016/j.agwat.2025.109347

Probst, P., Wright, M., & Boulesteix, A.-L. (2019). Hyperparameters and Tuning Strategies for Random Forest. *Wiley Interdisciplinary Reviews. Data Mining and Knowledge Discovery*, *9*(3), e1301. DOI: 10.1002/widm.1301

Schulz, O., Busche, H., & Benbouziane, A. (2008). Decadal precipitation variances and reservoir inflow in the semi-arid Upper Drâa basin (South-Eastern Morocco). In *Climatic changes and water resources in the Middle East and North Africa* (pp. 165–178). Springer Berlin Heidelberg.

Singh, J., Sandhu, S. S., Singh, D., & Hadda, M. S. (2017). Soil Management to Optimize Water in Rice-Wheat Cropping. In Lichtfouse, E. (Ed.), *Sustainable Agriculture Reviews* (pp. 253–279). Springer International Publishing., DOI: 10.1007/978-3-319-48006-0_8

Snoek, J., Larochelle, H., & Adams, R. P. 2012. “Practical Bayesian Optimization of Machine Learning Algorithms.” *Advances in Neural Information Processing Systems (NeurIPS)* 25: 2951–59 https://papers.nips.cc/paper_files/paper/2012/hash/05311655a15b75fab86956663e1819cd-Abstract.html

Wan, Z., Hook, S., & Hulley, G. (2021). *MODIS/Terra Land Surface Temperature/Emissivity 8-Day L3 Global 1km SIN Grid V061* [Data set]. NASA Land Processes Distributed Active Archive Center. Date Accessed: 2025-10-09DOI: 10.5067/MODIS/MOD11A2.061

Wan, Z., Hook, S., & Hulley, G. (2021). *MODIS/Terra Land Surface Temperature/Emissivity 8-Day L3 Global 1km SIN Grid V061* [Data set]. NASA Land Processes Distributed Active Archive Center. DOI: 10.5067/MODIS/MOD11A2.061

Wang, J., Liu, Q., & Zhao, P. (2023). Multi-Source Data Integration for Soil Moisture Estimation Using Machine Learning. *Remote Sensing*, *15*(5), 1123. DOI: 10.3390/rs15174214

Wang, Y., Shi, L., Hu, Y., Hu, X., Song, W., & Wang, L. (2024). A comprehensive study of deep learning for soil moisture prediction. *Hydrology and Earth System Sciences*, *28*(4), 917–943. DOI: 10.5194/hess-28-917-2024

Willmott, C. J. (1981). On the Validation of Models. *Physical Geography*, *2*(2), 184–194. DOI: 10.1080/02723646.1981.10642213

Yu, T., & Zhu, H. (2020). *Hyper-parameter optimization: A review of algorithms and applications. arXiv preprint arXiv:2003.05689.* https://doi.org//arXiv.2003.05689DOI: 10.48550

Zhang, K., Hu, J., & Xu, L. (2023). Ensemble Machine Learning Approaches for Daily Surface Soil Moisture Estimation. *Geoscientific Model Development*, *16*, 5825–5843.

# Chapter 8
# Integration of Machine Learning and AI With Geospatial Analysis

**Oladapo Muyiwa Oladoja**
https://orcid.org/0000-0002-0766-2248
*Abiola Ajimobi Technical University, Ibadan, Nigeria*

**Latifat Adebisi Abimbola**
https://orcid.org/0009-0000-2924-8235
*Abiola Ajimobi Technical University, Ibadan, Nigeria*

**Elizabeth Ayodele Oladejo**
*Abiola Ajimobi Technical University, Ibadan, Nigeria*

## ABSTRACT

*The aim of the chapter is to provide a more comprehensive vision of how machine learning (ML) and artificial intelligence (AI) technologies can be used together with geospatial analysis to help solve modern environmental problems. This study shows that spatial data, remote sensing, and powerful computational techniques can be used to transform environmental monitoring, prediction, and decision-making processes using the power of ML/AI-enhanced geospatial tools. Demonstrated via hands-on case studies based on wind speed data of Lagos, Nigeria, it was shown how several ML algorithms, such as Random Forest (RF), Naïve Bayes (NB), XGBoost (XGB), Linear Regression (LR) and Neural Networks (NN), can be utilised to recognise and predict patterns in the environment. The models were validated using Root Mean Square Errors (RMSE) and Mean Absolute Error (MAE).*

DOI: 10.4018/979-8-3373-6608-1.ch008

## INTRODUCTION

The rapid growth of coastal megacities in West Africa presents unprecedented challenges to environmental management and sustainable development. Several challenges can be observed by examining the case of Lagos, Nigeria's economic capital and one of the world's most rapidly developing urban centres, with a population of over 17 million people that grows by 3.5% per year (World Population Review). Local atmospheric dynamics, particularly wind patterns, have become increasingly important in renewable energy planning, air quality measures, urban climate control, and climate change adaptation measures (Oyewole & Aro, 2018).

Conventional methods in meteorological analysis may be part of the solution, though they usually fail to recognise the detailed interplay between urban morphology, coastal effects, and the atmosphere that act to influence wind circulation patterns in tropical coastal metropolises (Odekanle *et al,* 2022). The combination of Geographic Information Systems (GIS), remote sensing, machine learning (ML), and artificial intelligence (AI) presents new opportunities to enhance various capabilities for understanding environmental systems and inform evidence-based decisions that promote sustainable urban development. AI and ML have taken centre stage in geospatial analysis to handle this complexity (Dristas & Trigka, 2025).

Remote sensing has entered the big data era thanks to the quick development of sensor technologies, lower satellite deployment costs, and broad IoT adoption. Massive volumes, high acquisition frequencies, format heterogeneity, and high dimensionality set modern geospatial data apart from traditional datasets. These traits present substantial storage, processing, and integration challenges in addition to opening doors for deeper insights. Deep learning algorithms, including convolutional neural networks (CNNs) to analyse images and recurrent neural networks (RNNs/LSTMs) to learn time-varying dynamics, have increased the capacity to identify more complex spatial and temporal behaviours. In addition, there are new developments in cloud computing (e.g., Google Earth Engine, AWS, Azure) and edge computing, which have democratised access to high-performance analytics, making it possible to run near real-time applications like disaster response and precision agriculture.

Despite these advances, several challenges remain. Concerns such as data quality, interoperability, algorithm bias, model interpretability, and ethical considerations must be addressed to achieve reliable and socially responsible results (Siddique, 2024). Furthermore, successful implementation involves interdisciplinary co-operation, which involves data science, spatial analysis, and domain-specific expertise. New opportunities have already arisen to provide secure, scalable, and real-time geospatial intelligence through emerging technologies, such as deep learning, Internet of Things (IoT) sensors, and blockchain, which are likely to promote this integration further. Geospatial Artificial Intelligence (GeoAI) is a relatively recent paradigm that has been developed over the last few years, aiming to integrate spatial analysis with AI and ML (Sharifi & Mahdipour, 2024). The method is based on high-performance computing, spatial big data, and advanced deep learning architectures to identify meaningful patterns, classify and automate classification as well as enhance predictive modelling in geospatial studies (Obisesan & Oladoja, 2022). As a case in point, deep neural networks, convolutional models, and recurrent architectures have been applied to problems such as land cover classification, vegetation monitoring, terrain mapping, and disaster response. Combining the multispectral, hyperspectral and synthetic aperture radar (SAR) data with ML has also maximised precision and extended capability to more powerful and real-time uses in a wide range of applications like agriculture, forestry, urban planning, and environmental survey (Oladoja *et al.*, 2023).

GeoAI is based on the assets of AI and spatial science. ML algorithms make it possible to automate, detect patterns, and make predictive models, and GIS allows the contextualisation and visualisation of that knowledge. This has seen GeoAI be used in various fields, including urban planning, smart cities, climate change adaptation, transportation, public health, disaster risk reduction, and tourism management (Fauzi, 2024). As an example, ML algorithms can be used to classify land-use, track environmental variations, and inform urban development, whereas AI-based GIS applications can improve analyses in real-time and policy development based on data. Geospatial analysis combined with ML and AI has already demonstrated revolutionary potential. Applications include public health mapping, biodiversity conservation, environmental monitoring, disaster response, and land cover classification and urban growth modelling (Kolluru *et al.*, 2024). For instance, by utilising spatial, spectral, and temporal data, ML-based methods improve the detection of deforestation, forecast urban sprawl, optimise agricultural practices, and model disease risk. These methods support real-time decision-making, increase scalability, decrease human error, and increase analytical efficiency.

A revolutionary strategy for tackling the escalating problems of urbanisation, environmental management, and sustainable development is the combination of geospatial technologies with artificial intelligence (AI) and machine learning (ML) (Odekanle *et al.,* 2022; Oladoja *et al.*, 2023). Spatial analysis and decision-making have long been aided by conventional geospatial tools like remote sensing and Geographic Information Systems (GIS). But as high-resolution Earth observation data, open geospatial datasets, and big data platforms proliferate, traditional analytical techniques are no longer adequate to manage the scope, intricacy, and dynamic nature of contemporary geospatial issues (Obisesan & Oladoja, 2022).

Recent developments show how enhanced pattern recognition, feature extraction, predictive modelling, and decision support are just a few ways that AI and ML can improve geospatial analysis. In rapidly urbanising cities like Bangkok, for example, population mapping demonstrates how combining open geospatial data, remote sensing, and machine learning-based classification can overcome data scarcity and generate fine-scale, building-level estimates for resilience and urban planning (Maneepong *et al.,* 2025). These methods enhance environmental monitoring, infrastructure planning, and disaster preparedness in addition to improving demographic insights. In a similar vein, AI-powered geospatial decision support systems (DSS) are showing promise across a range of fields, such as disaster risk reduction, urban management, and natural resource conservation. These systems optimise resource allocation, support climate adaptation strategies, and facilitate evidence-based policymaking by utilising machine learning, deep learning, and computer vision.

The combination of AI and geospatial analysis is not without its difficulties, despite its potential. Wider adoption is still severely hampered by issues with data quality, interoperability, algorithmic bias, interpretability, and ethics. Other challenges in many developing regions include the high cost of proprietary datasets, inadequate infrastructure, and out-of-date census data. Open data initiatives, explainable AI, interdisciplinary cooperation, and strong governance frameworks are necessary to address these issues and guarantee fair and reliable results.

All things considered, a paradigm shift from descriptive to predictive and prescriptive environmental intelligence systems is represented by the convergence of AI, ML, and geospatial technologies. It makes scalable solutions for resource management, climate resilience, and sustainable urbanisation possible as well as real-time monitoring and adaptive management. This combination of computational intelligence and human expertise opens the door to accomplishing the Sustainable Development Goals (SDGs) of the UN and creating more resilient, inclusive, and sustainable societies.

This study demonstrates the feasibility of integrating geospatial-AI technology with the Lagos (Ikeja) wind speed dataset (2000-2011) from the Nigerian Meteorological Agency. Integrating ground measurements with satellite-based environmental parameters, and comparing land surface temperature (LST), Normalised Difference Vegetation Index (NDVI), cloud coverage, and topographical information, one can form a hypothesised overall model of the environment that spans both past meteorological science and future computational methodologies. The intended study will answer two important questions:

1. How to perform hand-in-glove implementation of machine learning algorithms to forecast wind speed patterns with the help of multi-source geospatial data?
2. How do integrated geospatial-AI have the potential to shape urban development and adaptation planning?

# GEOSPATIAL ANALYSIS

## Spatial Context and Geographical Setting

Geospatial technologies, specifically Geographic Information Systems (GIS) and remote sensing, have been and continue to play a key role in the analysis of spatial patterns, detection of changes in the environment, and evidence-based decisions (Singha *et al.,* 2024). Historically, GIS has offered the means through which spatial data can be captured, stored, manipulated, and displayed, and has been used in a variety of fields including agriculture, urban planning, disaster management and monitoring of natural resources. Nonetheless, the increasing size and data complexity of geospatial data have revealed the constraints of traditional approaches to data analysis in dealing with high-dimensional, heterogeneous and real-time data. Remote sensing has progressed to such heights, beyond mere aerial photography, to include complex satellite constellations, unmanned aerial vehicles (UAVs), on land Internet of Things (IoT) sensors, and provides detailed multispectral and hyperspectral information on the monitoring of systems of the earth (Farooqi *et al.,* 2025). Such technologies have expanded tremendously in use in agricultural, forestry, climate monitoring, urban planning, and disaster management. The increasing demand for workable solutions that integrate AI into geospatial workflows is demonstrated by the growing commercialisation of GeoAI, which includes toolboxes within popular GIS platforms. These advancements show that the combination of AI and geospatial analysis represents a paradigm shift in spatial science rather than just a technical breakthrough. Researchers and practitioners can better address complex societal issues like climate change, disaster risk reduction, food security, and sustainable urbanisation by combining AI-driven predictive capabilities with geospatial context.

Even with its quick development, there are still issues with combining AI and geographic analysis. Although many AI models continue to underuse these geospatial features, traditional spatial methods place a strong emphasis on spatial dependence, heterogeneity, and autocorrelation. Furthermore, problems with data quality, interoperability, AI model interpretability, and ethical issues like privacy and bias still prevent wider adoption. However, the combination of machine learning, artificial intelligence, and geographic analysis is changing how complicated spatial issues are handled. Cloud computing, big data analytics, geocomputation, and other emerging technologies are enhancing GeoAI's capabilities, and interdisciplinary cooperation between computer scientists, geographers, urban planners, and policymakers is guaranteeing its practicality. In a world where data is becoming more and more important, GeoAI has enormous potential to promote sustainable development, increase urban resilience, and improve quality of life by fusing predictive intelligence with spatial knowledge.

Ikeja meteorological station (6.5244°N, 3.3792°E) gives a strategically located observation point of the dynamics of the atmosphere in Lagos. The station is 15km inland at 39 meters above sea level along the Atlantic coast, which gives the station the level of transition between coastal and continental values to local winds. This geographic location is important in gathering the land-sea breeze circulation, which prevails on the diurnal winds of Lagos. GIS analysis shows that the station depicts the suburban-urban fringe of Lagos, consisting of mixed land use typology, which entails residential locations, business districts, and residual vegetation (Gilbert & Shi, 2024). The 15 km coastline distance factors the station within the most favourable range to monitor penetration of the sea breeze inland, and the relatively low station altitude provides minimal orographic mixing in the determination of the wind measurements (Babalola *et al.,* 2025).

Figure 1 illustrates the spatial data infrastructure and the absolute significance of geographic situation in environmental AI perspectives. Its coordinates (6.5244°N, 3.3792°E) put it in the suburban interior of Lagos. It is relatively low in elevation, of 39m above sea level, as is characteristic of the Lagos plateau. The inland distance of the coast is 15km, which is a considerable length in order to comprehend the land-sea breeze influence. Ikeja stands as a strategic location of the transition between urban and rural zones of Lagos. Inland wind, which is captured by the station, is unique compared to the coastal measurements (Spatial Representativeness). The distance of 15 km to the coast can be used to measure urban heat island effects on wind (Urban Gradient Analysis), and the 39m elevation may be used to determine a baseline to make required adjustments to atmospheric pressure (Topographic Influence). ML training is conducted on point measurements that are an indication of a broader regional wind.

*Figure 1. GIS Spatial Context*

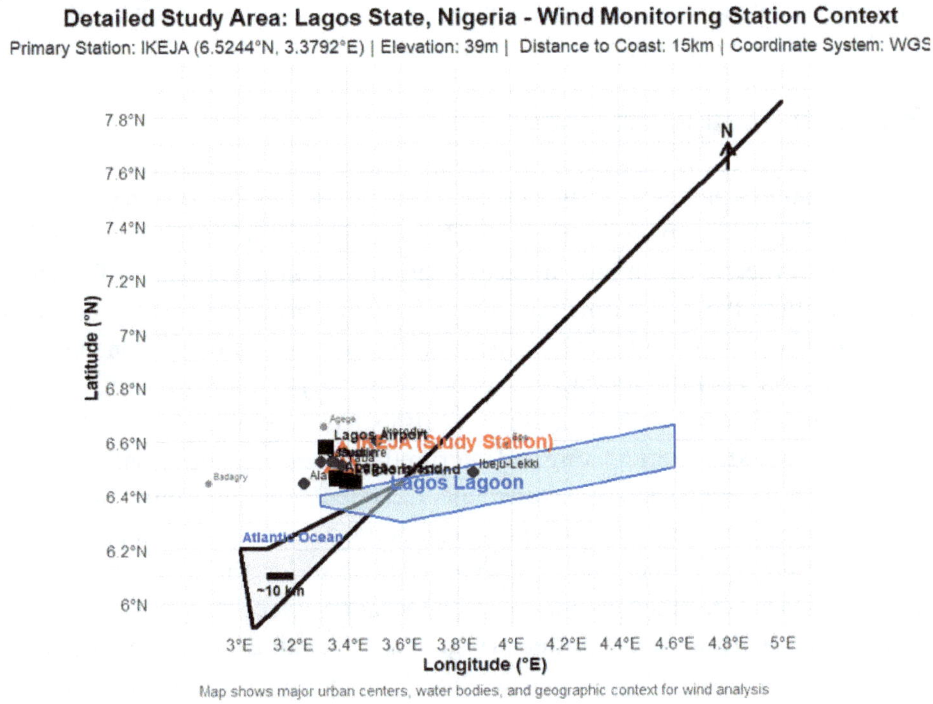


## Multiscale Spatial Analysis

Composite spatial datasets facilitate the characterisation of the environment in various scales (He *et al.*, 2024). Terrain context Digital Elevation Models (DEMs) can be used to reveal that Lagos has a relatively flat topography and a small level of elevation at various points that affect local wind flow patterns (Gilbert & Shi, 2024). Based on satellite imagery, land use categorisation determines the trend of urban heat island development and its association with the variation of wind speed (Yu *et al.*, 2024).

The distance analysis techniques compute accurate values between the meteorological station and important geographic sites, such as the coast, the main water bodies, and cities. These spatial measures are basic predictor variables for machine learning algorithms, to find geographic regulation of atmospheric processes (Awoseyi *et al.*, 2022; Dadashpoor *et al.*, 2024; Wang *et al.*, 2025). Emerging problems related to the rise of air pollution have created a myriad of challenges in human society, and therefore, exploration of an effective approach in the context of forecasting air pollutant concentrations has become extremely important (Zhang *et al.*, 2024). The main weaknesses of the current investigation involve disregard of the non-Euclidean nature of the site distribution of data and a severe dependence of interest on space and time of the dispersion process of pollutants.

## Temporal-Spatial Integration

Wind patterns analysed by time series indicated definite seasonal patterns in line with the West Africa monsoon pattern. The combination of time- and space-based analysis helps detect which geographical location affects the seasonal shifts of winds, such as curbing the dry Harmattan season (December-February) of north-eastern winds of Lagos, in the northern hemisphere, and shifting to the wet monsoon period (June-September) of the southwestern zone in the city.

In the world of big data, time series analysis has developed as one of the most important methods of anticipating and interpreting time-series information (Oladoja *et al.*, 2023). A sequence of measurements recorded at uniform intervals (of time) is called time series data. Both of those data points tend to be correlated and allow one to find patterns, trends, and seasonality. Time series analysis and forecasting are vast, with applications in modelling and weather forecasting.

Temporal analysis is exemplified in Figure 2, calculating GIS stations of the Ikeja area (6.5244°N, 3.3792°E). Raw wind speed data at two timepoints per 24 hours (09:00 and 15:00) are visualised along with moving averages (7-day and 30-day) that smooth shorter-term variations to indicate long-term trends and high sampling frequency (time resolution) observations over an 11-year period with seasonal changes and inter-annual variability. The prevailing wind speeds are within the 0-8 m/s, and there are definite trends involving higher speeds in a specific season. The moving averages show the circling cycles that the raw data conceals, and the large variance on a daily and seasonal basis of the wind patterns.

*Figure 2. Lagos Wind Speed Time Series Analysis*

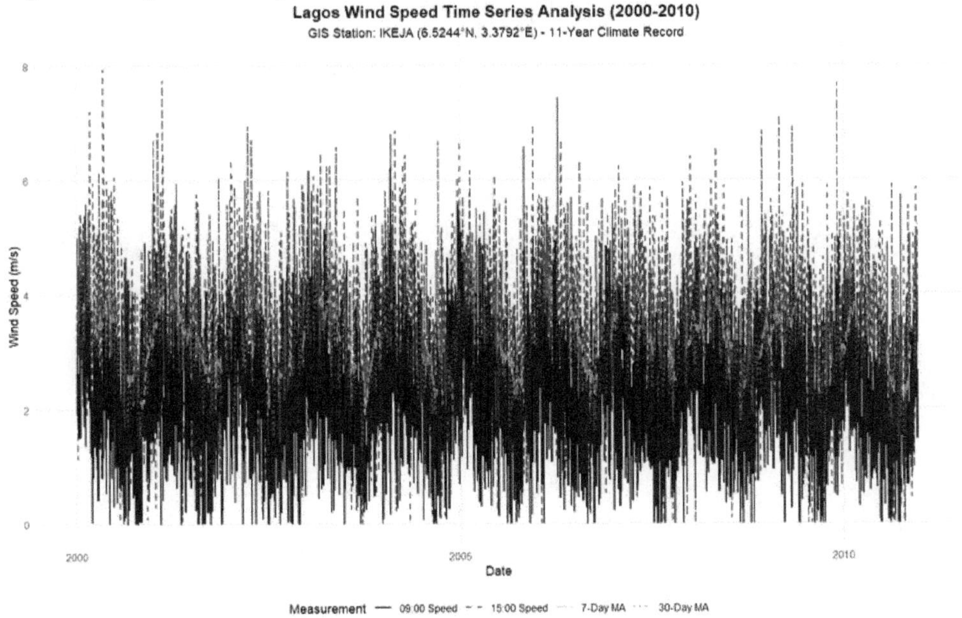


Wind rose analysis shows the directional preferences of the wind flow, where 26.2% of the prevalent direction is in southwest, indicating monsoon activity and 18.8% in northeast, indicating the presence of Harmattan. A combination of this directional analysis with speed distributions yields important information to select sites of renewable energy generation, as well as in urban ventilation planning.

Figure 3 is an advanced GIS-based tool to assess wind resources, named polar visualisation, that is essential to renewable energy planning in Lagos. They are dominated by the southwest winds 26.2% of the time, which is characteristic of the coastal monsoon weather condition in Lagos. The seasonal impact of the Harmattan is shown in northeast winds (18.8%). Mean speeds average 3.0-3.6m/s, with an increased speed as a rule associated with the directions SW and NE and displayed on the coloured gradient. The 8-sector analysis shows that there are very strong directional preferences which are essential in the siting of wind turbines. A specific geographic reference is given by Station coordinates (Ikeja). Direction analysis takes into consideration local topography and adjacent topography by sea. Data are used to show long-term climatological trends during the planning of energy infrastructure.

*Figure 3. Wind Rose Analysis for Energy Applications*

## MACHINE LEARNING

### Feature Engineering and Data Preparation

Environmental geospatial analysis deals with intricate correlations of atmospheric processes with geographical factors (Oladoja & Adegoke, 2023). To this end, the feature engineering exercise of machine learning applications was accomplished with caution. The set of features comprises meteorological factors (wind speed at 09:00 and 15:00, atmospheric pressure, humidity, temperature), satellite parameters (NDVI, LST, cloud cover) and geographic characteristics (elevation, distance to coast, urban heat island intensity).

The quality control is a part of data pre-processing in outlier detection, imputation of missing values, and alignment of very-long time series when synthesising multi-source data. Standardisation and normalisation methods make compatible variable measurements that were conducted at different scales and units. The feature correlation analysis finds the possible clusters of multicollinearities and directs the variable choice to the best model of performance

Figure 4 shows the correlation between meteorological and geospatial variables, that is, the similarity of the values, which is a feature engineering in machine learning models. Wind Speed variables possess great consistency across measurement occasions. The pressure of the atmosphere is proportional to the Land Surface Temperature (LST). Plants grow well in wet areas, and such atmospheric physics and clouds mitigate the surface temperatures because of the urban growth, decreasing vegetation, and cloud cover. The results indicate little correlation between the Distance to Coast and most of the variables, which indicates the local dominance of meteorology. Correlation of topographic influence includes the fact that elevation is perfectly correlated with local climate. The correlation matrix informs the selection of features in wind energy (coordinating the model by use of LST, pressure, and humidity), urban heat island (harnessing NDVI, distance to coast, and elevation) and air quality (coordinating wind patterns with urban heat and vegetation indices).

*Figure 4. Machine Learning Feature Correlation Heatmap*

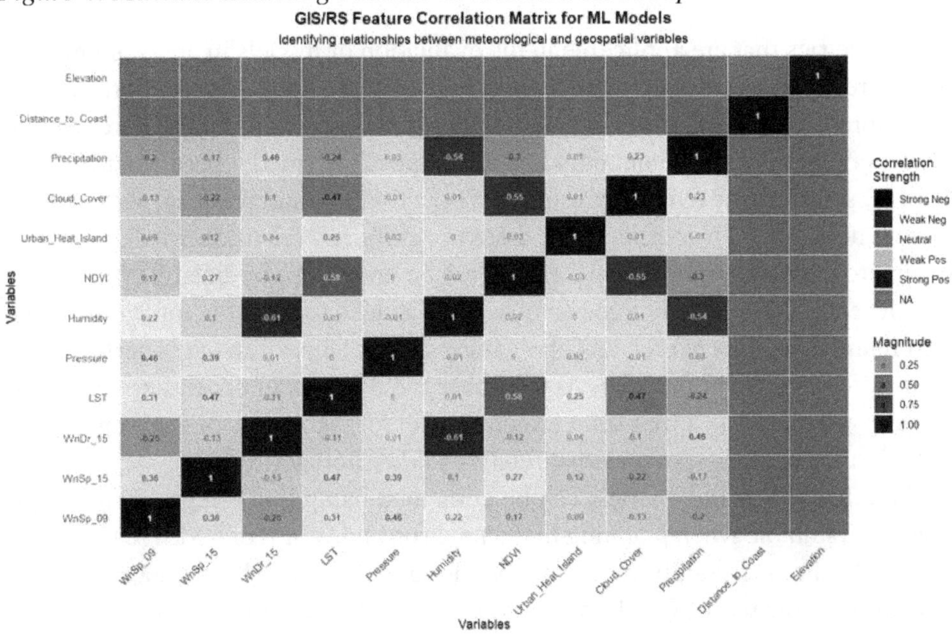


This study analyses how contemporary environmental science combines historical weather analysis with the most advanced AI/ML-based approaches, generating potent instruments for sustainable urbanisation in fast-growing and coastal cities, like Lagos. The combination of wind resource assessment with in-depth analysis of environmental variables provides the basis for evidence-based policy formulation and infrastructure development.

## Model Selection and Implementation

Five machine learning algorithms were applied to consider different facets of wind speed prediction in the study area. To consider linear relationships, Linear Regression (LR) was used, to check non-linear relations and feature prioritisation criterion, Random Forest (RF) was applied, to consider complex patterns, Neural Networks (NN) were implemented, to consider gradient-boosted ensemble, XGBoost (XGB) and Naive Bayes (NB) were implemented and used to compare results with others.

Strategies of cross-validation accommodate the temporal autocorrelation of environmental time series data. This is implemented following an 80%-20% train-test split and temporal hold-out validation so that models should be able to generalise into the future, but not merely optimise over the past. Grid search methods are used in hyperparameter optimisation to find the best model.

## Performance Evaluation and Framework

Two metrics that are applicable in the evaluation of models in areas of environmental prediction are also applied. Root Mean Square Error (RMSE) reports prediction accuracy as a number in original units of wind speed (m/s), therefore making a direct comparison to the quantity of interest possible (model performance). The Mean Absolute Error (MAE) presents an effective measurement of performance which is not affected by outliers. The assessment framework acknowledges the fact that environmental systems have a characteristic level of unpredictability, and even quite decent levels of predictive performance can lead to valuable information and interpretation in terms of practical applications. The comparison of the performance of the different algorithms helps us to choose the best model depending on the usage case, whether it is a real-time forecast, where it takes time or long-term trend prediction, where it is less sensitive relative to the short-term.

The reliable baseline models can be computed using linear regression, simple models are suitable when planning in the long run and values of RMSE can be used to quantify the uncertainty in energy predictions. Quality control checks can be made through the use of Random Forest. It has an alert in case flag predictions are outside normal RMSE limits. In order to adapt, retraining on growing datasets is required. This study employs a new integration of the relationships between NDVI and wind speed. The impact of the Urban heat Island is quantified in the wind patterns. Comparing multiple algorithms in the geospatial prediction, along with scalability in development country applications.

The ML linear regression performs as the best model with the lowest RMSE and MAE (Table 1). This means the model captures the intricacies in the linearity among the features. The ensemble learning RF performance follows the LR to ascertain several complexities involved in the erratic nature of wind speed. Convoluted NN and XGB perform badly as compared to LR and RF to ascertain that not all geospatial problems need sophisticated or advanced ML models to handle. The performance of NB shows that the problem at hand is not a classification problem but a regression task.

*Table 1. Comprehensive Machine Learning Model Comparison for Wind Speed Prediction in Lagos*

| Model | Type | RMSE | MAE |
|---|---|---|---|
| **Linear Regression** | Regression | 1.234 | 0.9790 |
| **Random Forest** | Regression | 1.256 | 0.9962 |
| **Neural Network** | Regression | 1.286 | 1.0202 |
| **XGBoost** | Regression | 1.293 | 1.0219 |
| **Naive Bayes** | Classification | 2.666 | 1.8806 |

**Note:** RMSE = Root Mean Square Error (lower is better); MAE = Mean Absolute Error (lower is better); Best performing values and poor performing values are highlighted.

## ARTIFICIAL INTELLIGENCE

### Intelligence Pattern Recognition and Automated Decision Support

The use of artificial intelligence methods goes beyond the traditional statistical methods, allowing the establishment of complicated patterns in environmental data, and the findings may not be otherwise obvious (Adegoke *et al.*, 2023). Neural networks exhibit the ability to learn non-linear associations among various factors in the environment at the same time, and thus may represent the emergent properties of the coupled land surface and atmospheric system.

In this application, deep learning architectures demonstrate moderate performance and exemplify the possibilities of more advanced pattern recognition as the size of the datasets increases. The possibility of the neural network to automatically determine and extract features of raw environmental data is a great step compared to the existing methods of having to manually specify the features.

To find the best conditions for many environmental applications, AI-based analysis allows them to be performed automatically. There is a need to incorporate wind speed forecasts into other environmental parameters to facilitate intelligently made decisions concerning renewable energy production, air quality and modelling of urban spaces. The ensemble learning methods use a combination of predictions made by multiple algorithms in order to give robust predictions having the uncertainty quantified. This element of an ensemble resembles the human expert. It contemplates numerous lines of evidence before environmental conclusions are made.

### Adaptive Learning Systems

Machine learning algorithms portray the ability to drive towards continuous enhancement with the emergence of new information. Such adaptive capacity is especially worthwhile when dealing with environmental monitoring systems which need to adapt to shifting climatic conditions and the trends of urban development. The framework allows a gradual learning, thus enabling the models to keep revising their lists of environmental relationships as Lagos evolves and as regional climates might change in response to global climate change.

## INTEGRATION OF MACHINE LEARNING AND AI

### Multi-Algorithm Ensemble Approach

The combination of various ML algorithms develops a complete prediction system that considers the advantages of the diverse techniques. LR gives an understandable set of initial forecasts with clean interaction between the coefficients, whereas RF models the non-linearity of interactions and gives a hierarchy of feature relevance. NNs are sophisticated in pattern recognition, and the ensemble techniques merge various views in making strong predictions.

This multi-algorithm method illustrates one of the fundamental principles of environment AI applications, that is, there cannot be an algorithm that is best suited to all scenarios. The combined framework enables the choice of algorithms depending on the use cases, availability of data, and the accuracy that is necessary in prediction.

### Geospatial-AI Workflow Integration

With the seamless fusion of GIS analysis and machine learning, there can be a production of a strong workflow in environmental evaluation. These processes in geographic data pre-processing, spatial analysis and predictive modelling are not distinct analytical steps, but they work as an integrated entity. With this combination, it will be able to perform automatic processing of satellite images, online addition of weather information, as well as the ever-changing forecast of the environment.

The workflow in this study shows that it scales to both one-station analysis and city-wide or regional environmental monitoring systems. Future trends of monitoring systems will be able to load more data sources, newly developed geographic variables, and extended temporal assignments into the framework.

## Environmental Intelligence System

Joint application of geospatial analysis, machine learning, and domain knowledge spells environmental intelligence systems that are able to assist in complicated decision-making. These systems are able to consider various environmental factors at the same time, evaluate trade-offs between various management options and give a quantification of the uncertainty in order to be able to evaluate risk.

The merger will allow the creation of early warning systems in environmental hazards, optimisation algorithms on the deployment of renewable energy and adaptive management in urban development under a changing environment.

## Model Performance Analysis

The overall model comparison opens a great deal of knowledge concerning the use of machine learning in environmental prediction. Linear regression has shown the overall optimal performance (RMSE = 1.234 m/s), indicating the primary substantiation that described fundamental physical relationships between environmental variables and the wind speed are linear in this coastal urban setting.

Random Forest also achieved reasonably well as a next-best alternative (RMSE = 1.256 m/s) and demonstrated the ability to measure non-linear relationships and also evade overfitting. The performance of both linear regression and Random Forest being comparable is indicative of the fact that this term relationship between geospatial variables and wind speed is chiefly linear, to the extent that there are not much of non-linear interactions within the existing data.

The more advanced algorithms, such as Neural Networks (RMSE = 1.286 m/s) and XGBoost (RMSE = 1.293 m/s), showed moderate outcomes, which means that the complexity of the model does not always enhance the correctness in the particular task in an environmental environment. This observation is a testament to the need to associate models and their complexity with the nature of data used and the sizes of samples available.

Figure 5 shows the evaluation of the regression model employing Root Mean Squared Error, which is an important indicator of continuous environmental forecasts. Linear models are good when the relations between geospatial variables and wind speed are more linear. It shows that simple models are effective at times than complex models, require less computation to monitor the environment in real time and are simple to interpret by the policy makers and environmental managers. Ensemble methods (Random Forest) have good results, where nonlinear patterns are captured without overfitting. It demonstrates the capacity to explain spatial heterogeneity in environmental data, has the capacity to deal with mixed data types (continuous satellite data, categorical land use), has the capacity to list the importance of the features in environmental interpretation and is resistant to the outlier that usually abounds in environmental datasets. Deep learning (Neural Network) performs at a relatively low level, which may be because of the training data. The Naive Bayes is a poor performer because it does not work on regression but on classification.

*Figure 5. RMSE Comparison Across ML Models*

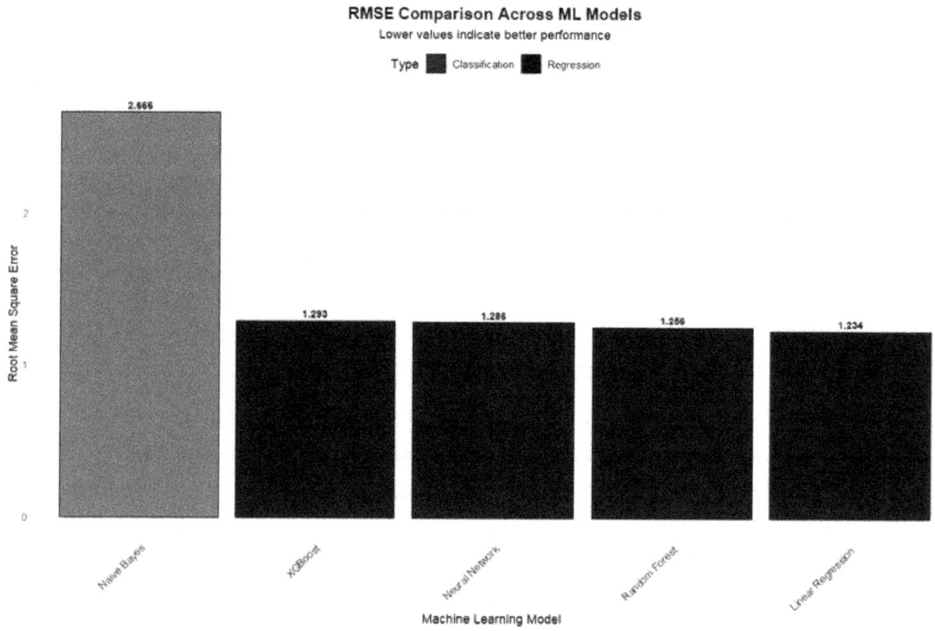


## Environmental Control on Variability

The identification of the main environmental controls on the wind speed patterns observed in Lagos was undertaken in the conducted study. Wind speed is positively correlated with the land surface temperature, which is attributed to the fact that warm air tends to force the circular circulation patterns in a given area. The increase in temperature brought about a pressure gradient that favours the intensity of wind flow, especially during the heating cycle in the day.

Vegetation indices (NDVI) portray the combination of intricate relationships with wind patterns, where a moderate level of vegetation is linked with the best wind conditions. Such correspondence denotes the trade-off between the surface roughness effects, which either increase turbulent mixing or block the movement of the winds by their density and spacing patterns.

There are systematic influences of atmospheric pressure and temperature that reflect on the speed of wind, such that strong winds of up to 5.2 m/s are recorded in high-pressure, high-temperature conditions. Such a tendency represents the yearly cycle of monsoon circulation and local thermal forcing that supports Lagos in the wind climate.

Figure 6 is an intricate visualisation that illustrates the factorial analysis of environmental controls, which is ideal to be used in determining the importance of features in ML models. The progressive pattern demonstrates an orderly amplitude of wind velocities with the combinations of pressure and temperature and portrays seasonal and daily atmospheric condition gradients. It is in the high vegetation where the highest wind speed is observed, which depicts augmented turbulence and assortment. The effect of vegetation preserved to all conditions of atmospherics and most exceptional stimulation is witnessed in the middle-to-high pressures/temperatures category. Such integration of multi-scale data, such as the interaction of point measurements (weather station) and regional environmental gradients (vegetation, pressure patterns) is typical of the contemporary environmental analysis. This study shows how environmental modelling directly applies to renewable energy, urban planning, and climate adaptation, and therefore, it proves to be an exemplary study of translational environmental science.

*Figure 6. Environmental Controls on Wind Speed*

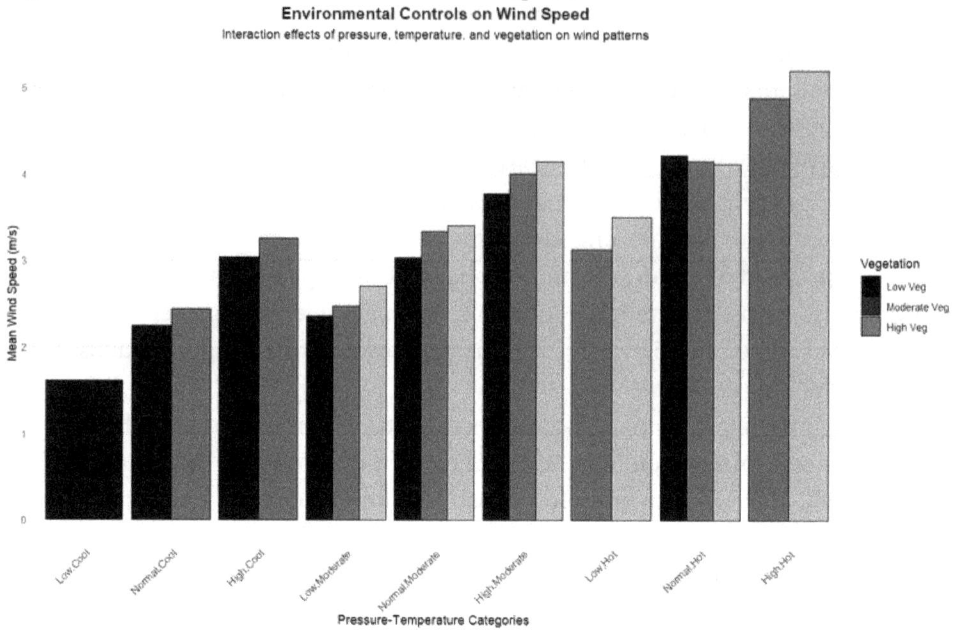


## Seasonal and Directional Patterns

These studies, based on wind rose analysis, show a clear predominance of flow in the southwest direction (26.2%) during the wet season and prominent northeast flow (18.8%) during the dry Harmattan season. These patterns of direction are vital in learning how pollutants are dispersed, ventilation in cities, and renewable sources of energy, depending on the seasons.

This use of remote sensing data and meteorological data demonstrates the effect of vegetation phenology on the patterns of seasonal winds. In times of elevated vegetation effects (high NDVI), increased surface roughness results in more turbulent flow that may be desirable in wind energy uses, but makes the dispersion of water pollution more difficult to model

The illustration in Figure 7 shows the integration of multispectral remote sensing because it incorporates ground-surface meteorological measurement (wind velocity bars), near-infrared satellite-derived vegetation coverage (NDVI) as the line and seasonal grouping using colour association. It has increased wind velocity and moderate vegetation during the dry harmattan season, high wind velocity throughout and vegetation starting to turn green in the pre-wet season. The wet monsoon and transition period have lower wind speeds and vegetation growth peaks, and there are increasing winds during the post-wet seasons.

*Figure 7. Seasonal Wind Patterns with Remote Sensing Integration*

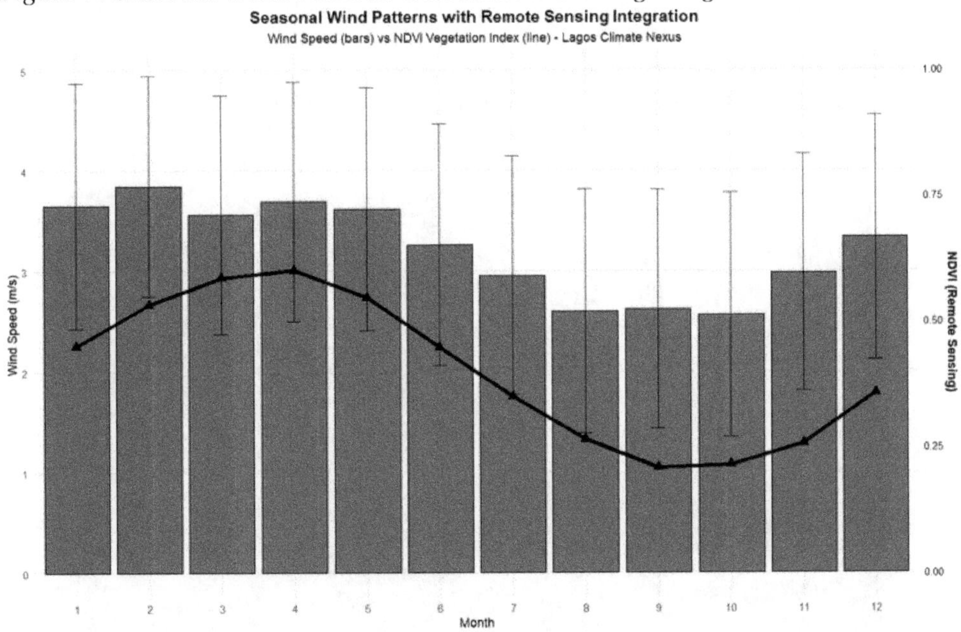


The integration demonstrates the direct influence on the environment of Lagos by West African monsoon systems. Inversion correlation between the level of vegetation and wind (NDVI) shows the pattern of seasonal moisture changes to influence productivity in the ecosystem. The accumulation of pollutants in lower wind speeds could occur during the rainy season. Knowledge about the peak winds may be used to maximise the natural ventilation. Wind-associated patterns in vegetation can yield insights into the location of green infrastructure. The NDVI combination indicates how the weather conditions trigger plant phenology, which is highly important when it comes to agricultural timing and irrigation scheduling, ecosystem persistence and preservation of biodiversity. The existence of high wind speed periods as determined in the analysis assists in forecasting the occurrence of dust storm events in the harmattan season, the possible occurrence of wind damage during the transitional times and flooding when accompanied by seasonal rainfall data.

Figure 8 illustrates long-term forecasting analytics and trending of the wind energy in Lagos. The range of wind power density has a high seasonal influence. The trend line is alarming, with the value falling from 31 W/m$^2$ (2000) to 23 W/m$^2$ (2011). Obvious patterns in a year, which are around 45-50 W/m$^2$ maximum and 10-15 W/

m² minimum. Inter-annual variability exhibits significant year-to-year differences in peak magnitude. The trend of decline of wind power has various implications. The decline in wind deposits disdains the goals of sustainable power and indicates the necessity of renewable energy planning. In addition, the diversification of the renewable energy portfolio and long-term Infrastructure Investment is needed.

*Figure 8. Wind Power Potential Assessment with Trend Detection*

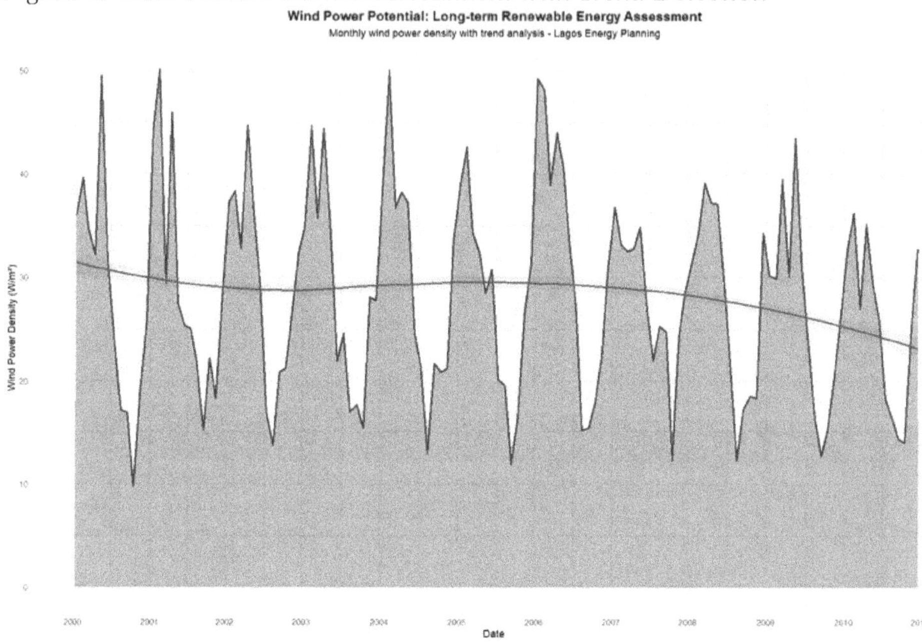


Exploratory data analysis in this scatter plot matrix displayed in Figure 6 is the visualisation that illustrates correlations of environmental factors evaluated by satellite observations and wind speed, that is, variables and their relation to one another. In cloud cover against wind speed (top left), there is a minor correlation between high cloud cover and low speeds of wind speeds. A high variability demands a complicated setting in the atmosphere, and cloud cover conveys a predictor variable in the wind forecasting models in ML Relevance. In the case of LST against wind speed (top right), there is a positive correlation between the temperature and strong winds and thermal gradients cause the movement of wind. In NDVI against wind speed (bottom left), high vegetation index correlates with moderate wind increments, vegetation affects local wind formation, and variations in NDVI have an impact on wind stream properties. On urban heat island (UHI) against wind speed (bottom right), there is a weak association between UHI intensity and wind speed. That is relevant to the comprehension of atmospheric meet-up dynamics on the city scale.

The wind-temperature relationships allow pollutant dispersion predictions, how the LSTs and winds interact regarding cooling urban regions to reduce urban heat, and how NDVI-wind correlations define where to place vegetation to generate climate advantages.

*Figure 9. Remote Sensing Integration Analysis*

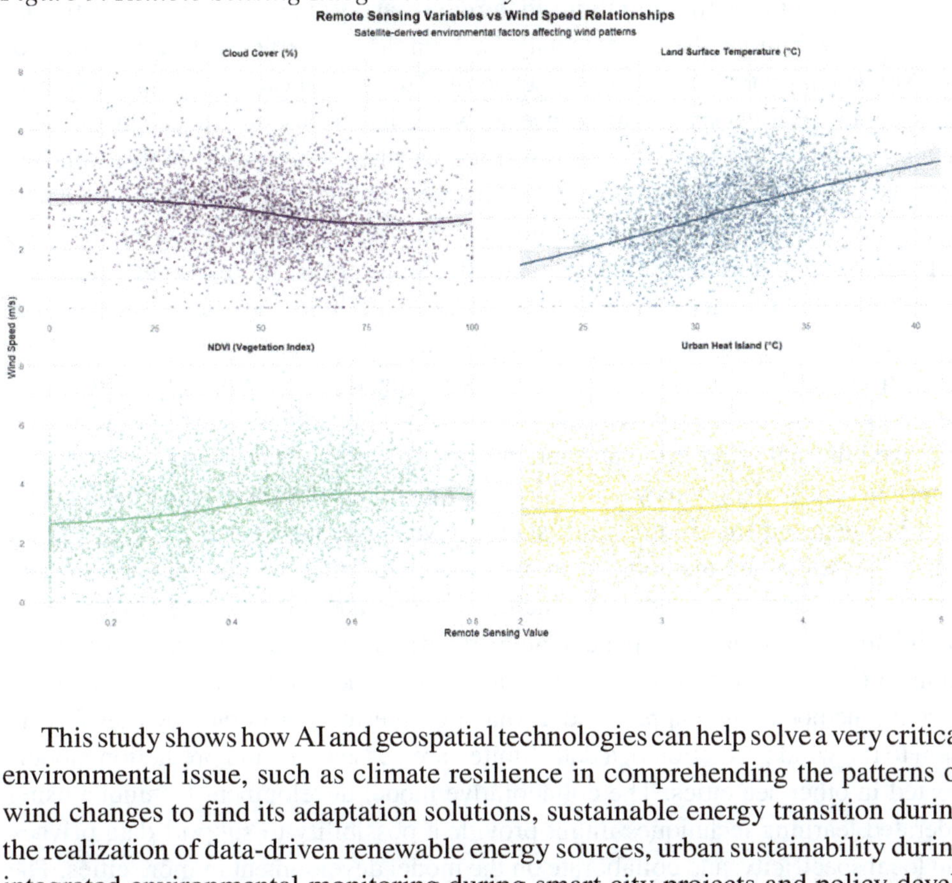


This study shows how AI and geospatial technologies can help solve a very critical environmental issue, such as climate resilience in comprehending the patterns of wind changes to find its adaptation solutions, sustainable energy transition during the realization of data-driven renewable energy sources, urban sustainability during integrated environmental monitoring during smart city projects and policy development during evidence-based environmental management operations. Long-term trend analysis and multi-source remote sensing data offer a useful tool in critical decisions, particularly on sustainable development in the context of the fast-growing coastal-urban region of Lagos, by environmental scientists, city planners, and policy makers.

This predictability of the simple linear models indicates that simple atmospheric physics forms the basis of wind patterns in Lagos, since simple concepts of meteorology should be used to make predictions which are more likely to be accurate compared to using complex algorithms. The result has significant implications in practice with operational environmental monitoring systems, where the main concern is on interpretability and computational efficiency.

## LIMITATIONS AND FUTURE RECOMMENDATIONS

There are a number of limitations in the data that confine the model performance and generalizability in the present analysis. A single meteorological station does not have sufficient spatial coverage in a megacity of the size of Lagos and may therefore fail to capture critical spatial differences in winds in various areas of the metropolis. In future studies, it is possible to add more weather stations and high-resolution wind data that would help to take into consideration the spatial heterogeneity. The time series is 11-year temporal and hence big enough to recognise climatic shifts and variability, but not complete to depict the long-term climate changes and variables. Longer temporal coverage would enhance the Understanding of inter-annual and decadal wind pattern changes and especially in the planning of climate change adaptation. The limitations of satellite data with regard to the temporal resolution of measurements limit the possibilities to record sub-daily processes in the atmosphere. Improved diurnal wind cycle and interaction with urban heat island development pattern may be understood by using higher temporal resolution remote sensing data.

It is suggested that future work should investigate the idea of state-of-the-art artificial intelligence methods, such as physics-informed neural networks, where atmospheric equations are directly included in the machine learning model. Good candidates to use temporal dependencies in wind patterns better are Long Short-Term Memory (LSTM) networks and other recurrent neural network architectures. Transfer learning methods might apply models that are superior in accuracy by transferring them to the wind prediction in a data-limited area based on wind prediction models created in other sea cities. The collaborative model development technique using federated learning techniques might provide a possibility to support data privacy and local specificity, and collaborate on the model development in more cities. The existing real-time environmental monitoring and forecasting systems might be assisted by the improved infrastructural environment, like cloud computing infrastructure and edge computing devices. Interfaces and support tools in decision making that are user-friendly may aid in their adoption by urban planners, environmental management and policy makers. Mobility integration with the mobile apps and the information systems in the public could allow residents of Lagos to gain real-time environmental data to promote individual decision-making and resilience-building in the community.

## CONCLUSION

Using Lagos as a case study, this study has shown how machine learning and artificial intelligence can be successfully combined with geospatial analysis to address environmental issues in coastal areas that are rapidly becoming more urbanised. The analysis identified important factors influencing wind variation, including vegetation cover, seasonal circulation, and land surface temperature, by fusing satellite remote sensing with meteorological data collected on the ground. Planning for renewable energy, managing air quality, and developing climate adaptation plans can all benefit from these insights. Ensemble approaches demonstrated that, although sophisticated algorithms such as gradient boosting and neural networks perform well, simpler models frequently yield more reliable and understandable outcomes for environmental applications.

Beyond its applicability locally, the framework created here can be scaled to other megacities and adjusted to a variety of climatic and geographic conditions. By making evidence-based planning for sustainable cities, renewable energy development, and climate change mitigation possible, it offers a route towards creating climate-resilient urban systems that are in line with the UN Sustainable Development Goals.

In order to enable quicker responses to changing environmental conditions, future work should concentrate on extending the framework to include real-time data streams, citizen science inputs, and adaptive learning systems. Predictive accuracy will be improved while addressing concerns of transparency and trust by incorporating explainable AI models, cloud-based computing, and higher-resolution sensors. In coastal megacities around the world, this will increase the ability of environmental governance systems to handle the mounting stresses of urbanisation and climate change.

## REFERENCES


Adegoke, T. M., Oladoja, O. M., Bashiru, S. O., Mustapha, A. A., Aderupatan, D. E., & Nzei, L. C. (2023). Topp-Leone Inverse Gompertz Distribution: Properties and Different Estimation Techniques and Applications. *Pakistan Journal of Statistics, 39*(4).

Awoseyi, A. A., Akinsola, J. E. T., Oladoja, O. M., Adeagbo, M., & Adebowale, O. O. (2022). Hybridisation of decision tree algorithm using sequencing predictive model for COVID-19. In *Emerging Technologies for Combatting Pandemics* (pp. 201–221). Auerbach Publications. DOI: 10.1201/9781003324447-10

Babalola, A. S., Adeogun, A. O., Thabet, H. S., TagEldin, R. A., Oyeniyi, T., Adekunle, O., Izekor, R., Adetunji, O., Olalekan, O., Omotayo, A., Abiodun, O., Adediran, A. D., Adekeye, T., Adegbola, A. O., Isaac, C., Okoko, P. O., & Harwood, J. F. (2025). Geospatial modelling of geographical spread of *Aedes* species, in relation to climatic and topographical factors in Lagos State, Nigeria. *PLoS Neglected Tropical Diseases, 19*(2), e0012860. DOI: 10.1371/journal.pntd.0012860 PMID: 39913605

Dadashpoor, H., Khaleghinia, A., & Shabrang, A. (2024). Explaining the role of land use Changes on land surface temperature in an arid and semi-arid metropolitan area with multi-scale spatial regression analysis. *Environmental Monitoring and Assessment, 196*(2), 124. DOI: 10.1007/s10661-023-12241-2 PMID: 38195837

Farooqi, S. M., Kanwal, A., Zaman-ul-Haq, M., Saqib, Z., Akhtar, N., Tariq, A., Abdullah-Al-Wadud, M., Mubbin, M., & Bokhari, S. A. (2025). Integrating Geo-AI with RS & GIS for comprehensive assessments of urban land cover transformations and integrated responses. *Environmental Earth Sciences, 84*(1), 3. DOI: 10.1007/s12665-024-12005-2

Fauzi, C. (2024). Review Geospatial Artificial Intelligence (Geo-AI): Implementation of Machine Learning on Urban Planning. *Proceedings of the International Conference on Applied Science and Technology on Engineering Science 2023 (iCAST-ES 2023)*. 311-329. https://doi.org/DOI: 10.2991/978-94-6463-364-1_30

Gilbert, K. M., & Shi, Y. (2024). Quantitatively analysing the driving factors of urban spatial evolution in Lagos (2000-2020). *OAlib, 11*(3), 1–22. DOI: 10.4236/oalib.1111303

He, Y., Tu, B., Liu, B., Chen, Y., Li, J., & Plaza, A. (2024). Hybrid multi-scale spatial-spectral Transformer for hyperspectral image classification. *IEEE Transactions on Geoscience and Remote Sensing, 62*, 1–18. Advance online publication. DOI: 10.1109/TGRS.2024.3443662

Kolluru, V., Mungara, S., Chintakunta, A. N., Kolluru, L., & Telaganeni, C. S. (2024). Geospatial Intelligence Enhancement Using Advanced Data Science and Machine Learning: A Systematic Literature Review. http://dx.doi.org/DOI: 10.5121/csit.2024.141502

Maneepong, K., Yamanotera, R., Akiyama, Y., Miyazaki, H., Miyazawa, S., & Akiyama, C. M. (2025). Towards High-Resolution Population Mapping: Leveraging Open Data, Remote Sensing, and AI for Geospatial Analysis in Developing Country Cities—A Case Study of Bangkok. *Remote Sensing (Basel)*, *17*(7), 1204. DOI: 10.3390/rs17071204

Obisesan, K. O., & Oladoja, O. M. (2022). On Normal Process of Diffusion Equation in Monitoring Carbon Monoxide Concentrations in Nigeria. *International Journal of Statistical Distributions and Applications.*, *8*(2), 24–29. DOI: 10.11648/j.ijsd.20220802.11

Odekanle, E. L., Fakinle, B. S., Odejobi, O. J., Akangbe, O. E., Sonibare, J. A., Akeredolu, F. A., & Oladoja, O. M. (2022). COVID-19 induced restriction in developing countries and its impacts on pollution load: A case study of Lagos megacity. *Heliyon*, *8*(8), e10402. DOI: 10.1016/j.heliyon.2022.e10402 PMID: 36065213

Oladoja, O. M., & Adegoke, T. M. (2023). Detection of Cervical Cancer Risk Factors in Venezuela Using the Decision Tree Algorithm. *Reliability: Theory & Applications, 18*(3 74), 492-500.

Oladoja, O. M., Obisesan, K. O., Adebayo, S. A., & Adegoke, T. M. (2023). On the Bayesian modeling of road accidents due to vehicle type in Oyo State, Nigeria. *Reliability: Theory & Applications, 18*(2 73), 223-232.

Oyewole, J. A., & Aro, T. O. (2018). Wind speed pattern in Nigeria (a case study of some coastal and inland areas). *Journal of Applied Science & Environmental Management*, *22*(1), 119–123. DOI: 10.4314/jasem.v22i1.22

Sharifi, A., & Mahdipour, H. (2024). Utilising geospatial artificial intelligence for remote sensing applications. *Environmental Earth Sciences*, *83*(9), 274. DOI: 10.1007/s12665-024-11584-4

Siddique, I. (2024). Machine learning empowered geographic information systems: Advancing Spatial analysis and decision making. https://doi.org/.DOI: 10.30574/wjarr.2024.22.1.1200

Singha, C., Rana, V. K., Pham, Q. B., Nguyen, D. C., & Łupikasza, E. (2024). Integrating machine learning and geospatial data analysis for comprehensive flood hazard assessment. *Environmental Science and Pollution Research International, 31*(35), 48497–48522. DOI: 10.1007/s11356-024-34286-7 PMID: 39030454

Wang, Z., Li, D., Jiang, R., & Okumura, M. (2025). *Continuous sign language recognition with multi-scale spatial-temporal feature enhancement.* IEEE., DOI: 10.1109/ACCESS.2025.3526330

Yu, Y., Zhang, Y., Cheng, Z., Song, Z., & Tang, C. (2024). Multi-scale spatial pyramid attention mechanism for image recognition: An effective approach. *Engineering Applications of Artificial Intelligence, 133*, 108261. DOI: 10.1016/j.engappai.2024.108261

Zhang, B., Chen, W., Li, M. Z., Guo, X., Zheng, Z., & Yang, R. (2024). MGAtt-LSTM: A multi-Scale spatial correlation prediction model of PM2.5 concentration based on multi-graph attention. *Environmental Modelling & Software, 179*, 106095. DOI: 10.1016/j.envsoft.2024.106095

## KEY TERMS AND DEFINITIONS

**Artificial Intelligence:** This is a Computer science discipline that has been approached more broadly to develop systems that are capable of performing all functions which traditionally demand human intelligence/abilities, like reasoning, learning, perception and decision making.

**Deep Learning:** This is a generated part of machine learning that applies a layered form of artificial neural networks (hence deep) to model and interpret intricate trends in data. It is specifically good at image recognition and natural language processing.

**Ensemble Learning Method:** An ensemble learning method involves using the outcome of one or more models and a different combination of models that are used to make a stronger and more accurate final forecast. The idea is that several models, when combined, usually outperform an individual model.

**Environmental Monitoring:** This is the monitoring of environmental conditions and changes systematically with time. This involves monitoring of the quality of air and water, climate factors, biodiversity, pollution and the wellness of ecological systems through diverse sensors, sampling, and data collection.

**Extreme Gradient Boosting (XGBoost):** This is a sophisticated machine learning algorithm that trains models one upon another, and new models update inaccuracies within the previous models. It works very well with structured data and competitive machine learning.

**Geographic Information Science (GIScience):** This is the field of study examining the theories, methods, and techniques of computation on which Geographic Information Systems (GIS) rely. It includes geographical analysis, map making, remote sensing and geographic data processing algorithms.

**Geospatial Analysis:** This is the procedure of analysing and interpreting data that has a geographic or spatial element. It entails the examination of trends, patterns, and relationships in data that could be identified with geographical locations on the face of the earth in tools such as GIS (Geographic Information Systems).

**Geospatial Tools:** These are technologies and software programs that geographic data was collected, stored, analysed, and visualised. The examples are GIS software (such as ArcGIS, QGIS), GPS, satellite imagery applications, mapping applications, and spatial databases. The tools are used to process location-based information to be used in different applications.

**Linear Regression:** This is a statistical technique which approximates the connection between an independent variable or variables and a dependent variable by fitting a linear equation to collected results. It can be applied in forecasting and in grouping the correlation of variables.

**Machine Learning:** This is a branch of artificial intelligence capable of teaching computers to output decisions according to data on the basis rather than having a specific command of every action. Pattern recognition and experience enable systems to attain better performance.

**Naïve Bayes:** This is a random learning based on the Bayesian theorem, where it is assumed that the features are independent of one another. Although it makes a municipally naive assumption, it is precisely effective for the classification tasks, especially the text classification and spam filtering.

**Neural Networks:** This is a bio-inspired computer-based network modelled after neural networks in the body; networks of nodes (neurons) that process information. Through training, they can get to know complex patterns and relationships in data.

**Normalised Difference Vegetation Index (NDVI):** This is a form of satellite imagery where a numerical measure scores the health and density of vegetation. NDVI images have a -1 to +1 range, into which higher positive values reflect healthier, denser vegetation. It is determined by the contrast in the near IR and the red-light reflectance.

**Random Forest:** This is a machine learning algorithm which is an ensemble that constructs several decision trees to produce the predictions and combines them. Its reliability and capacity to process huge amounts of data containing a large number of characteristics are known.

**Remote Sensing:** This is the study of acquiring details about the Earth's surface and atmosphere remotely, usually through satellites, aeroplanes or drones that employ equipment to monitor electromagnetic radiation.

**Root Mean Squared Error (RMSE):** This is a statistical parameter applied to appraise the effectiveness of predictions or models. It figures the square root of the mean of squared deviations between forecasted and genuine values, and it gives how good a model is functioning.

**Temporal Spatial Integration:** This is the integration of both the time-based (temporal) and location-based (spatial) data analysis. Under this approach, the variations of the phenomena both in space and time are analysed, and different researchers can get the concept of dynamic patterns and processes.

**The Land Surface Temperature (LST):** This is the thermal infrared sensors, usually satellites, that measure the surface of our Earth. LST is not the same thing as air temperature, and it plays a key role in urban heat island-related research, climate change, farming, and environmental surveillance.

**Topography:** This is the careful survey and analysis of the surface features of the earth, such as surface elevation, gradients or slopes, land forms as well as terrain characteristics. Topographic data refers to the three-dimensional form of the surface nearby and is vital in the determination of fluid movement, erosion processes and land use designs.

**Windrose Analysis:** This is a graphical instrument that shows the data regarding the direction and the measurement of the wind speed in a certain location at a given time. It depicts how often the wind is coming from various directions and the speed with which it comes, which is usually in a circular diagram and looks like a compass rose.

**Windspeed:** This is the speed of erratic movement of air molecules measured in meters per second (m/s), kilometres per hour (km/h) or miles per hour (mph). It is a basic meteorological variable in weather forecasting, renewable energy planning and climatic research.

# Chapter 9
# Contribution of DSAS and GIS Tools in Shoreline Evolution:
## The Case of Mehdia

**Mohamed Drissi**
https://orcid.org/0009-0007-4653-5616
*Ibn Tofail University, Morocco*

**Mouhcine Batchi**
https://orcid.org/0000-0002-0915-6592
*Department of Geography, Ibn Tofail University, Morocco*

**Said Laaribya**
*Department of Geography, Ibn Tofail University, Morocco*

## ABSTRACT

*The present study manipulates satellite imagery (Landsat TM+ and ETM+) and GIS tools to analyze the changes in the Mehdia coastline between 1990 and 2023. We estimated shoreline change rates, particularly the End Point Rate (EPR) index, using a diachronic approach and the DSAS 5.0 model in ArcGIS. Results show that the coastline has significantly retreated, particularly between 1990 and 2005 and 2005 and 2023, with erosion concentrated in the central and northern sectors. Across the estuary, the overall trend shows an average retreat of –3.25 m/year. In addition to human actions including the construction of dams and dike systems, urbanization, tourism, and dredging, this vulnerability is also linked to natural forces like storms and climate change.By affecting sediment dynamics, these factors have accelerated*

DOI: 10.4018/979-8-3373-6608-1.ch009

*beach loss, particularly in the north. The findings highlight the critical need for sustained monitoring and resilient coastal management to prevent present erosion and anticipate future environmental challenges.*

## 1. INTRODUCTION

Nearly 60% of the world's population lives within 60 kilometers of the coast, according to reports from international organizations like the UICN. (2010), Nearly half of the world's population, or about 3.8 billion people, live within 150 kilometers of coastlines(Neumann et al., 2015). As a result, over two-thirds of sandy coasts are experiencing, with rate varying from 1cm /year to 10m/ year (Pilkey & Hume, 2001). In the study by Luijendijk et al. (2018), just 20% of sandy or gravelly coastlines are stable, and only 10% are advancing, while 70% are retreating.

The coast characterized by great environmental diversity, such as sandy or rocky beaches, dune ridges, cliffs, lagoons, and estuaries, is mostly dynamic and susceptible to often forces hydrodynamic and meteorological influences (Snoussi, 2020).

The Morocco's coastline holds a significant place in the nation territory, stretching over than 3500 km along both the Atlantic and Mediterranean coasts. It is a major center for multiple infrastructures and economic activities, and is host to over half of Morocco's population(CESE, 2021).

However, the coastline is subject to both natural processes (currents, swell, tides, waves, storms, etc.) and anthropogenic impacts (infrastructure, replenishment operations, etc.), which affect the shoreline causing rapid changes that are particularly marked by dangers, erosion, and inundation (Mallett et al., 2012).

Climate change directly affects the sea level by accelerating up its elevation, with an annual rate that more than doubled between 1993 (0,18 cm/an) and 2023 (0,42 cm/an)(NASA, 2024).

The coastal zone is an extremely delicate ecosystem, and highly stressed environment, undergoing a natural phenomenon of retreat of the coastline that that impacts a significant portion of the Moroccan shores. As with the Atlantic coastal cities, the coastline of Mahdia has changed over the years as a result of rising sea levels, a rise in the frequency of extreme weather events, and increased anthropogenic pressures. These changes in the coastline can sometimes pose a threat to nearby infrastructure, making Mehdia's shore highly vulnerable and under constant pressure. Therefore, it is essential to monitor shoreline changes over time and across different areas.

Given this, the primary goal of this research is to monitor and investigate the variations in Mehdia's shoreline from 1990 to 2023, in order to identify its overall trends and the key factors driving these changes. For this reason, our study aims to guide policymakers and coastal managers to develop successful management and adaption strategies in response of Morocco's increasing issues with coastal erosion.

## 2. MATERIALS AND METHODS

### 2.1 Study Area

The study area is situated along the Atlantic façade of Gharb (Ennouali et al., 2023). It is located between 34°16' and 34°14' N latitude, and 6°38' and 6°41' W longitude in Rabat-Salé-Kénitra region, near the city of Kénitra and 30 km northeast of the capital, Rabat. It extends between oued Sebou in the north and Sidi Boughaba beach in the west and the biological reserve of Sidi Boughaba in the south-west.

The coastline exist in the western Mehdia city, its bay extend about 3 km in a straight line, between the estuary of Oued Sebou and Sidi Boughaba Beach.

According to Laassilia. (2013) this study area is sandy beach with a slope about 3% to 5%, a shorter coastline less than 150 meters, protected by an aeolian dunes that extend 3 to 5 km and have an height from 5 to 15 m (Hakkou & Maanan, 2023). The coastal erosion in this area is a result of human activities and effects of climat change(Ennouali et al., 2023; Tafraouti, 2025).

*Figure 1. Location map of the study area*

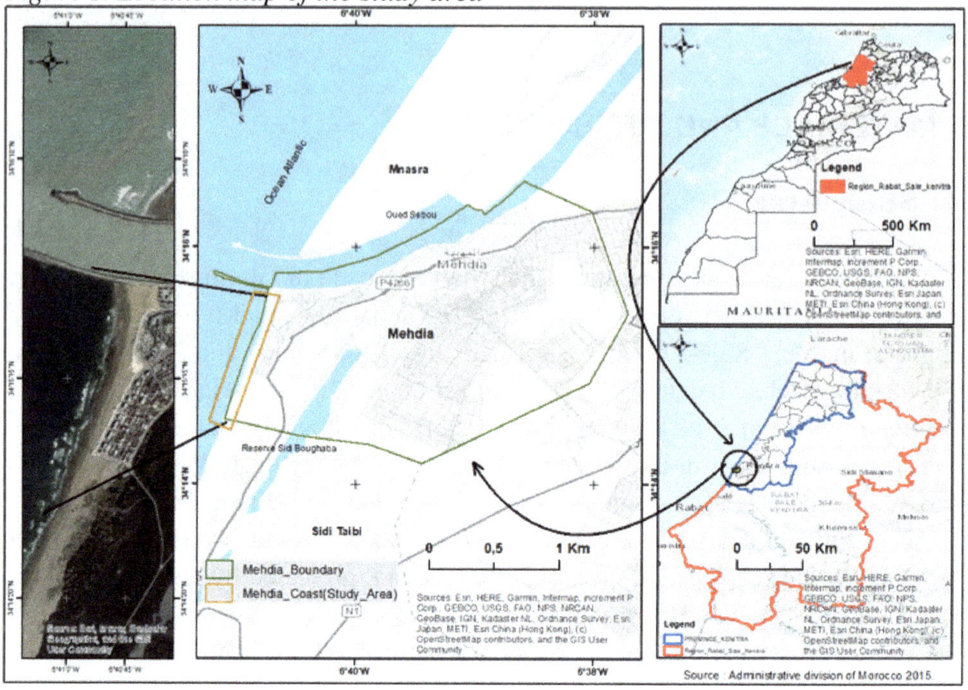


## 2.2 Methodology

The methodology of this study is multidisciplinary, and it aims to analyze the trends of coastal erosion by determining the most impacted zone along of this area between 1990 and 2023, using satellite images and different tools of GIS.

### 2.1.1 Geospatial Tools Context for Coastal Studies

The Digital Shoreline Analysis System (DSAS) is one of the best specialized analytic tools, its creation by USGS, make a significant revolution in Geographic Information Systems particularly in coastal studies, this tool allows to calculate an accurate estimations of accretion and erosion rates, and has become a standard technique for evaluation coastal change (Himmelstoss et al., 2021).

DSAS tools have the ability to analyze diver data sets efficiently and precisely, which can identify, and visualize trends and coast dynamics.

However, there are different techniques that can offer more effectives results in coastal studies and remote sensing. For example, the integration of the high-resolution satellite images with machine learning provide an automated coastline extraction and enhances the comprehension of coastal processes. Still have limitations, such as the integration of bathymetric data for a comprehensive modeling of long-term predictive.

### 2.1.2 Application of the Methodology to the Study

The methodology of our study based on different techniques and tools of geomantic and remote sensing, such as ENVI, ARCGIS, DSAS (Digital Shoreline Analysis System), Google Earth, and other software.

Firstly, we were reducing noise from Satellite images and improving images quality. Then, we were applying an atmospheric correction to eliminate adverse atmospheric effects.

This comprehensive methodology is based on a rigorous approach to study the evolution of the coastline at Mehdia. It incorporates the use of state-of-the-art techniques and in-depth analysis to achieve meaningful results.

The purpose of this study is to estimate the trend of evolution of the Mehdia line in the reference period 1990-2023 using satellite images, and to make the analysis by the DSAS tool.

The analysis of the shoreline evolution will proceed according to the following scenario:

*Figure 2. Scenario for the Progression of the Study on the Evolution of the Mehdia Coastline*

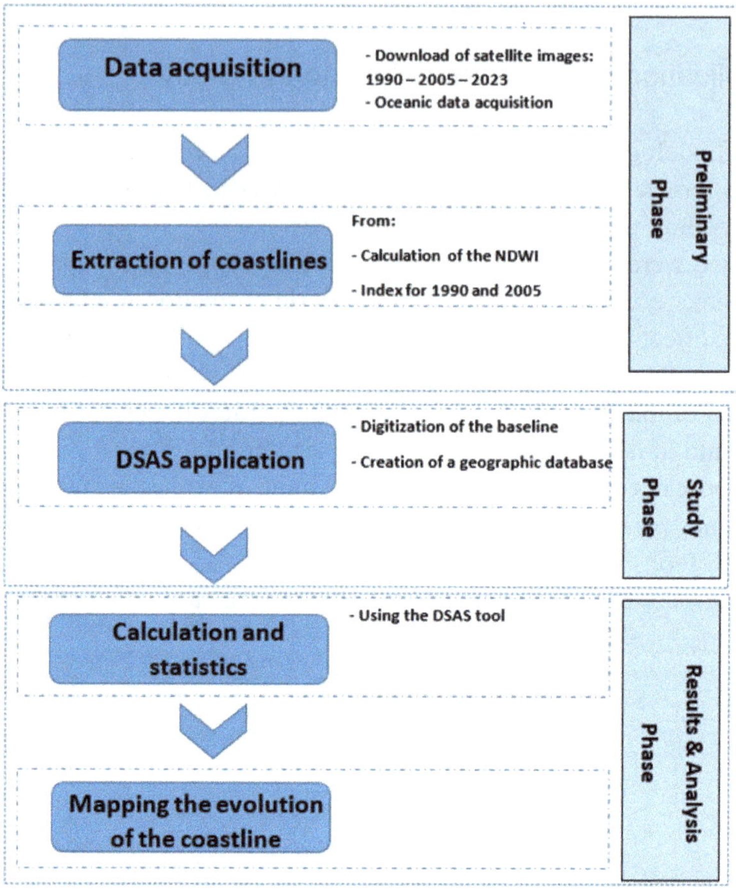


- **Data acquisition:**

To investigate this study, satellite images and ocean data, including high tide periods, are required.

- <u>The satellite images:</u>

The diachronic analysis of the Mehdia coastline evolution is based on a comparison of three satellite images from 1990, 2005, and 2023 (see table below).

These images were freely downloaded from the EarthExplorer website.

To reduce inaccuracies caused by water level fluctuations, satellite images must be collected during similar seasons and under nearly equivalent tidal conditions for the purpose of comparative study. As indicated by recent research employing the DSAS plugin for quantitative coastal change analysis, this method assures consistency in shoreline detection(Thior et al., 2021).

*Table 1. Periods and acquisition dates of satellite images.*

| Period | Acquisition Date | Satellite | Sensor ID |
|---|---|---|---|
| 1990 | 1990-01-14 | $LANDSAT_5$ | MSS |
| 2005 | 2005-02-16 | $LANDSAT_7$ | ETM |
| 2023 | 2023-01-25 | $LANDSAT_8$ | OLI TIRS |

- Oceanic data:

The tide data obtained from the Puertos.es website and was extracted from SIMAR point 1052038, which is the nearest point to Mehdia.

From puertos.es, the high tides corresponded on days bellow:

*Table 2. High tide data according to the Puertos website*

| year | month | day | hour | significant Hight (m) | Direction ° | Pic period (s) |
|---|---|---|---|---|---|---|
| 1990 | 1 | 31 | 3 | 5.9 | 310 | 17.86 |
| 2005 | 1 | 5 | 7 | 5.36 | 269 | 12.92 |
| 2023 | 1 | 18 | 10 | 4.22 | 317 | 14.66 |

For the year 2023, high tides were observed more clearly from Google Earth using the date and time options of the monitoring tool.

- **Coastline extraction**
    - Choice of the reference line

In similar studies, it is difficult to automatically delimit shorelines from medium-resolution satellite images (Maiti & Bhattacharya, 2009). For sandy coasts, the beach backslope is considered the best indicator (Robin, 2002), it represents the boundary where the upper part meets the lower section of the beach.

The high tide line, also known as the beach foreshore, was used as the reference line for the coastline in this study, it indicates the highest point attained by the tide on the intertidal zone during the entire duration of slack tide. This reference line is crucial for accurately monitoring and analyzing the evolution of the shoreline over time. According to (Crowell et al., 1991), it is the line that separates the dry areas, which appear light gray to white, from the moist parts which typically appear dark gray. According to Crowell et al. (1991) and Boak & Turner (2005), the line of rivage can be identified by the spectral contrast between the dry (generally more clear) and humid (generally more sombre) zones on satellite images.

*Figure 3. The high tide line*

- **Coastline digitalization**

In this study, the shoreline was determinate and extracted by using Google Earth Pro and GIS tools.

For the satellite images taken in 1990 and 2005, the shoreline was detected by determining the normalized difference water index (NDWI):

NDWI = (Greenband – InfraRedband)/ (Greenband + InfraRedband)

This index allow detecting the boundary between the wet surface and the dry surface.

The result of the NDWI equation is positive values for water features and negative (or zero) values for soil and land vegetation.

The NDWI values correspond to the following ranges:

- 0.2 – 1.0 – Water surface,
- – 0.2 – Flooding, moisture,
- -0.3 – 0.0 – Moderate drought, non-water surfaces,
- -1.0 – -0.3 – Drought, non-water surfaces.

In our case, the coastline (high water line) is the boundary between the wet surface (its value is between 0 and 0.2) and the dry surface (its value is between -0.3 and 0.0).

After establishing the boundary, we digitized the coastline using GIS tools.

However, for the image 2023, the reference line was digitized using Google earth pro.

- **Application of DSAS for Studying Shoreline Evolution**

The changes in the coastline's evolution at Mehdia Beach were measured using the Digital Shoreline Analysis System (DSAS) version 5.1 (Thieler et al., 2009).

This system allow representing statistically and graphically of the reference line dynamics along transects perpendicular to the shore.

Furthermore, before applying DSAS tools, it recommended firstly to organize all required data in a personal geodatabase. In addition, calculating of the rate of shoreline retreat or advancement, transects must be generated from a reference line.

In this study, we used The End Point Rate (EPR), to estimate coastal variations, dividing the number of years, by the distance between the line's positions at the start and the end of the period.

To determine the shoreline change rate (m/y), all the transect's width was calculated. The negative EPR values indicates a rate of erosion, while the positive values indicates a rate of accumulation. All transects are launched at a user-specified interval perpendicular to the baseline. the coastal intersections along this baseline permit to obtain the rate of change statistics (Hakkou & Maanan, 2023).

In this study, 119 transects measuring 500 meters in length and 20 meters distant were created by the DSAS program and oriented perpendicular to the reference line along the Mehdia coast.

## 3. RESULTS

To make correct conclusions and determine trends so as to better understand the period of decline, the study results were projected on maps representing both periods: 1990-2005 and 2005-2023. The study area (Mahdia beach) was also divided into three sections: South Mahdia, Central Mahdia and North Mahdia.

## 3.1 Overall Trends (1990-2023)

In general, this period combines the 1990–2005 and 2005–2023 periods and is characterized by a general decline in the littoral. However, the rate of its decline varies depending on the beach's zones.

The northern zone is severely eroded, especially in the areas close to close to the estuary, where a rate of 5 meters per year is observed.

At the central zone, the rate of retreat is returning, with speeds of up to 5.67 metres per year in the northern part of the zone, while the rest of this zone and the southern area of the beach show a moderate to slight decline, ranging from 4 metres per year to 0.71 metres per year (see figure below).

The most important statistics determined by the DSAS tool during this period are: Average EPR (End Point Rate) rates expressed in m/year:

- Total number of transects: 114
- The average rate: -3.25725445478715

*Figure 4. Summary of EPR rates between 1990 and 2023*

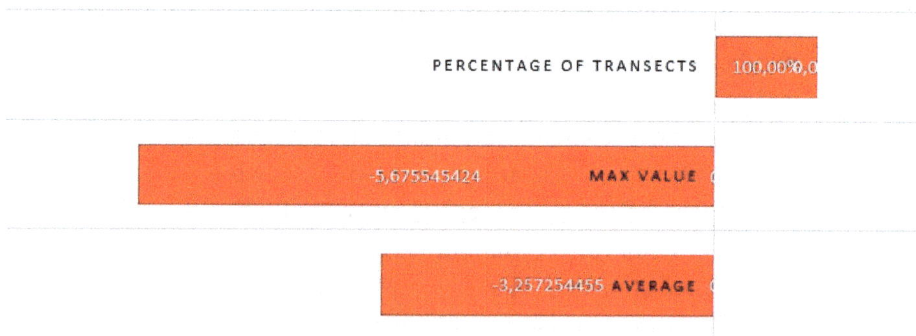


| | average | max value | percentage of transects |
|---|---|---|---|
| Erosion | -3,257254455 | -5,675545424 | 100,00% |
| Accretion | 0 | 0 | 0,00% |

*Figure 5. Evolution of Mehdia's coastline between 1990 and 2023*

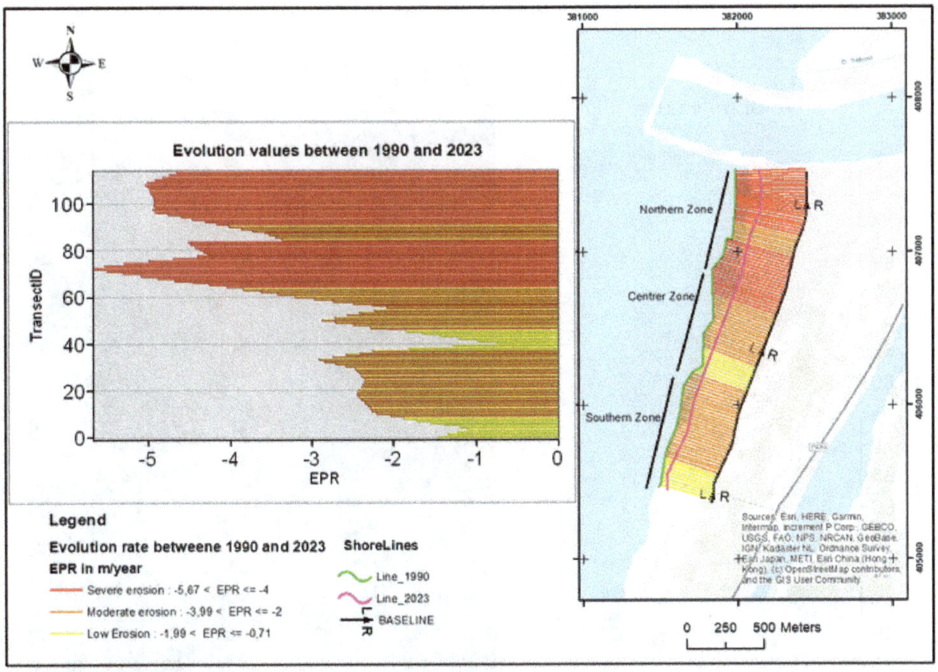


## 3.2 Spatial Patterns Analysis (North/Central/South)

During the period between 1990 and 2005 (Figure 4), the northern portion of Mehdia beach suffered significant erosion, with an average rate of nearly 2.66 meters per year. However, the rate of decline in the middle region, which can reach 6.63 m/year, is far greater than that in the northern portion of this range.

In the southern zone, there is a considerable accretion, an average of 1.17 m/year, and this is probably due to the great role played by the boundary dune, which compensates for the sediment lost as a result of the currents generated by the breakup of the waves.

During 2005-2023 period (Figure 5), strong erosion occurs in the northern zone near to the estuary of the Oued Sebou, reaching 6.94 m per year. In contrast, in the central and southern regions, erosion rates vary from moderate to low, between 6 m/year and 2.45 m/year .

*Figure 6. Evolution of Mehdia's coastline between 1990 and 2005*

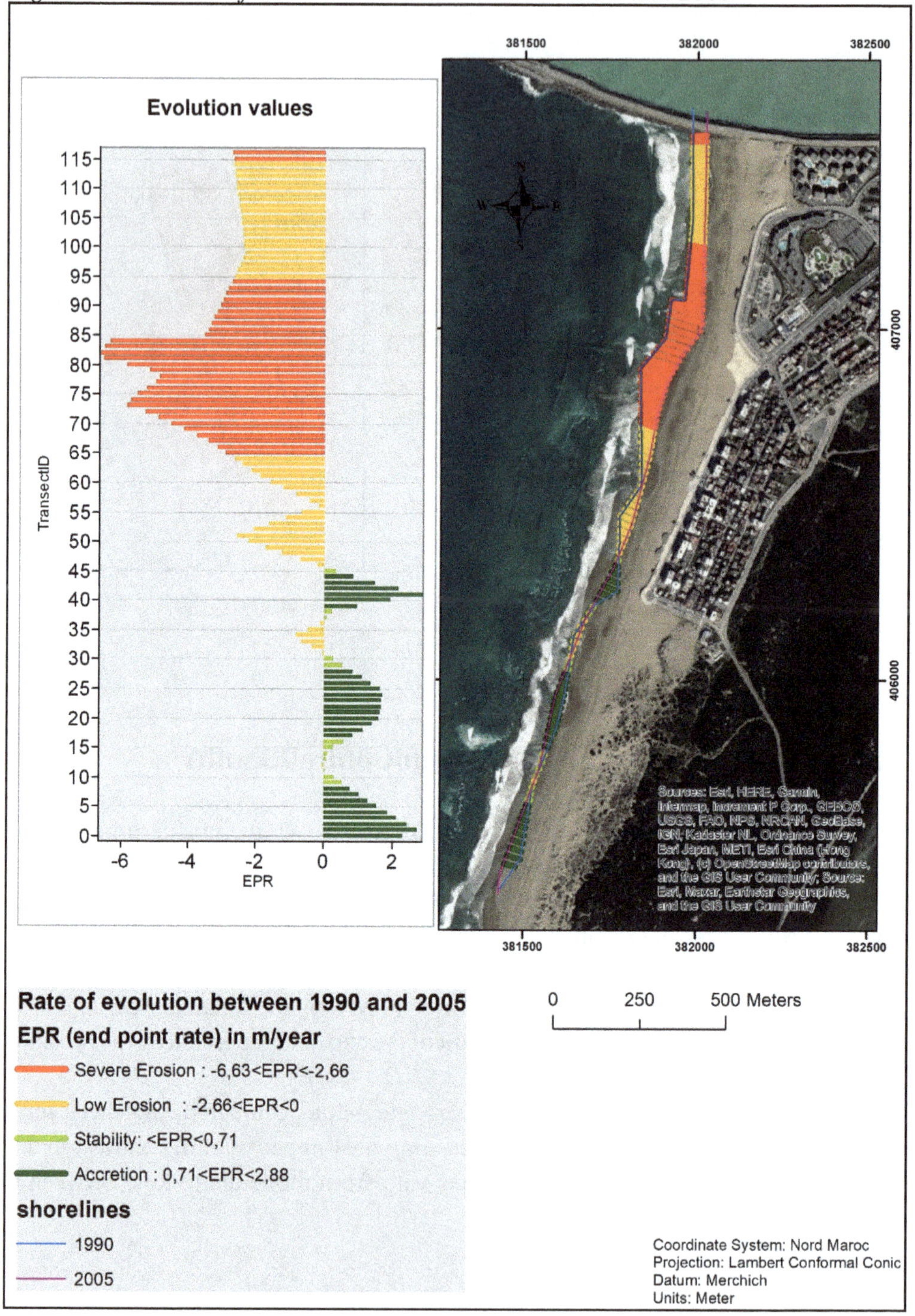

*Figure 7. Evolution of Mehdia's coastline between 2005 and 2023*

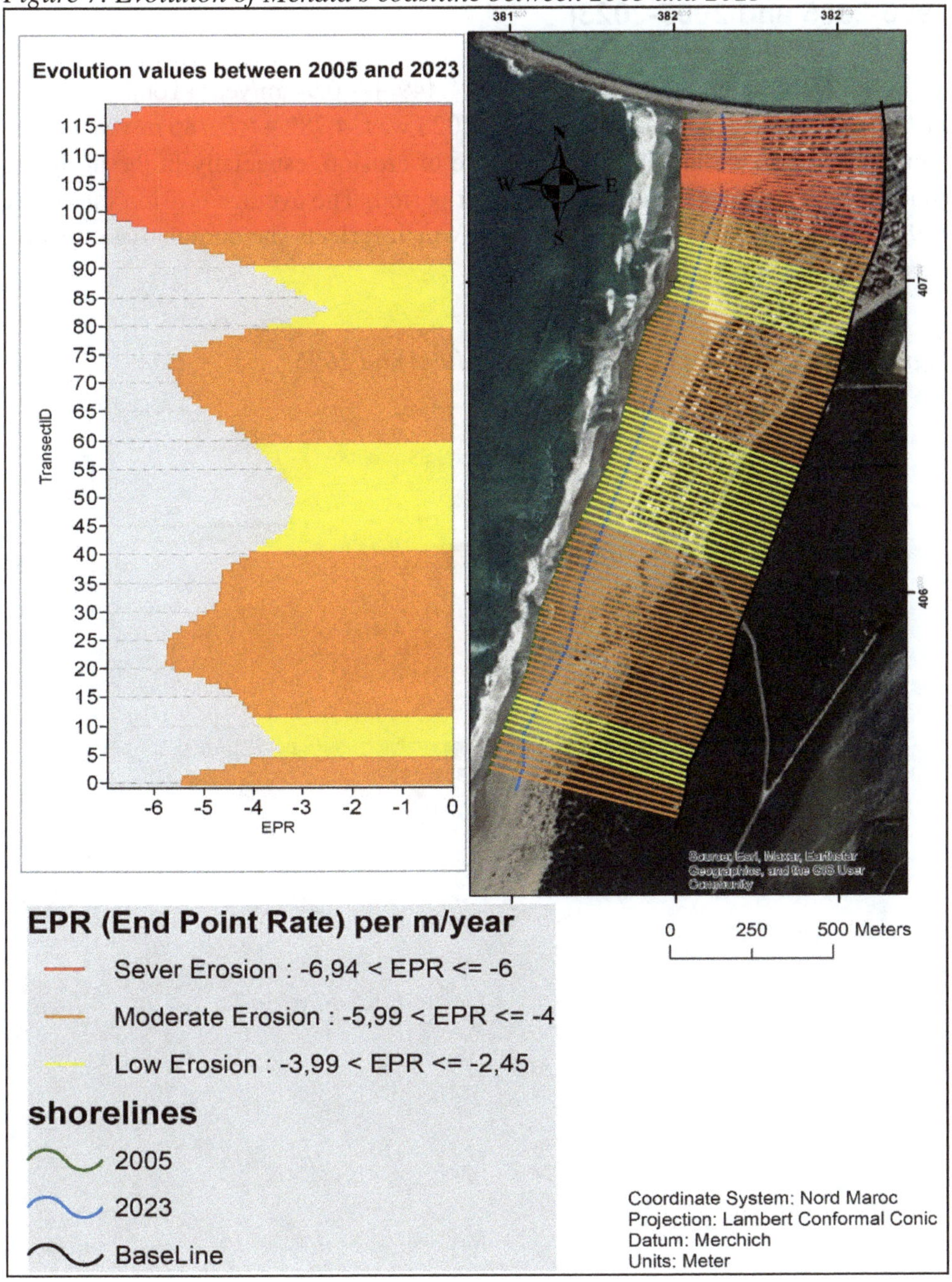

### 3.3 Temporal Evolution Patterns (Comparing 1990-2005 and 2005-2023)

The average EPR for 1990–2005 was -1.449 +/- 0.94 m/year (Total transects: 117), however the average EPR for 2005–2023 was -4.782 +/- 0.786 m/year (Total transects: 119). This indicates that the rate of erosion, especially in the northern zone, significantly accelerated in the latter period. The average erosion rate from 1990 to 2023 is -3.257 +/- 0.429 m/year, which reflects the cumulative effect of these temporal variations.

*Figure 8. Summary of EPR rates between 1990 and 2005*

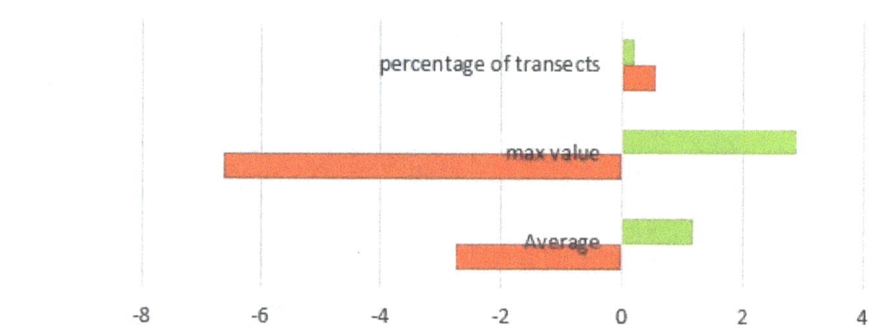


| | Average | max value | percentage of transects |
|---|---|---|---|
| Accretion | 1,178075492 | 2,888553274 | 19,66% |
| Erosion | -2,763210936 | -6,631127217 | 55,56% |

*Figure 9. Summary of EPR rates between 2005 and 2023*

| | Average | max value | percentage of transects |
|---|---|---|---|
| erosion | -4,782363179 | -6,942699119 | 100,00% |
| accretion | 0 | 0 | 0,00% |

## 3.4 Driving Mechanisms Analysis

**Spatial Variability:** A combination of natural and anthropogenic factors contribute to the pronounced temporal and spatial variations in erosion rates observed along the Mehdia shoreline.

The severe erosion in the northern area, which is adjacent to the Sebou River estuary, is primarily driven by the area's direct exposure to dominating wave regime and most importantly to the significant reduction in sediment supply from the river. River dam construction is mostly responsible for this decline, which is estimated to be over 90% (Ahizoun et al., 2010; Snoussi et al., 2002), decreasing the coast of its natural sediment accumulation. Furthermore, erosion drifting is probably made severe by port infrastructure, which interferes with longshore sediment transport.

The accretion observed in the southern zone during the first period was most likely caused by the protective function of the boundary dunes, which preserved sand and momentarily stopped sediment loss. The increase in erosion in this zone during the second period suggests that this natural retaining ability was eventually reduced.

**Temporal Acceleration:** An increase in the force of climatic factors, such as more intense and frequent winter storms(Fox-Kemper et al., 2021), can be attributed to the notable acceleration of erosion between the two periods(from -1.449 to -4.782 m/year). Reducing the supply of sediment du of these factors, and the growing of anthropogenic pressures affecting the coastline and made the coastal system even less resilient and more unstable.

## 4. DISCUSSION

According to the results of our study, the average rate of erosion is (-3.25 m/yr). This result aligns with previous studies, such as (Hakkou et al., 2018), who defined a rate of (-3,2 m/yr) for the years (1936–2014), and (Blais & Akhloufi, 2025), who indicates a rate of (-3,4 m/yr) for the years (1990–2023) . This similarity indicates a persistent coastal degradation, which may have been more accentuated in recent decades. this dynamic coastal sector may considered in the perspective of the larger region. For example, Sidi Taibi, in particular, shows noticeably more significant erosional intensity, with retreat at a rate of (-4.37 m/yr) (Ennouali et al., 2023).

The morphology of Mehdia coast (sandy beach, low topographic gradient), which increase sensitivity to winter storm impacts and the effects of climate change in general, may be the principal cause of this tendency . This vulnerability pattern has been founded in sandy coastal systems around the world (Forzieri et al., 2016). The projections of Current climate predict that severe storms are becoming more frequent and more intense, potentially increasing littoral vulnerability in the upcoming decades *(Fox-Kemper et al., 2021)*. According to recent studies, Morocco's sea level is expected to rise significantly, increasing the risk of erosion and coastal submersion(Chtioui et al., 2024; GIEC, 2021).

These changes have important socioeconomic consequences although they are often underestimated(Tafraouti, 2025). For instance, coastal erosion immediately damages vital infrastructure, including buildings, roads, and hotels, leading to considerable repair or relocation costs for local communities and authorities (World Bank, 2024b). Furthermore, the continual decline of coastal areas, which are crucial to the economy, puts the local revenue of communities that rely on the tourism industry at risk(World Bank, 2024a).

According to the alarming results from this study and previous research, significant and systematic coastal erosion would result from an extended absence of proactive and integrated coastal management. It is crucial to implement these conclusions into practice by applying the current legal framework, such as Law 81-12 on the coast, which aims to ensure the sustainable management of Moroccan coastal areas.

However, due to the study's limitations of our research, these results are still approximations, even while the Landsat (30 m) imagery are perfectly appropriate for a large-scale analysis, their spatial resolution may result some inaccuracy in precisely determining the coastline. Furthermore, the chronological difference between the precise times of satellite acquisition and high tides periods, also compromises the accuracy of the results.

Additionally, the absence of recent bathymetric data presents a gap in the modeling of submerged sedimentary transports and the dynamic equilibrium of the coast.

The limited number of temporal points (1990, 2005, 2023) is another important limitation. consequently, periods of more severe erosion or short recovery phases may be hidden by the estimated average erosion rates..

Effectively, there are many perspectives could be considered to improve the precision and applicability of future studies. Using the high-resolution satellite images, such as Sentinel-2 (10m), Pléiades (0,5 m), or drones, would enable a more precise detection of the coastline. In addition, it would also be crucial to integrate the spatial analysis with topographical and bathymetric data in order to better understand the interaction between land and ocean processes. Moreover, there may be enormous opportunity for improving shoreline prediction and spatiotemporal land cover analysis through the combination of AI, remote sensing, and sophisticated statistical models. For areas like the Mehdia coast, coastal managers can use these technologies to predict future changes, evaluate risks, and create robust adaptation plans by implementing an integrated framework, as proposed by Laaribya and Alaoui. (2025) . Future work should concentrate on enhancing model accessibility, interoperability, and coherence with policy requirements to guarantee real-world impact.

In the end, the results of this study highlight how urgently Mehdia Beach and the Moroccan coastline require proactive, integrated coastal management policies. Our investigation leads us to make the following important recommendations:

- Improved Data Collection and Monitoring: Continuous and high-frequency monitoring of shoreline changes utilizing sophisticated remote sensing techniques (e.g., Sentinel-2, drones) is essential given the indicated temporal data gaps. Additionally, using bathymetric data would offer a more thorough comprehension of the dynamics of sediment transport.
- Comprehension Based on Process: The causes of local variations in erosion, such as local hydrodynamic conditions, sediment budgets, and the impact of anthropogenic structures, need to be investigated further in future studies. This will make it possible to create interventions that are more focused and successful.
- Integrated Coastal Zone Management:Morocco's Law 81-12 on coastal protection should be completely implemented, encouraging a comprehensive strategy that incorporates scientific discoveries into policies related to environmental protection, urban development, and land-use planning. This includes promoting nature-based solutions and controlling building along the coast.
- Socio-Economic Vulnerability Assessment: In order to inform decision-making and efficiently distribute resources for adaption measures, comprehensive evaluations of socio-economic impacts are required, including dam-

age to infrastructure, relocation expenses, and effects on activities (tourist, fisheries).

- Engagement of Organizations and Public Awareness: To promote coordinated efforts and ensure the long-term resilience of coastal areas, it is essential to inform stakeholders, policymakers, and local communities about the risks of coastal erosion and the importance of sustainable practices.

## 5. CONCLUSION

The evolution of Mehdia Beach's coastline from 1990 to 2023 was the main topic of this study. Through a diachronic study based on photo-interpretation, we were able to measure the shoreline's dynamics and answer the research question, "How does the coastline of Mehdia change between 1990 and 2023?"

The results of this study clearly demonstrate that the rate of coastal erosion has dramatically accelerated over the past 30 years, moving from a moderate process to a tendency of rapid erosion. Significant human factors that interact with natural forces, like as the vulnerability of the beach's morphology and the effects of climate change, include urbanization, the development of port infrastructure, and a reduction of sediment supply as a result of dam construction.

Our study area is an example that indicates a global issue, the climate change and sea level rise, apply more pressures on sandy beaches across the globe, which is affecting the resilience of coastal ecosystems.

This needs an effective and planned management strategy that includes regular scientific monitoring, the protection of natural areas, and strict supervision of coastal artificialization. The effective implementation of legislative laws such as Law 81-12 on coastal sustainability in conjunction with current adaptation strategies is an efficient way that can prevent this progression, sustainably preserve the Mehdia coastline, and contribute to global efforts for the protection of coastal zones.

## REFERENCES


Blais, M.-A., & Akhloufi, M. A. (2025). Advances in Remote Sensing and Deep Learning in Coastal Boundary Extraction for Erosion Monitoring. *Geomatics, 5*(1). *Scopus., 5*(1), 9. Advance online publication. DOI: 10.3390/geomatics5010009

Boak, E. H., & Turner, I. L. (2005). Shoreline definition and detection : A review. *Journal of Coastal Research, 21*(4), 688–703. DOI: 10.2112/03-0071.1

CESE. (2021). *Le développement durable du littoral marocain : Enjeux et perspectives [Sustainable development of the Moroccan coast : Challenges and prospects]*. https://www.cese.ma

Chtioui, T., Hakkou, M., Aangri, A., El Hassani, F., El Mostafa, Z., & Benmohammadi, A. (2024). Assessing coastal flood risk under extreme events and sea level rise in the Casablanca-Mohammedia coastline (Morocco). *Natural Hazards, 120*(12), 10727–10752. DOI: 10.1007/s11069-024-06624-y

Crowell, M., Leatherman, S. P., & Buckley, M. K. (1991). Historical shoreline change : Error analysis and mapping accuracy. *Journal of Coastal Research*, •••, 839–852.

Ennouali, Z., Fannassi, Y., Lahssini, G., Benmohammadi, A., & Masria, A. (2023). Mapping Coastal vulnerability using machine learning algorithms : A case study at North coastline of Sebou estuary, Morocco. *Regional Studies in Marine Science, 60*, 102829. DOI: 10.1016/j.rsma.2023.102829

Forzieri, G., Feyen, L., Russo, S., Vousdoukas, M., Alfieri, L., Outten, S., Migliavacca, M., Bianchi, A., Rojas, R., & Cid, A. (2016). Multi-hazard assessment in Europe under climate change. *Climatic Change, 137*(1-2), 105–119. DOI: 10.1007/s10584-016-1661-x

Fox-Kemper, B., Hewitt, H. T., Xiao, C., Aðalgeirsdóttir, G., Drijfhout, S. S., Edwards, T. L., Golledge, N. R., Hemer, M., Kopp, R. E., Krinner, G., Mix, A., Notz, D., Nowicki, S., Nurhati, I. S., Ruiz, L., Sallée, J.-B., Slangen, A. B. A., & Yu, Y. (2021). Ocean, cryosphere, and sea level change. In Masson-Delmotte, V., Zhai, P., Pirani, A., Connors, S. L., Péan, C., Berger, S., Caud, N., Chen, Y., Goldfarb, L., Gomis, M. I., Huang, M., Leitzell, K., Lonnoy, E., Matthews, J. B. R., Maycock, T. K., Waterfield, T., Yelekçi, Ö., Yu, R., & Zhou, B. (Eds.), *Climate Change 2021 : The Physical Science Basis. Contribution of Working Group I to the Sixth Assessment Report of the Intergovernmental Panel on Climate Change* (pp. 1211–1362). Cambridge University Press., DOI: 10.1017/9781009157896.001

GIEC. (2021). AR6 [*The Physical Science Basis. Chapitre : Sea Level Rise ] [projections citées par la Banque Mondiale pour le Maroc] [.* https://climateknowledgeportal.worldbank.org/country/morocco/sea-level-projections]. *Climatic Change, 2021*, •••.

Hakkou, M., & Maanan, M. (2023). *Évaluation multi-décennale des changements du littoral à l ' aide d ' outils géospatiaux et de calculs automatiques sur la côte de Kenitra, au Maroc*. 232-239.

Hakkou, M., Maanan, M., Belrhaba, T., El Khalidi, K., El Ouai, D., & Benmohammadi, A. (2018). Multi-decadal assessment of shoreline changes using geospatial tools and automatic computation in Kenitra coast, Morocco. *Ocean and Coastal Management, 163*, 232–239. DOI: 10.1016/j.ocecoaman.2018.07.003

Himmelstoss, E. A., Henderson, R. E., Kratzmann, M. G., & Farris, A. S. (2021). Digital Shoreline Analysis System (DSAS) version 5.1 user guide. In *Open-File Report* (Nos. 2021-1091). U.S. Geological Survey. DOI: 10.3133/ofr20211091

Laaribya, S., & Alaoui, A. (2025). Spatio-Temporal Trends and Climate Change Impacts on Land Cover Dynamics in Talassemtane National Park (Morocco) Using Artificial Intelligence and Google Earth Engine. *Journal of Sustainable Forestry, 44*(6), 289–318. DOI: 10.1080/10549811.2025.2516072

Laassilia, O. (2013). *Approche multidisciplinaire de suivi de la dynamique marine du littoral de Mohammedia*. Université Cadi Ayyad,Faculté des Sciences et Techniques Marrakech.

Luijendijk, A., Hagenaars, G., Ranasinghe, R., Baart, F., Donchyts, G., & Aarninkhof, S. (2018). The State of the World's Beaches. *Scientific Reports, 8*(1), 1–11. DOI: 10.1038/s41598-018-24630-6 PMID: 29311619

Maiti, S., & Bhattacharya, A. K. (2009). Shoreline change analysis and its application to prediction : A remote sensing and statistics based approach. *Marine Geology, 257*(1-4), 11–23. DOI: 10.1016/j.margeo.2008.10.006

Mallett, R., Hagen-Zanker, J., Slater, R., & Duvendack, M. (2012). The benefits and challenges of using systematic reviews in international development research. *Journal of Development Effectiveness, 4*(3), 445–455. DOI: 10.1080/19439342.2012.711342

NASA. (2024). *NASA Analysis Sees Spike in 2023 Global Sea Level Due to El Niño*. NASA Sea Level Change Portal. https://sealevel.nasa.gov/news/270/nasa-analysis-sees-spike-in-2023-global-sea-level-due-to-el-nino/

Neumann, B., Vafeidis, A. T., Zimmermann, J., & Nicholls, R. J. (2015). Future Coastal Population Growth and Exposure to Sea-Level Rise and Coastal Flooding—A Global Assessment. *PLoS One, 10*(3), e0118571. DOI: 10.1371/journal.pone.0118571 PMID: 25760037

Pilkey, O. H., & Hume, T. (2001). The shoreline erosion problem : Lessons from the past. *Water and Atmosphere*, *9*(2), 22–23.

Robin, M. (2002). *Télédétection et modélisation du trait de côte et de sa cinématique. Le Littoral: Regards, Pratiques et Savoirs: études offertes à F.* Verger, Presses de l'Ecole Nationale Supérieure.

Snoussi, M. (2020). *Diagnostic des zones côtières de la Région de Rabat-Salé-Kenitra*. 1-70.

Tafraouti, M. (2025, janvier 9). *The Erosion of the Mediterranean Coast of Morocco : Reality and Implications | Earth Journalism Network*. https://earthjournalism.net/stories/the-erosion-of-the-mediterranean-coast-of-morocco-reality-and-implications

Thieler, E. R., Himmelstoss, E. A., Zichichi, J. L., & Ergul, A. (2009). *The digital, Shoreline Analysis System (DSAS) version 4.0 An ArcGIS extension for calculating shoreline change. (Ver. 4.3.4730, April 2012). US. Geological Survey Open-File Report 2008-1278, accessed June 2012. En ligne :* Https://pubs.er.usgs.gov/publicat. https://doi.org/DOI: 10.3133

Thior, M., ASy, M. A., Cisse, I., Dieye, E. H. B., Sane, T., Ba, B. D., Solly, B., & Descroix, L. (2021). A cartographic approach of the evolution of the coastline of the Casamance estuary. *Mappemonde, 131*, 0-22. DOI: 10.4000/mappemonde.5939

UICN. (2010). *Blue Planet : Coastal Zones. Union Internationale pour la Conservation de la Nature*. https://www.iucn.org/resources/report/blue-planet-coastal-zones

World Bank. (2024a). *Climate Change impacts on the Blue Economy in Morocco : Prospects for jobs in Coastal Tourism. Maghreb Technical Note*(12).

World Bank. (2024b). *Disappearing coasts in the Maghreb : Coastal erosion and its costs*. World Bank. https://www.worldbank.org/en/country/morocco/publication/disappearing-coasts-in-the-maghreb-coastal-erosion-and-its-costs

# Chapter 10
# Geo-Spatial Perspectives on Urban Sprawl and Environmental Sustainability

**Sitesh Kumar Singh**
https://orcid.org/0000-0002-7108-0808
*College of Engineering, National University of Science and Technology, Muscat, Oman*

**Jaishri Gothania**
https://orcid.org/0000-0002-2656-642X
*Lingaya's Vidyapeeth, Faridabad, India*

## ABSTRACT

*Urban sprawl has a great question mark over the issues of environmental sustainability, infrastructure, and urban resilience. This chapter explains how geospatial technologies (remote sensing, GIS, and spatial modeling) can be used to monitor and manage developments in cities. It discusses the ecological, social, and infrastructural effects of sprawl, taking Muscat, Oman, as an illustrative example, and outlines development-planning concepts of compact cities, smart growth, and eco-friendly modes of transport. Cities can optimize planning needs between development and ecosystem conservation through the incorporation of spatial intelligence in governance and policy frameworks. The chapter highlights the planning and participatory strategies as suitable in ensuring a sustainable, resilient, and livable urban environment in accordance with global sustainability objectives.*

DOI: 10.4018/979-8-3373-6608-1.ch010

# 1. INTRODUCTION

The urban sprawl is nowhere as different as a challenge to sustainable development in the 21st century. Cities are often referred to as the reflection of the ideas of progress-developing economies, new innovations, and availability of social mobility-their uncontrolled expansion has resulted in numerous, adverse environmental and social effects. Uncontrolled sprawl separates and discontinuities the landscape, occupies agricultural land, compromises habitats and increases the ecological footprint and infrastructural challenges of urban or regional systems. Moreover, urban sprawl is also related to higher amounts of carbon intensity compared to urban efficiency models (Muniz & Dominguez, 2020).

By the year 2050, nearly 70 percent of the world population is expected to live in cities, and this will subject ecosystems, water resources, and carbon sinks to enormous pressure (Angel, 2023). In a bid to overcome such response, geo-spatial technologies, which include remote sensing, Geographic Information System (GIS) and spatial modeling have emerged as very important in the measurement, analysis and management of sprawling. Technology offers spatially detailed, temporally shifting information of growth trends, environmental implications and scenario-based future.

## 1.1 Urban Expansion and the Sustainability Challenge

An implicit structural contradiction of urban expansion is also present; or, in other words, the necessity to increase the city size on the one hand, and the necessity to be sustainable on the other hand. Urban sprawl leads to economic dynamism but the patterns of sprawling development, that is, low density, long commuting and bizarre zoning impose undue burdens on land, energy and ecological systems.

And it is convincingly proved that big urban areas cause significantly bigger per-capita greenhouse emissions and consumption of resources in comparison to cities with a compact structure and transit orientation (Muniz & Dominguez, 2020). It also affects the fast-urbanizing countries, including India, where the development, which is not controlled, is invading arable and wild lands, threatening the food security of populations and exposing them to the climate risks of high temperatures and floods (Das et al., 2020). The contrast between growth and environmental degradation reveals exactly what is missing in the urban growth patterns i.e. smarter patterns.

## 1.2 Defining Urban Sprawl in the Context of Global Urbanization

Urban sprawl is sparse and discontinuous since it is usually patchy and inaccessible to infrastructure leading to the reliance on automobiles and poor use of land.

The process of sprawl development began as urbanization in North America and Europe but the creation of informal settlements, peri-urban in most of the cities in Asia and Africa, which are non-serviced and subjected to unfavorable ecological circumstances is the mode of informal development (Angel, 2023).

It is important to note that urban sprawl is not an entirely spatial phenomenon, but socio-economic phenomenon. This implies that effective measurement frameworks of sound are a prerequisite that facilitates effective assessment and control of sprawl. In this example, geo-spatial analysis based on multi-temporal land use/land cover (LULC) detection and morphological measures provide the opportunity to quantify portions of the spatial patterns and fragmentation, which can be used to respond to planning (Zhu et al., 2022).

## 1.3 The Role of Geo-Spatial Perspectives

The estimation and the process of urban sprawling mitigation have changed with the geo-spatial technologies. Remote sensing is capable of detecting LULC change over time, GIS can be applied to compute metrics including fragmentation, density and connectivity and spatial modeling (e.g. cellular automata, scenario-based forecasting) can be applied in simulating future urban examples under varying policy conditions (Amoah & Gorsevski, 2025).

With their help, it is possible to identify areas of unsustainable growth, to evaluate trade-offs between development and ecological preservation, and to visualize of the situations e.g. compact vs. sprawling urban trajectory, which are relevant in making evidence-based planning (Zhang et al., 2021). Combination of advanced AI, volume data, and predictive analytics with spatial analysis would elevate the abilities of providing real-time data and long-term plan to help urban sustainability beyond any uncertainty, and geo-spatial intelligence is a facilitator that cannot be ignored in changing the lifestyle in urban areas.

*Figure 1. Conceptual Framework of Urban Sprawl and Environmental Sustainability*

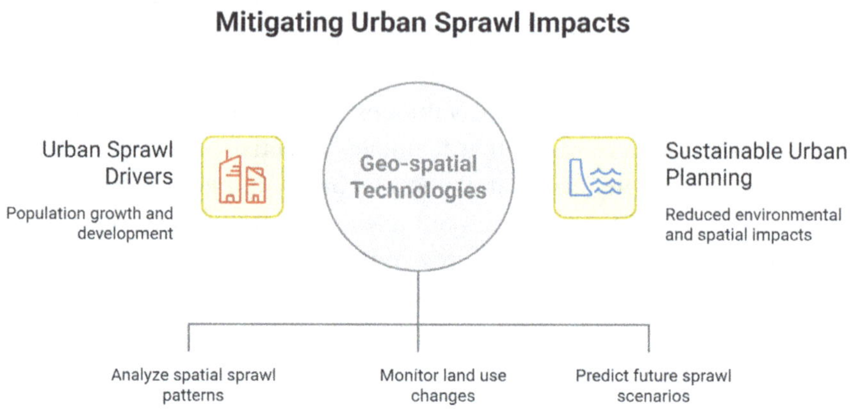


Figure 1 reveals the abstract connections between the drivers of urban sprawl, spatial consequences and environmental effects related to sprawling. It suggests that geo-spatial technologies (GIS, remote sensing and spatial modeling) contribute to analysis and monitoring that enables shaping sustainable urban planning so that it will overcome adverse effects of ecological degradation and provide sustainable, compact and environmentally friendly cities.

## 2. CONCEPTUAL FRAMEWORK

Another phenomenon is that of urban sprawl, but it is a multi-layered and multi-faceted phenomenon, and it has much more complicated interpretations than the straightforward extension of the limits of cities. It is about interaction in space, social, economic and environmental, as each dimension affects and is impacted by another. The characteristics of urban sprawling constitute low-density and scattered development with inefficient land patterns spread over large areas of land at the peripheries of the already existing urban settlements (Pradhan et al., 2024). Socially, it can intensify segregation, reduce the proximity to vital amenities and affect the social cohesiveness of communities especially as people spread out over larger geographical distances. Urban sprawl has an impact on the cost of infrastructure, the need of transportation and local economies, and in most cases, the cost of having a sprawling environment increases government spending, through road networks, utilities and service provision, and at the same time, lowering the efficiency of resource distribution. Environmentally the sprawling development led to fragmenting of the habitats, greater greenhouse gas pollution as people will now have to travel longer distances to commute and scarce natural resources like water and land are put under pressure.

It is not insignificant that the conceptual framework, which is supposed to bring numerous theoretical approaches and empirical models into the scope of understanding urban sprawl purely, is required. Such a framework considers the processes that lead to the development of the phenomenon of sprawl including an increase in population size, economic development, policy change and cultural shifts coupled with examination of its effects to sustainability. Through the study of the interconnections between city development and environmental, social and economic sustainability, researchers and city developers will have the capacity to develop strategies, which will eliminate the adverse effects, instead, increasing compact, efficient and more robust urban development. Besides, the connection of the research on sprawl and Sustainable Development Goals (SDGs) created by the United Nations (UN-Habitat, 2023) provides a global opportunity to address the issues of the sustainable cities and communities (SDG 11), climate action (SDG 13), and responsible consumption and production (SDG 12). This holistic practice will ensure that the development of cities will be in tandem with the realization of sustainability, which will provide sustainability in the long-term to the surroundings environment, the social and economic environment.

## 2.1 Theories and Models of Urban Growth and Sprawl

City development theories have been evolving continuously, trying to assure more knowledge about the process, and trends of the development of cities, stepping a quite primitive early paradigm towards more complex ones. First-generation models such as the concentric zone model, formulated by Burgess (1925) were simply models of how cities grew, and they assumed that cities could be considered as growing in rings with the central business district at the center. Although this input existed, the models were likely to ignore numerous realities of urban development that were manifested in multi-focal places of activity, multi-economic-societal diversity and influence of regional and global economies. More recent theories have attempted to expand on these shortcomings, by taking into consideration additional concepts, such as edge cities, post-suburban development, polycentric urban regions etc., to more effectively highlight the dynamic and multi-dimensional nature of urban sprawl. The approaches value the fact that the cities are no longer constructed around a central location but rather, cities may consist of a variety of centralities with various functional, economic and social roles.

Among the innovations made in the sphere of knowledge development of urban growth, there is the fact that spatial measures have been implemented in the methodology of modeling. The recent tools such as: Geographic Information Systems (GIS), remote sensing and spatial modelling can also be used to quantify and simulate urban sprawl and facilitate the researcher to track how land utilization, density and urban forms have changed over time. Simultaneously, the GIS-based analysis as new applications can be applied to identify the low-density development areas, fragmented urban forms, and over-dependence on cars to offer empirical-grounded support to theoretical methodologies (Mansour et al., 2025).

The theoretical methods can be equated to spatial analytics so that the planners and policymakers can develop composite strategies to cope with the process of urban development at the sustainable stage. Part of these measures may involve sustaining compact cities, fixing up of public transportation, and preserving green spaces therefore preventing impacts on the environment and supporting long run resiliency of cities. In conclusion, such theory-technology synthesis provides a viable platform on which to address the contemporary challenges of urbanization to meet a bigger sustainability agenda.

## 2.2 Linkages Between Urban Expansion and Environmental Sustainability

The relationship between urban development and environmental sustainability is in its turn complex, and it is a complex network, which is founded on the tradeoff relationship that is known to exist between the benefits of urbanization and its impact on the environment. The urban development process can stimulate economic growth due to its ability to enable its people to be employed, better infrastructure, and have more viable services required i.e. education, available healthcare facilities and transport. These benefits are, however, normally accompanied by enormous environmental trade-offs where urban sprawl occurs in an uncontrolled or unplanned manner. Unregulated urban sprawls will result in the overconsumption of the available land and strain on the available natural resources to an extent of increasing pollution that comes at the cost of sustainable development (Dehghani et al., 2025).

The empirical effects of the sprawling form of urbanism on the environment have been reported many times. Low density development, which is dispersed, tends to increase the use of energy per capita particularly due to increased transportation, personalized motorized vehicles and poor land use. This eventually leads to an increment in greenhouse emission and creates climate change. Along with that, the urban sprawl typically entails the infringement of ecologically sensitive areas and the following habitat fragmentation and biodiversity erosion, as well as the disruption of the ecological services. Forests, natural landscapes, and wetlands could be transformed into cities and result in the decreased ecological resilience and be vulnerable to environmental hazards, including floods and heatwaves.

The impacts of urban sprawl are unevenly spread and are severely controlled with reference to the geographic, socio-economic and governance parameters. Due to the presence of semi-arid regions, the process of sprawl may further exacerbate the shortage of water, and in tropical regions, the process of deforestation may be triggered by the presence of sprawl. That does not imply that such governance structure will not have more deplorable environmental effects, but where the governance structure looks after the proactive land use policies and sustainable development plans then the effects of this will be averted. To the extent that they need to convey the correct response to these context-specific processes, such as compact development, green areas, and sustainable mobility, can address the mismatch between global urban growth and sustainable development and attain global goals, such as SDG 11 and SDG 13 are therefore critical in informing a successful response to them (OECD, 2023).

## 2.3 Relevance to Sustainable Development Goals (SDGs)

There are several Sustainable Development Goals (SDGs) that are directly connected with urban sprawls, i.e., SDG 11 (Sustainable Cities and Communities) (UN-Habitat, 2023) and SDG 13 (Climate Action) which directly states that urban development should be followed on sustainability grounds. The SDG 11 implementation emphasizes the establishment of inclusive, secure, resilient, and environmentally friendly cities, the significant aspects of which are related to proper land management and equal access to governmental services, to protecting the natural and urban environment. The deficient control of urban development is detrimental in achieving these objectives since it promotes low density development, an experience associated with high cost of delivering infrastructures and a shortage of green spaces that ensure ecological stability and human welfare. The disjointed urban development is also constraining the ability to get access to essential services, adds to the social differences, and creates difficulty in the governance mechanisms which undermine the sensible objective of developing sustainable and resilient urban communities (UN-Habitat, 2023).

Adaptation trends in urban development also influence SDG 13 that aims to address climate action. The sprawling cities are characterized by increased per capita energy use, and emission of greenhouse gases because of the long commuting distances, the use of individual transportation systems, and energy-intensive infrastructure. Also, the sprawling urban environments are mostly unable to cope, in case of floods, heat waves and hurricane surge. The cities are made more vulnerable by weak planning and a lack of green infrastructure and reduce the resilience of the cities to environmental shocks.

To address these challenges in its support of SDG 11 and SDG 13, integrated urban planning plans are required. Small-scale, mixed-use urban forms can reduce land use and enhance accessibility and investment in transport systems, non-motorized urban mobility, and green infrastructure can increase sustainability and resilience. With these strategies incorporated into the local and regional planning, cities will have the capacity to strike a balance between the growing urban centers and the environment, social equality, and climate change resiliency to ultimately form sustainable urban development in line with the global community sustainability goals.

Figure 2 shows the relationship between drivers of urban growth; the sprawl patterns and consequences of development on the environment and how sustainable interventions of planning can be used to stem the adverse impact. It is theoretically inclined in its combination with realistic style of portraying a framework of how to analyses and manage urban sprawl towards compact, resilient and environmentally sustainable urban centers.

*Figure 2. Urban Growth and Sustainability Nexus Model*

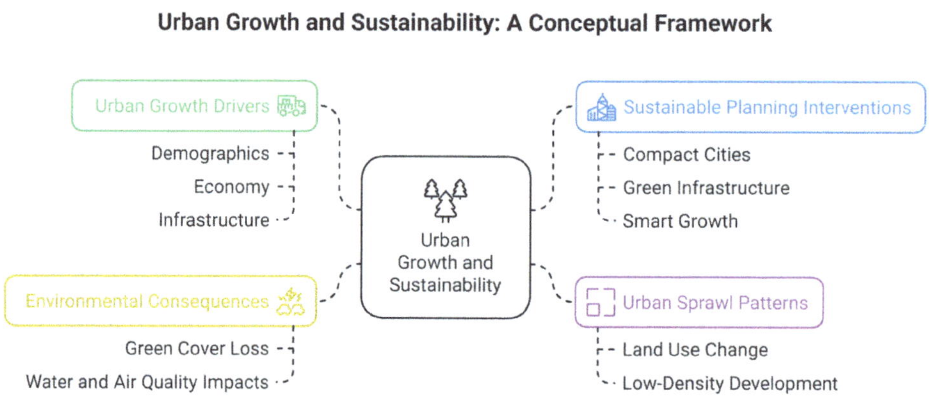

## 3. GEO-SPATIAL APPROACHES TO ANALYZING URBAN SPRAWL

Another significant issue to the planning of sustainable cities is urban sprawl which is considered in most instances to be low-density and automobile-based development. This scattered pattern usually results in poor use of land resources, energy wastage and use of greenhouse gases, overloading of the public infrastructure and provision of services. The control, monitoring, and mitigation of the effects of sprawl involves accuracy, analysis and forecasting, which geo-spatial technologies have the potential of supporting. Remote sensing can be utilized, e.g. to observe LULC changes over time and thus to trace the patterns of the cities expanding, encroachments on natural habitats, and environmental degradation (Seyam et al., 2023).

GIS supplements remote sensing because it allows integration, visualization, and analysis of spatial information. Urban morphology, density patterns, and connectivity between neighborhoods and transportation networks can be discovered through GIS by the urban planners. Spatial modeling can fully enhance these functions in that it can be used to model future urban developments, which is based on the demographic trends, land availability, policy measures and landscape constraints. Incorporated among models like cellular automata, agent based modeling and logistic regression are used to provide predictive insights useful in sustainable land- use planning and policies (Sudhira et al., 2003).

All these geo-spatial technologies enable evidence-based planning through the identification of high-risk areas, optimal investment in infrastructure and compact, resilient and ecologically conscious city construction. The same should be incorporated into the urban management strategy to restrict the adverse features of sprawl and organize its growth in combination with sustainability goals, such as SDG 11 (Sustainable Cities and Communities) and SDG 13 (Climate Action) (Goal 13: Climate Action - the Global Goals, 2024).

### 3.1 Remote Sensing Applications for LULC Change Detection

Remote sensing has been found to be an indispensable tool when it comes to monitoring land utilization and land deployment (LULC) changes, providing important data to the way urban sprawling is moving (Seyam et al., 2023). The satellite images, e.g. Landsat, Sentinel and MODIS inform the researcher and planners with spatial and temporal indicators of the urbanization, through the delivery of multi-temporal and spatially explicit data on the land cover, which are essential in tracking the alteration of built-up areas, the cover of vegetation, water bodies and other land covers that can guide a comprehensive assessment of the impacts of the urbanization process on the environment.

Recent developments in machine learning algorithms have made them more accurate and efficient mostly in the LULC classification. One such advanced application of the remote sensing data is the land cover classification and time change analysis via the algorithm, which is often random forest, support vector machines (SVM) and deep learning on the remote sensing data (Bayable et al., 2025). These methods are particularly useful in the activities of handling very huge volumes of data, identifying small trends, and enhancing the quality of estimating urban sprawling. The classification process, which was identified to be the human aspect and mistakes, could be faster and automated through machine learning techniques to provide stable and scalable solutions in terms of surveilling an ever-dense city.

Lahane et al. (2025) as an example applied machine learning to analyze modifications in LULC in a rapidly growing area with the help of remote sensing. They observed major expansion in build-up and a loss in agricultural and green lands, hence showing the effects of urban sprawl to environment. This demonstrates the significance of integrating remote sensing and other advanced methodologies of analysis in informed urban planning, good land management as well as policy formulation that are geared towards addressing the bad consequences of sprawling.

## 3.2 GIS-Based Spatial Metrics and Urban Morphology Analysis

The techniques of spatial model have provided a precious resource in the simulation of sprawl, projection, thereby rendering the techniques the best available tools that can provide planners and policymakers with the much-needed foresight in managing the challenge of sprawl in a sustainable manner (Sudhira et al., 2003). Cellular Automata (CA), Markov Chains and Agent-Based models (ABM) have been employed to represent the complexity of urban processes by incorporating multifactor affects that comprise socio-economic motives, population growth, land suitability, transportation infrastructure and policy. The models can recreate the current city forms in the urban areas as well as forecast the patterns of growth in the future that makes it possible to plan and make decisions based on scenarios.

Spatial patterns are represented by the Cellular Automata models, and existence of neighborhood effects are represented by the interaction between the land parcels in the Cellular Automata models which model urban growth. This is supplemented with ABM models which assume that the behavior of the individual agents like households or developers is modelled and evaluated about their net contribution to urban form.

The practicability of such methods has been revealed in the studies of Zhang et al. (2025) who applied the CA-Markov model to anticipate urban growth in one of the fastest urban centers. The trends were also reflected in the predictions of the model that confirms the efficiency of the spatial modeling in the sphere of urban planning. The development in recent years has also led to the increased predictive capacities as machine learning is implemented. Spatial models can also be extended to deep learning algorithms that would enable them to process small-scale spatial datasets and detect non-linear tendencies and increase the accuracy of the prediction. Such a form of integration will provide the planners with sophisticated planning tools to forecast and plan urban development, ideal land use, and reduce the consequences of the sprawl, the consequences of environmental and socio-economic degradation.

Figure 3 reveals the geo-spatial based geo-spatial workflow applied in the study of urban sprawl by use of remote sensing, GIS and spatial models. It demonstrates the manipulation of input data to determine the land use changes, quantify the urban morphology in terms of the city form, predictive capabilities of the urban morphology estimates, and make the supportable environmentally friendly urban development planning possible.

*Figure 3. Geo-Spatial Analysis Workflow for Urban Sprawl*

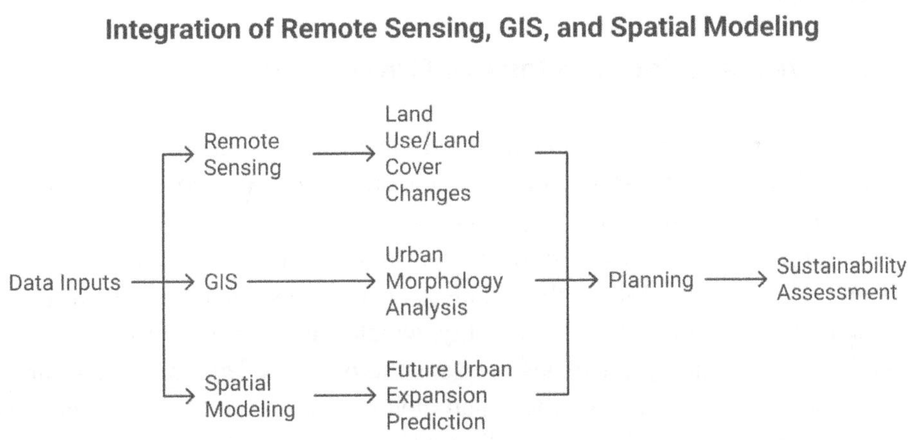


## 4. ENVIRONMENTAL IMPLICATIONS OF URBAN SPRAWL

The effect of the urban sprawl is multi-sided and consequently multi-faceted since the effects of urban sprawl extend way beyond the spatial and social effects of the growing cities. Among the most noticeable implications is the depreciation of land and the loss of green cover whereby low-density development is likely to epilate part of the forests, agricultural and open lands leading to the breaking of the habitat and the reduction of biodiversity. The vegetative destruction also lowers the purification services of the air, temperature, and soil stabilization services offered in the ecosystem.

It also affects water resources, as urbanization causes the increase of impervious surfaces, alteration of the natural drainage pattern, and decreasing the groundwater recharge. The transformations contribute to the further possibility of floods, water pollution, and water shortages in the rapidly developing cities. Even the air quality and the health of the ecosystem is stifled, as sprawling cities will result in increased emissions during vehicles, industry and pollution as well as garbage.

Finally, the implication of urban sprawl on carbon emission and climatic effects is great. Increased transportation demands and energy intensive infrastructure and dispersed settlements contribute to the aggravation of greenhouse gas emissions. Such environmental concerns require active and comprehensive urban planning that focuses on intensive building, green infrastructure-based sustainable utilization, and conservation of resources to reduce the adverse environmental and weather effects of sprawling.

## 4.1 Land Degradation and loss of Green Cover

The issue of urban sprawl can be asserted to be one of the major causes of land degradation because in most instances, the uncontrolled expansion of the built-up territories will engulf and enter the productive agricultural land, forests, wetlands, and other ecologically sensitive areas. The nature is displaced, the natural soil structure is altered, therefore making the land less fertile and subject to erosion; it also alters the hydrological cycles in the locality, which can attract scarcity and floods. Breaking up of the green spaces also lessens the burden of green space in urban areas and hence the capacity of the urban environment to endure environmental disruptions such as air purification, cooling, carbon sequestration, and flood control (Seto et al., 2012).

A body of evidence on these changes has been provided by GIS and remote sensing-based studies which can quantitatively assess LULC changes with time. The Indian population is concerned with the ecological imbalance issues and food security because of the irreversible conversion of agricultural lands into urban areas because of the rapid urbanization (Singh et al., 2024). In addition, vegetation cover leads to the intensification of urban heat island effect that increases the local temperatures and strains the micro climatic conditions, energy consumption, and health in communities. Such problems require comprehensive planning and preservation of green infrastructure to obtain ecological sustainability.

## 4.2 Impacts on Water Resources, Air Quality, and Ecosystems

The urban sprawl that is happening haphazardly is putting strain on the water resources and its ability to sustain itself could be exceeded at any time. The development of constructed surfaces such as roads, pavements and roof-tops inhibit the groundwater recharge and increase the issue of surface run off and the flood hazards. This run-off usually contaminates the degrading water quality of rivers and lakes and presents a threat to aquatic communities. The disturbance of the natural water systems through over-pumping of rivers, lakes, and ground water sources to meet urban demand also has adverse impacts on the river ecosystem, wetlands, and agricultural activities (Shaamala et al., 2025).

The air quality of Minnesota city in which the issue of sprawling is ongoing is also very poor. The greater use of Personal vehicles, and the consequent longer commutes augment the emissions of PM, nitrogen oxides and greenhouse gases. Research has demonstrated that large urban agglomerations often surpass the allowable air pollution level and produce a higher percentage of respiratory and cardiovascular diseases in the population (Marshall et al., 2009). Further destruction of biodiversity is through human urbanization into previously environmentally sensitive zones which fragments the ecosystems and reduces their resilience. All the environmental pressures put an increased emphasis on sustainable urban planning, compact development, and green infrastructure to reduce the impacts of water stress, air stress, and ecological stress.

## 4.3 Climate Change and Carbon Footprint of Sprawling Cities

Sprawling urbanization contributes immensely to climate change mainly due to the consumption level of energy and the high level of carbon emissions. Automobile-dependent and poorly densified developments require more energy to move around, heat, cool, and maintain lengthy infrastructure networks, whereby more energy-thirsty sprawling cities are drawn in comparison to compact urban patterns (Ewing & Hamidi, 2015). Also, the transformation of greenery to construction spot lessens the extent of carbon storage, further intensifying the carbon imprints and global heating.

Recent literature draws attention to the fact that climate change should be introduced in urban planning and design as it is a necessity. It is found that in the absence of effective policies to address sprawl, the emissions per person could be 30 percent higher after 20 years in many cities (Seto et al., 2012). Green infrastructure, mixed-use/transit-oriented development, the development of improved public transportation to connect suburbs and adjacent cities, as well as optimization of urban densities is critical to energy demand and emission reduction. Not only do these strategies provide a reduction in the carbon footprint of sprawl, but it also enhances resiliency within cities to help support adaptation to climate change and achieve the long-term viability of urban areas.

## 5. CASE STUDY: URBAN SPRAWL IN MUSCAT, OMAN

The city of Muscat in Oman has witnessed significant urbanization in recent years due to the consumption of population growth, economic growth of the country, and the increased in housing and commercial construction. This rate of high growth has transformed the land use that has led to transformation of green land into residential areas, business and industries. Expansion in the infrastructure creation, including roads, utilities and services, have been able to match the size of the population to support the size of the increased loads, however at the expense of ecological niches and open spaces. The environmental connotation of this expansion is the displacement, increase in energy and water waste and exposure to urban heat and flooding, which makes urban planning activities sustainable. The main indicators of urban sprawls are summarized in Table 1 and they include population density, land consumption, coverage of the impervious surface, commuting patterns, and availability of green spaces. These measures assist in evaluating environmental sustainability, determining trade-offs between development and environmental impact and inform urban planning policies, including compact development, green infrastructure and transit-based policies.

*Table 1. Comparative Indicators of Urban Sprawl and Environmental Sustainability*

| Indicator / Metric | Description | Measurement / Unit | Impact on Sustainability | Policy Implication |
|---|---|---|---|---|
| **Population Density** | Residents per unit area | Persons/km$^2$ | Low density → higher land consumption, fragmented urban form | Encourage compact, high-density development |
| **Land Consumption Rate** | Rate of urban land expansion | km$^2$/year | High → habitat loss, ecosystem disruption | Implement urban growth boundaries |
| **Impervious Surface Coverage** | Proportion of non-permeable surfaces | % of urban area | High → increased runoff, flood risk | Promote green infrastructure & permeable surfaces |

continued on following page

*Table 1. Continued*

| Indicator / Metric | Description | Measurement / Unit | Impact on Sustainability | Policy Implication |
|---|---|---|---|---|
| **Commuting Distance / Car Dependency** | Average travel distance & vehicle reliance | km/day or % trips by car | High → $CO_2$ emissions, energy inefficiency | Improve public transit, TOD |
| **Green Space Availability** | Parks, forests, and natural areas per capita | $m^2$/person | Low → reduced ecological services, lower urban resilience | Integrate urban green planning |

## 5.1 Urban Expansion Patterns

Cities have caused this through their urban sprawl such that the city of Muscat has changed tremendously to become one and integrated urban fabric as opposed to a series of individual settlements. The city has expanded tremendously through these forces that encompass high population growth rate, economic diversification, the government policies that were strategic to develop housing, commercial and infrastructure development. The outcomes of these drivers are far-ranging in terms of the change of land use where residential, commercial and industrial land becomes an upwards trend occupying agricultural lands, natural vegetation and other green spaces. Remote sensing and GIS-based studies were also significant in providing clear maps of change in land cover, the extent of conversion, and urbanization problems of Muscat (Fadda et al., 2019).

The brightest example of such development was the Al Khoud district of western Muscat that can be seen as an example of tendencies typical of the sprawling metropolis. The construction of this project started in 2007 and the total residential plot of the project, which is currently in the development process, is approximately 2,000 together with roads, utility, schools, and commercial units. The area is a typical example of the suburban sprawl of Muscat in the form of sprawling low-density, auto-oriented development. The dependence on individual transport along with a low-density population is an indication of the difficulty associated with efficiency in social controls, environmental feasibility, and provision of services. An analysis of an area such as Al Khoud can be used to inform urban planners on how to govern growth, maximize land use, and encourage better practices of development as far as being environmentally sustainable is concerned in Muscat (Kader & Mahmoud, 2019).

## 5.2 Infrastructure Development and Transportation

The high rate of urbanization in Muscat has fueled massive investments in infrastructure to support rising population and economic activities. Some of the main projects, in this respect, are the Muscat Expressway, six-lane highway to be used to connect several districts of the city. This major arterial road has enabled free flow of people, goods and services and thus it has facilitated residential and commercial development in sprawling urban areas. The expressway has also brought urban sprawl and improved access to different job opportunities, colleges, and shopping centers, through enhanced connectivity between the outlying areas of the city and the city center and now it is projected for further expansion (Nair, 2024).

These advantages notwithstanding the growth of road networks have deepened the dependence on personal motor cars that have led to cementing car-oriented cities. This has led to traffic jams, high levels of emissions of pollutants and various health- related issues such as Asthmatic attacks and inactive living. To counter these adverse effects, Muscat must urgently stimulate the development of transport-related solutions that are sustainable, such as adequate transport infrastructures, walking and bicycle lanes, and sensitive footpath networks. Incorporating these options in the planning of the cities will help in reducing environmental pressures, creating efficient mobility and an increase in the quality of life of the inhabitants besides facilitating sustainable urban expansion.

## 5.3 Environmental Impacts

The fast development of a city has resulted in severe environmental impact, as natural environment and its ecosystems are extensively modified. As a result of the growth of residential, commercial, and industrial zones, natural grounds are lost, which is accompanied by a loss of biodiversity and destabilization of ecosystem activities. This has been concentrated in coastal regions especially, which have been encroached by the cities, affecting marine systems by lowering the quality of water, and putting some species of their dependent ecosystems under threat. Anthropogenic cover has also upset natural hydrological processes as the land has been covered with impervious surfaces, including roadways, pavements, and construction. Any flooding due to heavy rain is increased and sources of freshwater are reduced since there is less groundwater recharge and even surface run off increases.

It also accelerates the urban heat island effect by contributing to a loss of vegetation and green spaces and adding to local temperature and cooling loads. Further, a lack of greenery decreases the city's ability to clean the air pollutants, which leads to a decrease in the air quality and related health outcomes. These environmental stressors are some of the reasons why green infrastructures, sustainable land-use planning, and ecosystem conservations should be incorporated in the urbanization planning in Muscat to curb the negative impacts of sprawl.

## 5.4 Policy and Planning Responses

The Omani government has attempted to address the infrastructural and environmental problems that occur because of urban sprawl by putting in place several strategic planning and policy interventions in a bid to propagate sustainable urban development Spatial Strategy. (2023). Sultan Haitham City in Seeb is one such development, a master-planned city that is meant to set an example of how healthy urban growth must be achieved. This project focuses on diverse land use, increasing building density, and the incorporation of green spaces, making it compact and pressuring less on natural resources and improving the life quality of the residents. The city integrates parks and recreation spaces as well as environmentally sustainable design features to counteract the growth of urbanization with environmental conservation (Bitwize, 2023).

Besides large-scale planning processes, it has been reported that space syntax analysis into the urban design process has proven beneficial to ensuring that cities become more walkable and less reliant on the utilization of private vehicles. Pedestrian networks can also support the connection and communication of transport that may reduce the threat of traffic jams and make the environment more accessible through enhancing the accessibility of residential, commercial and communal zones by maximizing the street networks, connectivity and walking routes. With the potential to decrease the adverse impacts of sprawling in urban areas, negative health impacts of individuals, and environmental sustainability in the long term, these solutions can amalgamate urban growth with other global sustainability goals such as SDG 11 and SDG 13.

# 6. POLICY AND PLANNING PERSPECTIVES

The problem of urban sprawl is one of the significant topics of sustainable development and predetermines the need of policy framework and integrated urban plans. Others among the major methods that have been utilized effectively include the use of spatial intelligence using Geographic Information Systems (GIS), remote sensing and spatial modeling in the monitoring of land use changes, study of the pattern of urban development and placement of evidence-based decisions. Increasing the use of mixed-use development in urban areas will be more compact to reduce the use of land, build more access, and rely on the use of vehicles. Besides, there is the question of governance that should be considered, i.e., the absence of consistent planning authorities, inappropriate regulations, and the lack of stakeholder engagement as one of the ways of ensuring the consistent growth towards sustainability. All these strategies lead to sustainable and resilient, efficient urban development.

Different urban planning interventions, including compact city form, urban growth boundary, transit-oriented design, and green infrastructure, have been identified to impact environmental sustainability in cities through reduction in land use, reduction in carbon emissions, and improvement in ecosystem services (Ewing & Hamidi, 2015; Burton, E., et al., 1996). These strategies are captured in Table 2 together with their effects.

*Table 2. Comparative Urban Planning Strategies and Their Sustainability Impacts*

| Strategy/Policy | Objective | Key Implementation Tools | Environmental Impact | Effectiveness/Remarks |
|---|---|---|---|---|
| **Compact City Development** | Reduce sprawl & improve density | Zoning, infill development | Less land consumption, lower emissions | High, when combined with TOD |
| **Urban Growth Boundaries (UGB)** | Control city expansion | GIS-based boundary mapping | Protects farmland & green spaces | Medium; requires enforcement |
| **Public Transit-Oriented Design** | Reduce car dependency | Bus/rail corridors, MaaS | Lower air pollution, reduced $CO_2$ | High in dense urban zones |
| **Green Infrastructure** | Enhance resilience & biodiversity | Parks, green roofs, urban forests | Improved ecosystem services | High, if adequately maintained |

## 6.1 Integrating Spatial Intelligence into Urban Planning Frameworks

The use of geo-spatial technologies and remote sensing, geographic information system (GIS), and spatial modeling are also rapidly becoming essential in the urban planning process in general and control of the urban sprawl in particular. Multi-temporal satellite data that could be provided by remote sensing would allow users to track the LULC changes, and identify which regions are fast urbanizing so that planners will more easily determine how development is impacting natural resources and ecosystems. The information mapped into GIS assists in quantifying urban morphologies and density distribution, among other infrastructural patterns so that a city or an area can be extensively evaluated. The spatial modeling also intensifies the planning by simulating the situation of the city growth, its further expansion, and defining the possible environmental and social effect of the decision, which is made (Costa et al., 2024).

By providing real-time data on the dynamics of the city development, the incorporation of the technologies into the framework of urban planning helps to conduct evidence-based decision-making and scenario-building. The spatial decision support systems (SDSS) offer interactive platforms to the planners which incorporate both the spatial data and the analytically driven models and interactive sources that support the policy and planning decisions. SDSS facilitates multi-criteria decision analysis to make the one-sided urban priorities counter-balanced e.g. economic development, social equity and environmental sustainability. In SDSS, as an example, it has been utilized to maximize land in fast-growing cities, model transportation networks, develop green infrastructure, and a sustainable development plan. Spatial intelligence will enable the urban planners in development of compact, robust and environment friendly urban forms, through identifying the high-risk regions and forecasting the catalogue of infrastructure requirements. These solutions are the major factors in mitigating the ill effects of urban sprawls and in supporting the realization of the growth of sustainable urbanization in accordance with global agendas such as SDG 11 (Sustainable Cities and Communities) and SDG 13 (Climate Action).

## 6.2 Strategies for Compact City Development and Smart Growth

The concept of smart growth and compact city development is gaining prominence and acceptance in the list of tools of mitigating the issue of urban sprawl. The strategies encourage high-density mixed-use planning involving housing, business, and leisure in compact affairs where individuals do not have to use cars since they can walk in their neighborhoods and thus use energy sparingly (Bibri et al., 2020). Compact types of urban sprawl are also environmentally friendlier with the safeguarding of the broader environment around natural and agricultural lands, and the quality of the air is nurtured. Green spaces and city parks are a natural city component, where services are provided in the ecology, recreational form, and decreasing the urban heat Island effects.

Urban growth boundaries (UGBs) may be considered as a political instrument that regulates the growth of a city to prevent uncontrolled sprawl. Urban planning in Oman is now adopted in the form of UGBs with bonuses on vertical affordances, infill developments as well as mixed-use development. The principles of Smart growth are also concerned with the application of the latest technologies in monitoring urban systems, sensor networks, and geographic information systems and making decisions based on the strategic plan guided by the data (Shrivastava and Sharma, 2011).

One of such strategies is sustainable transportation planning. TOD enhances compact city development within the higher transit areas through promoting mobility, car-independence, and greenhouse gas emission. Mobility-as-a-service (MaaS) systems progress in the convenience and effectiveness of public transport as well and can promote the change in mobility culture (behavioral transitions) particularly in regions surrounding the sprawl of cities (Singh and Jain, 2025). Together, the compact city and smart growth approaches provide city environments that are environmentally resilient, socially inclusive and economically productive.

## 6.3 Governance Challenges and Opportunities for Resilience

Genuine city management plays a crucial role in the achievement of sustainable city development, and particularly the challenge of urban sprawl. Fragile governance, and inconsistency in regulatory environments, where there is no coordination among municipal institutions can undermine any effort to infuse coherent policies in urban development leading to wasteful land use and an increase in the levels of environmental degradation and unbalanced service delivery (Zambrano-Monserrate, 2025). One of the measures that will assist in ensuring that governance is effective and the planning initiatives are in tandem with the goals of sustainable development in the long-term is empowering the institutions, enhancing collaboration among the stakeholders and using the data to support decision-making.

Adaptive governance belongs to the range of significant characteristics of the frameworks of the urban resilience, as it enables cities to respond to the alterations in the environment, society, and economics in a proactive way. It suggests a disposition of participatory processes of planning, cross sector interaction, and efficient monitoring and evaluation processes. With the help of the spatial intelligence- based on GIS, remote sensing, and spatial modelling, governments can find out where they are most vulnerable to sprawl, and how it can expand and distribute resources better (Wang et al., 2021).

Green, or environmentally friendly governance, also involves regulatory tools that encourage green buildings, green infrastructure and water sensitive urban design. The implementation of measures to encourage developers to use environmentally friendly constructions and the enforcement of the criteria of an environmental impact assessment when granting planning approvals also works to eliminate more of the negative results of urban sprawl. Collectively, these governance interventions enable the sustainable growth of cities that are contemporary, equitable, and ecologically viable and thus enable the cities to grow to meet the increasing demands, without compromising on the quality of environment and the quality of life.

# 7. CONCLUSION AND FUTURE DIRECTIONS

## 7.1 Conclusion

Urban sprawl is already one of the most urgent issues in modern urban planning, which affects the environment in terms of sustainability, general planning of the infrastructure, and social welfare. The fascinating aspects of urban development have been mentioned in the chapter highlighting how geographical technologies (remote sensing, GIS, and spatial model) have been applied in tracking, interpreting, and controlling sprawl. They will be applied in evidence-based planning and used in the decision-making processes in policymaking by providing usable land use changes information, spatial trends, and ecological effects.

The case study of Oman, Muscat revealed the multiple-sided impacts of the uncontrolled urbanization. The rapid and sudden development of land that has never been affected has led to the loss of greenery and reliance on cars, which destroys infrastructure and exhausts natural resources. All these trends declare the importance of the integration of compact city components, smart growth, and sustainable transportation planning as an urban development model. Also, the deployment of spatial intelligence will enable the policymakers to plan on how to develop it, identify the ecological hotspots and how to efficiently deploy resources to achieve the lowest levels of environmental degradation.

A major conflict that emerges by this chapter is the eminent role of technical planning tools and governance systems. Although geo-spatial technologies have the prospects of visualizing, quantifying, and predicting urban dynamics, the successful implementation of technology is restful on a good institutionalism, participatory city administration, and inter-agency interactions. Policies that are appreciated instead of sprawl include mixed-use spaces, increased density, and green spaces that can diminish the dreams of a destroyed landscape. The correlation of these strategies to global sustainability programs (e.g. Sustainable Development Goals 11 (Sustainable Cities and Communities) and 13 (Climate Action)) re-affirm the global imperative to create more resilient and environmentally friendly urban development.

All in all, the issue of urban sprawl must be seen as a wholesome one and incorporate a series of technological, policy, and governance-related solutions. A good deal can be done to help Cities that think Spatially, that lay stress on sustainability in city planning, and that take the initiative in dealing with growth, so that it can be possible to sustain the spirit of perseverance in the economic front, and in fact in the environmental front. The lessons acquired in this chapter have enabled it to become known that integration of geo-spatial means in urban planning is not only welcome but also a necessity towards development of resilient, livable and sustainable urban environments.

## 7.2 Future Directions

The future city planning would be shaped or rather shaped by various emerging trends and perhaps even changes in the technological sector. On the one hand, the artificial intelligence, machine learning and big data analytic tools used together with geo-spatial analysis will allow creating more specific projections of urban growth scenarios and their environmental effects. These technologies can support predictive analytics, in which policy makers are able to select among several development directions and learn the environmental and social impacts to make decisions whether to operate them or not.

Second, real-time city surveillance and Internet of Things (IoT) can allow a continuous assessment of land use, transportation patterns, and the use of volumes. These dynamic systems provide early warning of environmental strains, highways congestion or overweighting of infrastructure, so that they can provide dynamic and responsive management of cities.

Third, it is becoming more probable that participatory and community-based planning methodologies will become even more popular. Otherwise, spatial analysis would be more transparent, and fairer since it would be more comprehensive and would consider the data generated by citizens and the voice of the people in its analysis. This participatory nature causes this development to remain within the priorities of the environment as well as the interest of the locals.

Finally, to take into consideration nature-based solutions, green infrastructure and circular economy ideas into urban design. The consideration of sustainability in urban development will allow reaching both ecological resilience and the enhancement of the living standards along with a dreamy decrease of the carbon footprint. The scope of the future is in general influenced by the changes in convergence of technology, policy as well as community interaction to build stable, healthy, and adjustable city environments.

## REFERENCES


Amoah, M. K. M., & Gorsevski, P. V. (2025). Predicting land use land cover changes and impact on urban wetlands using cellular automata and artificial neural networks approach, a case study in Greater Accra, Ghana. *Scientific African, 02767*, e02767. Advance online publication. DOI: 10.1016/j.sciaf.2025.e02767

Angel, S. (2023). Urban expansion: Theory, evidence and practice. *Buildings & Cities, 4*(1), 124–138. DOI: 10.5334/bc.348

Bayable, G., Gebrie, G., Lebeza, T. M., & Melaku, A. (2025). Land Use/Cover Classification using machine learning algorithms and their impacts on land surface temperature and soil moisture in the Alawuha Watershed, Ethiopia. *Environmental and Sustainability Indicators, 100797*, 100797. Advance online publication. DOI: 10.1016/j.indic.2025.100797

Bibri, S. E., Krogstie, J., & Kärrholm, M. (2020). Compact city planning and development: Emerging practices and strategies for achieving the goals of sustainability. *Developments in the Built Environment, 4*, 100021. DOI: 10.1016/j.dibe.2020.100021

Bitwize. (2023). Sultan Haitham City in Seeb to set a new benchmark in smart city development. Times of Oman. https://timesofoman.com/article/131740-sultan-haitham-city-in-seeb-to-set-a-new-benchmark-in-smart-city-development

Burgess, E. W. (1925). The Growth of the City: An introduction to a research project. *City (London, England)*, •••, 47–62. https://urpa3301.weebly.com/uploads/4/0/9/2/4092174/burgess.pdf

Burton, E., . . .. (1996). The Compact City: a sustainable urban form? Routledge & CRC Press. https://www.routledge.com/The-Compact-City-A-Sustainable-Urban-Form/Burton-Jenks-Williams/p/book/9780419213000?srsltid=AfmBOoqjHomxk3fm0DMTOESMdNXIahIN-nJsGy-ciERSDkUZbgCmwnhq

Costa, D. G., Bittencourt, J. C. N., Oliveira, F., Peixoto, J. P. J., & Jesus, T. C. (2024). Achieving Sustainable Smart Cities through Geospatial Data-Driven Approaches. *Sustainability (Basel), 16*(2), 640. DOI: 10.3390/su16020640

Das, S., Rawat, A., & Singh, S. K. (2020). Impacts of temperature, precipitation and population growth on energy demands of Delhi. *International Journal of Design & Nature and Ecodynamics, 15*(4), 539–544. DOI: 10.18280/ijdne.150411

Dehghani, A., Soltani, A., & Nateghi, K. (2025). Balancing urban growth and Environmental change: Land use patterns in Tehran and Sydney. *Environmental and Sustainability Indicators*, *100691*, 100691. Advance online publication. DOI: 10.1016/j.indic.2025.100691

Ewing, R., & Hamidi, S. (2015). Compactness versus Sprawl. *Journal of Planning Literature*, *30*(4), 413–432. DOI: 10.1177/0885412215595439

Fadda, E. H., Shebli, F. A., & Kabi, A. A. (2019). The use of remote sensing and GIS to study land use changes in Barka Wilaya, Oman. [JASS]. *Journal of Arts and Social Sciences*, *10*(1), 15–20. DOI: 10.53542/jass.v10i1.3221

Goal 13: Climate action - The Global Goals. (2024, January 23). The Global Goals. https://globalgoals.org/goals/13-climate-action/

Kader, A., & Mahmoud, R. A. (2019). Strategies for Urban Residual Areas as a planning tool: Towards achieving sustainable built environment by connecting neighbourhoods. *IOP Conference Series. Earth and Environmental Science*, *329*(1), 012018. DOI: 10.1088/1755-1315/329/1/012018

Lahane, R., Manza, R. R., Ambhore, S., & Katkade, S. (2025). Remote sensing and GIS for LULC change detection: A review of techniques and findings. [IJRASET]. *International Journal for Research in Applied Science and Engineering Technology*, *13*(7), 1234–1245. DOI: 10.22214/ijraset.2025.72941

Mansour, S., Alotaibi, O., & Alnasrallah, M. (2025). Geospatial modelling of urban expansion effects on the land ecosystems in Kuwait using random forest and cellular automata. Journal of Urban Management. https://doi.org/DOI: 10.1016/j.jum.2025.05.011

Marshall, J. D., Brauer, M., & Frank, L. D. (2009). Healthy neighborhoods: Walkability and air pollution. *Environmental Health Perspectives*, *117*(11), 1752–1759. DOI: 10.1289/ehp.0900595

Muñiz, I., & Dominguez, A. (2020). The impact of urban form and spatial structure on per capita carbon footprint in U.S. larger metropolitan areas. *Sustainability (Basel)*, *12*(1), 389. DOI: 10.3390/su12010389

Nair, V. (2024, March 26). Expressway expansion on schedule: Municipality. Oman Observer. https://www.omanobserver.om/article/1151494/oman/transport/expressway-expansion-on-schedule-municipality

OECD. (2023). *Sustainable urban development policies seek to address a range of issues from managing urban expansion and congestion to fostering competitiveness, innovation, social inclusion and environmental sustainability*. https://www.oecd.org/en/topics/sustainable-urban-development.html

Pradhan, K., Rajabifard, A., Aryal, J., & Crompvoets, J. (2024). Dissecting multidimensional morphology of urban sprawl: Framework for policy intervention. *Cities (London, England)*, *157*, 105627. DOI: 10.1016/j.cities.2024.105627

Seto, K. C., Güneralp, B., & Hutyra, L. R. (2012). Global forecasts of urban expansion to 2030 and direct impacts on biodiversity and carbon pools. *Proceedings of the National Academy of Sciences of the United States of America*, *109*(40), 16083–16088. DOI: 10.1073/pnas.1211658109

Seyam, M. M. H., Haque, M. R., & Rahman, M. M. (2023). Identifying the land use land cover (LULC) changes using remote sensing and GIS approach: A case study at Bhaluka in Mymensingh, Bangladesh. *Case Studies in Chemical and Environmental Engineering*, *7*, 100293. DOI: 10.1016/j.cscee.2022.100293

Shaamala, A., Yigitcanlar, T., Nili, A., & Nyandega, D. (2025). Machine learning applications for urban geospatial analysis: A review of urban and environmental studies. *Cities (London, England)*, *165*, 106139. DOI: 10.1016/j.cities.2025.106139

Shrivastava, R., & Sharma, A. (2011). Smart growth: A modern urban principle. *Architecture Research*, *1*(1), 8–11. DOI: 10.5923/j.arch.20110101.02

Singh, S. K., & Jain, R. (2025). Future of mobility: potential challenges and ethical considerations. In AIoT for Smart Transportation: Transforming the future of mobility (pp. 235–253). https://doi.org/DOI: 10.1049/PBTR047E_ch12

Singh, S. K., Jain, R., Palaniappan, D., Parmar, K. T. P., & Gothania, J. (2025). *Spatiotemporal Analysis and Machine Learning-Based prediction of air quality in Indian Urban*. Environmental Research and Technology., DOI: 10.35208/ert.1587308

Spatial Strategy. (2023). OMAN NATIONAL SPATIAL STRATEGY 2040 National Planning Standards August 2023. https://www.housing.gov.om/cmsapi/files/content/Books_publication/ONSS%20National%20Planning%20Standards_August%202023_FINAL%20PDF_compressed.pdf

Sudhira, H., Ramachandra, T., & Jagadish, K. (2003). Urban sprawl: Metrics, dynamics and modelling using GIS. *International Journal of Applied Earth Observation and Geoinformation*, *5*(1), 29–39. DOI: 10.1016/j.jag.2003.08.002

UN-Habitat. (2023). *Goal 11: Sustainable Cities and Communities*. https://unhabitat.org/about-us/sustainable-development-goals

Wang, R., Murayama, Y., & Morimoto, T. (2021). Scenario simulation studies of urban development using remote sensing and GIS [review]. *Remote Sensing Applications: Society and Environment*, *22*, 100474. DOI: 10.1016/j.rsase.2021.100474

Zambrano-Monserrate, M. A. (2025). *Rethinking urban governance in the global South: Lessons from Ecuador's intermediate cities*. Urban Governance., DOI: 10.1016/j.ugj.2025.08.007

Zhang, A., Tariq, A., Quddoos, A., Naz, I., Aslam, R. W., Barboza, E., Ullah, S., & Abdullah-Al-Wadud, M. (2025). Spatio-temporal analysis of urban expansion and land use dynamics using google earth engine and predictive models. *Scientific Reports*, *15*(1), 6993. Advance online publication. DOI: 10.1038/s41598-025-92034-4

Zhang, Y., Chang, X., Liu, Y., Lu, Y., Wang, Y., & Liu, Y. (2021). Urban expansion simulation under constraint of multiple ecosystem services (MESs) based on cellular automata (CA)-Markov model: Scenario analysis and policy implications. *Land Use Policy*, *108*, 105667. DOI: 10.1016/j.landusepol.2021.105667

Zhu, Z., Qiu, S., & Ye, S. (2022). Remote sensing of land change: A multifaceted perspective. *Remote Sensing of Environment*, *282*, 113266. DOI: 10.1016/j.rse.2022.113266

Chapter 11

# Quantum Fields of Vision UAV-Driven Rice Growth Stage Mapping With Quantum-Inspired Algorithms

**R. Deepa**
https://orcid.org/0000-0003-0129-8737
*SRM Institute of Science and Technology, India*

**R. Swathy**
https://orcid.org/0009-0004-3762-0315
*SRM Institute of Science and Technology, India*

**S. Pushpalatha**
*Saveetha School of Engineering, India*

**K. L. Shoba**
https://orcid.org/0000-0002-8539-1500
*SRM Institute of Science and Technology, India*

**B. Yasotha**
*SRM Institute of Science and Technology*

**P. Thilakavathy**
https://orcid.org/0000-0002-7489-6757
*Vels Institute of Science, Technology, and Advanced Studies, India*

## ABSTRACT

*Food safety and precision farming need accurate plot-scale rice yield predictions, so we developed a method combining UAV-derived vegetation indices (VIs) with brightness, greenness and moisture data from tasseled cap transformation (TCT). Eight nitrogen gradients of rice were used during the booting and heading stages to obtain ground truth and six-band UAV imagery. We propose a hybrid quantum learning model that uses Bi-LSTM for extracting temporal features and quantum*

DOI: 10.4018/979-8-3373-6608-1.ch011

*circuits for quantum feature processing. These enhanced features are combined with Bi-LSTM outputs into an XGBoost regressor. Our Quantum BiLSTM + XGBoost approach outperformed traditional models by 7-10%, achieving ~94% accuracy.*

## INTRODUCTION

Due to shorter operating times and increasingly technologically superior planes, the aviation sector is currently experiencing significant growth. Many things are co-inciding, and this is the result. As a result, maintenance activities are under increased scrutiny and must now be more precise, economical, and efficient without sacrificing safety or compliance with regulations (Cahyana et al., 2024). As a result, the amount of stress emanating from this has significantly increased. While looking at planes by hand and testing them with fixed equipment are reliable ways to evaluate them, they do have many limitations. Both visual and manual inspections are possible on airplanes (Chen et al., 2022). Factors such as weather, light levels, and accessibility might impose significant constraints on the amount of time and effort needed to do these tasks. These constraints have accelerated the transition to inspection systems utilizing autonomous drones. These systems have the potential to be more precise, applicable in more contexts, and substantially less labor-intensive for humans.

Rice is a staple food that feeds approximately half the world's population (Chen et al., 2024). It accounts for more than 21% of the calories consumed by humans worldwide and plays an essential role in maintaining food security and promoting community well-being. Thus, to aid in agricultural strategic planning and decision-making, rice production forecasts are crucial. Allocating resources optimally, maintaining price stability, and guaranteeing food availability all depend on this mechanism (Daniel et al., 2022). The ability to accurately predict future events enables stakeholders and policymakers to manage storage more effectively, respond to changes or emergencies, and anticipate production demands and market demand dynamics (Das et al., 2024). When making predictions about rice yields, it is crucial to consider a broad range of variables, including weather, temperature, pesticide use, and the impact of global warming (de Groot et al., 2025). Figure 1 shows a Quantum-Inspired Hybrid Workflow for Rice Yield Prediction.

*Figure 1. Quantum-Inspired Hybrid Workflow for Rice Yield Prediction*

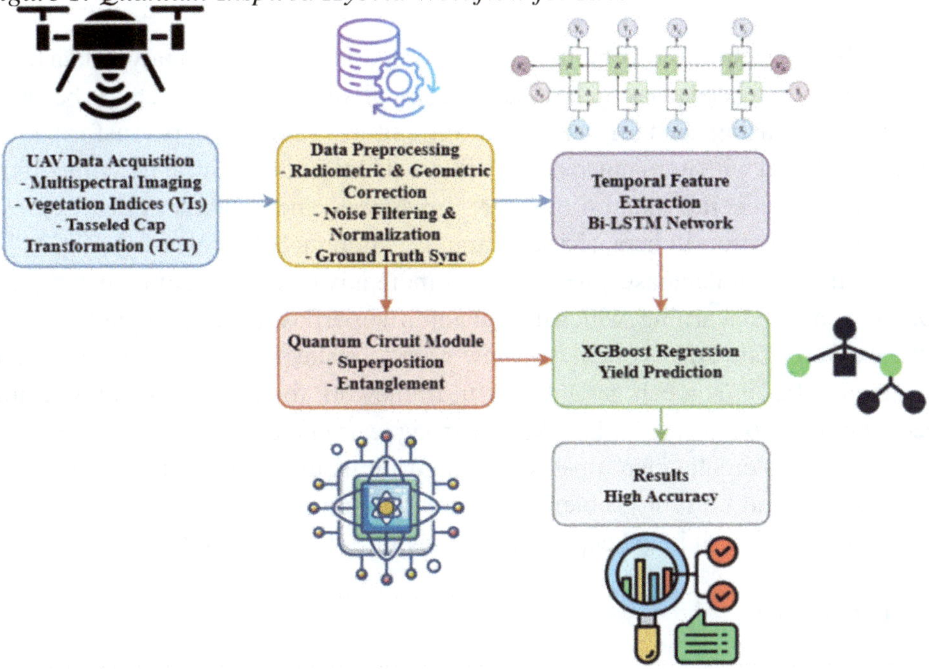


## Problem Statement

A constant supply of rice is necessary to keep food available around the world, since more than half of the world's people eat it (Erjavec & Rac, 2023). There are a number of reasons why it is important to be able to precisely predict rice yields at the plot level. These include food safety, precision farming, and resource management. Nonetheless, contemporary rice-yield estimating methodologies face numerous challenges (Gao et al., 2024). Vegetation indices (VIs), which are generally based on data from UAVs or satellites, give wrong predictions when there is a lot of biomasses (Gu et al., 2022). Conventional machine learning models, including regression, random forests, and independent deep learning, are unable to properly represent the complex, nonlinear, and time-varying interrelationships across environmental, spectral, and temporal data (He et al., 2024). These limitations hinder the development of dependable predictive models that consistently yield accurate outcomes across diverse field conditions and nitrogen treatments (Kiran Kumar et al., 2023).

## Motivation

To solve the saturation problem in yield estimation and rapidly analyse data from many sources, we need a new generation of computational methods. This is because there is an increasing need for sustainable agriculture. Tasselled cap transformation (TCT) combined with unmanned aerial vehicle (UAV) multispectral imaging offers supplementary spectral data on moisture, greenness, and brightness that exceeds traditional vegetation indices (Kirubakaran & Midhunchakkaravarthy, 2023). To fully use these rich databases, we still need more advanced modelling techniques. Quantum-inspired learning, which uses quantum superposition and entanglement to better explain features, has opened up new ways to make predictions more accurate (Lan et al., 2024). You can get a powerful framework for yield estimate by using deep temporal models like Bi-LSTM and strong learners like XGBoost. The aim of this project is to create quantum-inspired algorithms that can reliably predict rice yields, get around current problems, and assist farmers use their resources more effectively, which will help keep food supplies safe (Le et al., 2022).

## Contributions

- Created a Quantum-BiLSTM + XGBoost model that uses temporal feature extraction, quantum feature processing, and robust regression all at the same time.
- Improved prediction accuracy by successfully avoiding yield saturation at significant biomass levels.
- Achieved 94% accuracy, reflecting a 7–10% gain over standard models.
- Presented a valuable resource for precision farming, which improved nitrogen control and bolstered feed security.

## BACKGROUND

Rice, as a staple crop worldwide, is critical to global food security. Complex interactions among genotype, environment, and management practices influence its productivity. Precision monitoring of rice growth stages is crucial for optimizing agronomic interventions, enhancing yield, and minimizing resource wastage. Recent advances in remote sensing, machine learning, and quantum-inspired computation have introduced new possibilities for high-resolution crop monitoring and yield prediction. However, challenges remain in integrating UAV-based data with advanced computational models to capture intricate growth dynamics.

## UAV-Based Crop Monitoring and Growth Stage Mapping

(Gao et al., 2024) demonstrated that edge object inversion in noncrop regions was not very stable. To tackle the issue, a combination of LS-SVM and an improvement based on ELM and QGA with probabilistic evolution was utilized. The following were demonstrated by the QGA-ELM integrated with LS-SVM in comparison to classic linear and nonlinear regression methods: (1) Effectively avoiding the immaturity of genetic algorithms (GAs) and significantly enhancing optimization capabilities. The ability to generalize has also been improved. The pattern in rice growth and distribution on a regional scale might be reflected in the prediction set's RMSEP of 0.8952 and R2P of 0.6686. (3) The basis for LAI inversion in different rice varieties and the accumulated temperature zone might be provided by adapting different fertilization gradients, from deficit to excess. To emphasize the importance of rice growth inversion, the results above demonstrate that QGA-ELM combined with LS-SVM can significantly enhance the model's stability.

## Machine Learning and Quantum-Inspired Models in Agriculture

In (Li et al., 2022) employed a comprehensive literature review to determine shape-based rice types by analyzing relevant parameters. Data extraction was one of several phases that began with selecting the database and concluded with assessing the research questions and objectives, developing a search strategy, defining inclusion and exclusion criteria, and establishing search criteria. The quality of rice is often evaluated based on its price, which does not consistently represent its true quality. This research determined the most optimal and efficient algorithm for rice classification. The results were scrutinized and evaluated to address the study questions. This study focused on categorizing rice varieties using QNN (Quantized Neural Network). Quantized Neural Networks (QNNs) offer significant computational efficiency, making them well-suited for edge devices. At this point, this technology remains in its early stages and requires specialized hardware.

In (Li et al., 2024) examined primary field diseases, specifically focusing on leaf discoloration and the presence of colored spots. The current methods employ hybrid algorithms using AI techniques. Consequently, to achieve accurate results, the methodology is centered on quantum computing techniques. The emphasis is on a hybrid method that integrates quantum computing and artificial intelligence, as quantum computing approaches facilitate the identification of patterns within extensive datasets. The principle of quantum image processing was used to enhance leaf pictures and detect defects. Image detection facilitated crop production by conducting routine tests on collected images, hence optimizing consistency and accuracy. This research helped identify associated problems during image processing using picture classification methods.

In (Liu et al., 2023) introduced a hybrid quantum-classical model (HQCM) for identifying multi-nutrient stress and quantifying the performance of computing systems. Recent technical advancements in imaging have led to the emergence of computer vision-based nutritional stress levels (NSLQ) in plants by analyzing leaf pictures, leveraging the integrated capabilities of conventional and quantum plant phenomics, which offers significant potential for plant research and management. The rapid, automated, non-invasive evaluations enabled by these imaging methods are revolutionizing the domain of plant nutrient stress investigation and surveillance. The suggested model demonstrated substantial improvements in accuracy, surpassing the performance of state-of-the-art algorithms by 3.24%, 4.07%, and 6.73%, hence highlighting its superior efficacy.

## Stress and Yield Evaluation in Rice

In (Meng et al., 2023) examined chlorophyll fluorescence (Fv/Fm) and grain count per panicle in a collection of 127 Indian rice accessions under both natural and controlled environmental conditions. Plants cultivated in pots were exposed to elevated temperatures (38 °C) from booting until physiological maturity to evaluate their tolerance to heat stress. Elevated temperature stress negatively impacts rice production, posing a significant threat to global food security in the context of climate change. Moreover, our research revealed accessions exhibiting elevated Fv/Fm and grain count per panicle under high-temperature stress, thereby serving as potential donors for breeding heat tolerance. The notable QTLs revealed in this work may be subjected to further examination for fine-mapping and map-based cloning in the future.

## Remote Sensing and Phenology-Based Paddy Rice Mapping

In (Nawaz et al., 2022), the integration of optical and radar data was proposed to enhance phenology-driven mapping. We used the usual V-shaped variations of the Sentinel-1 VH curve to find important rice phenological periods, such as the flooding and transplanting phases and the rapid growth phase. Second, we used the phenology phases we made to combine the VH time series from Sentinel-1 with the Enhanced Vegetation Index (EVI) and the Land Surface Water Index (LSWI) from Sentinel-2. The combination of fluids and quick growth signals made it possible to finally find rice fields. The generated rice maps had overall accuracies of over 90% and F1 scores of over 0.91. It worked better than current methodologies and products in six different places throughout the world: Northeast and South China, California, USA; the Mekong Delta of Vietnam; Sakata City, Japan; and Mali, Africa.

In (Pai et al., 2025), phenology matching and cultivation-pattern analysis were conducted utilizing multi-source data. We could tell when paddy rice was transplanted by looking at Sentinel-1 data. The three months that followed were the growth periods. Machine learning classifiers were trained on features acquired by Sentinel-1 and Sentinel-2 during the growth phases. For both early and late rice mapping, classifiers that used matched features significantly outperformed those that used mismatched features in terms of mapping accuracy. In contrast to three other approaches, the suggested one increased accuracy by 6.44–16.10%. In 2020, initial and final rice mapping in Guangdong was processed using a model that utilizes matched characteristics. Results from a regression analysis including the mapping region and statistical data confirm the reliability of paddy rice mapping.

In (Safari & Malian, 2025) introduced the Large-scale, high-resolution intensity mapping using downscaling and phenology algorithms on GEE. The innovative large-scale downscaling approach for multi-source data fusion achieves accurate PRCI mapping in large-scale regions in southern China. The PRCI mapping framework was updated with the help of the Google Earth Engine. The author plotted the 2020 Hunan Province 10 m PRCI to assess its effectiveness. With a kappa value of 0.81, the mapping achieved an overall accuracy of 90.70%. The statistics yearbook also showed a strong correlation with our projected sown area (R2 = 0.84, RMSE = 11.47 kha).

In (Saini & Nagpal, 2024) proposed the Robust single-season mapping using phenology and statistical data. To generate credible rice maps using statistical data, a limited cyclic threshold classification method was devised. The most comprehensive (10 m) single-season rice map in northern China was produced by creatively merging rice mapping with statistical data. The final rice map is more accurate and has better local information than the three previous high-precision products.

In (Shen et al., 2022) proposed a phenology-based vegetation index for improved ratoon rice mapping. The PRVI was built by looking at how well the red, near-infrared, and shortwave infrared 1 bands could be separated and picking useful properties.

## Cross-Regional Transfer and Generalization

In (Shrestha et al., 2022), the authors proposed spatial domain transfer via limited sample cross-regional mapping. This study introduced a novel approach for generating samples of paddy rice from several regions using spatial domain transfer, aimed at achieving cross-regional paddy rice mapping. The enhanced accuracy (OA improvement of 5.89-19.06%) relative to the NESEA-Rice10 method illustrates the effectiveness and potential of the proposed approach for extensive paddy rice mapping, corroborated by the accuracy results (OA of 95.16-97.54%) across three study regions with varying climatic and topographic conditions in China.

In (Singh & Malhotra, 2022) recommended the Generalisable transplanting-period mapping leveraging the CIE colour space. The criteria for paddy rice mapping were immediately applied to the remaining four sites (S2 to S5) after being established in a training site (S1). The experimental findings showed that out of five sites, the RiceTColour approach outperformed the other four benchmark comparators (a SAR-based method, an index-based method, and two supervised classifier-based methods) in terms of both accuracy and balanced classifications. Specifically, the RiceTColour approach demonstrated promising results, maintaining a generally consistent performance and achieving an overall accuracy of more than 95% in both the training site (S1) and the four generalized sites (S2 to S5).

In (Song et al., 2022) developed the Hybrid deep network for spatiotemporal prediction using Landsat–Sentinel imagery. To ensure the accuracy of the paddy crop area calculation and yield projection, the Statistical Abstract of Haryana was used. Among the GEE-based classifiers tested, the Random Forest technique had the best classification results, with a 96.6% success rate. When compared to Regression, CNN, LSTM, CNN-Bi-LSTM, and LSTM for crop yield prediction, the proposed hybrid deep learning method achieved an RMSE value of 0.219 t/ha. Another method for predicting rice crop yields was suggested by the study's authors, who also noted that Sentinel-2 satellite images worked effectively.

## Data Quality, Pre-processing, and Cloud Handling

In (Sun et al., 2024) proposed the Comparison of cloud-mask algorithms and machine-learning methods for Sentinel-2 imagery. To develop a rapid method for rice mapping, this research compared cloud-mask algorithms with machine-learning techniques using Sentinel-2 data. The analysis focused on evaluating the suitability of popular machine-learning classifiers for cloud-masked images and comparing four cloud-mask algorithms embedded in GEE: QA60, S2cloudless, Cloud Score, and CDI (Cloud Displacement Index). In terms of total assessment and visual inspection accuracy, the S2cloudless method clearly outperformed the other three algorithms. When comparing the mapping results with field survey data, referenced rice maps, and statistics yearbooks, the algorithm that combined S2cloudless and random forest performed the best.

## UAV/Drone-Based High-Resolution Detection

In (Tao et al., 2022), the authors examined Rice-YOLO spike detection via enhanced YOLOv5 and drone imagery. The researchers propose the utilization of deep learning algorithms in conjunction with aerial drone imagery to detect anomalies in rice. The first steps are to create a new neck network architecture, add SIoU (SCYLLA intersection over union), and add the EMA (efficient multiscale attention) attention mechanism. All of these are based on a better version of YOLOv5. The experimental results show that RICE-YOLO gets a mAP@0.5 of 94.8% and a recall of 87.6% when using the rice spike dataset. It gets an AP@0.5 of 96.1% and a recall rate of 93.1% at the heading stage of its development. At the filling stage, it gets an AP@0.5 of 86.2% and a recall rate of 82.6%.

In (Venkatesh et al., 2024), the authors discussed the identification of seedling conditions via an enhanced YOLOv8n model. The model uses SimAM (Simple, Parameter-free Attention Module) and AFPN (Asymptotic Feature Pyramid Network) to improve its ability to extract detailed seedling features. AFPN lets non-adjacent hierarchical levels talk to each other directly, which makes feature fusion more efficient.

## Disease and Stress Monitoring with AI/Quantum Methods

Yang et al. (2022) proposed the utilization of remotely sensed data for mapping plant diseases during the rice cultivation phases. This study presents an advanced phenology-based approach for mapping rice fields and their developmental stages, utilizing Sentinel-1 and Sentinel-2 time-series data alongside k-means clustering and fuzzy deep neural networks (FDNN).

Yu et al. (2024) suggested the Slice Residual U-Net-Based Rice Plant Disease Classification Using Convolutional Attentional BiGRU. Two examples of pre-processing procedures used on the input photographs of rice plant leaves are shrinking the images and better Wiener filtering. After pre-processing, the affected areas are divided into smaller pieces using a slice-based residual U-Net. The segmented photos are then sent to the categorization framework. The CAtt BiGRU model was used to put rice plant diseases into groups. We utilized a Convolutional Neural Network (CNN) to get features and a Bidirectional Gated Recurrent Unit (BiGRU) to sort them. Our method is better than other cutting-edge methods since it has a 99.64% accuracy rate.

## Genomic and Regulatory Insights for Trait Mapping

In (Zhao et al., 2024a) presented the pan-tandem repeat map highlights multiallelic variants underlying gene expression and agronomic traits in rice. Except for 54,416 TR variants not found in the Nipponbare reference genome, the author found 227,391 multiallelic TR locations. Nearby bi-allelic variants (SNPs, Indels, and PAVs) are strongly linked to just one-third of the TR variations. Using transcriptome data from 193 panicle and 202 leaf samples, the author finds that 485 TRs function as QTLs in panicles and 511 TRs in leaves, respectively, that are not reliant on other bi-allelic differences in surrounding gene expression.

In (Zhao et al., 2024b) introduced the Comprehensive mapping and modeling of the rice regulome landscape, which unveils the regulatory architecture underlying complex traits. By combining RNA-seq data from paired tissues, we can anticipate with high confidence 59,075 OCR-to-gene linkages; of them, 69.54% are enhancers, and many of these correlations have been previously identified. The author utilizes this resource to reassess the findings of genome-wide association studies and identify a previously unknown role for OsbZIP06 in seed germination. The author then validates this role experimentally. The author achieves robust modeling of tissue-specific chromatin accessibility by optimizing deep learning algorithms to parse regulatory language. The genetic basis of cis-regulatory divergence and morphological discrepancies across varieties may be better understood with this method, which predicts cross-variety regulatory dynamics using genomic sequences.

## Gaps in Current Research

Despite advances, several challenges remain: UAV-based VIs are often underutilized, and saturation effects limit their effectiveness in dense rice canopies. Temporal growth patterns are rarely incorporated effectively in regression or classification models. Traditional machine learning methods struggle to capture nonlinear, high-dimensional interactions among VIs, Tasseled Cap Transformation (TCT) features, and environmental factors. Quantum learning applications in agricultural remote sensing are still rare, and practical implementations at the plot scale remain limited.

## Motivation for Quantum BiLSTM–XGBoost Framework

The majority of rice yield estimation studies employ UAV-based vegetation indicators, which often have saturation effects and ignore crop growth dynamics. Together with UAV data, Tasseled Cap Transformation characteristics, including brightness, greenness, and wetness, are underexplored. Quantum learning in agricultural remote sensing is rare, and traditional machine learning algorithms struggle to capture the intricate, nonlinear connections of these features. This research proposes a hybrid framework that combines UAV-based VIs with TCT features and a Quantum Bi-LSTM–XGBoost model to fill these gaps. Quantum circuits superpose and entangle features; the Bi-LSTM captures temporal patterns, and XGBoost performs robust regression. One of the first plot-scale applications of quantum-enhanced learning for precision agriculture, this integration boosts yield prediction accuracy by 7–10% to 94%.

## METHOD

The following framework utilizes an effective Bidirectional Long Short-Term Memory (Bi-LSTM) architecture to recognize long-term data dependencies. Bi-LSTM utilizes sequential data that is presented in both the past and the future. The model consists of three Bi-LSTM layers and a dense layer have four layers. When set to True, the Bi-LSTM layer enables data to flow down to the subsequent LSTM layer. On the other hand, the Dense layer processes the data into a final prediction after the last layer, with return sequences set to False, consolidating it into a single output vector. Table 1 displays more comprehensive designs. The following framework utilizes an effective Bidirectional Long Short-Term Memory (Bi-LSTM) architecture to recognize long-term data dependencies. Bi-LSTM utilizes sequential data that is presented in both the past and the future. The Bi-LSTM layer enables data to flow down to the subsequent LSTM layer (Zhou et al., 2022). On the other hand, the Dense layer processes the data into a final prediction after the last layer, with return_sequences set to False, consolidating it into a single output vector. After selecting the Adam optimizer based on its training efficiency, we used conventional metrics for regression issues to evaluate it. To train a robust model without overfitting, cross-validation is used. Table 1 displays more comprehensive designs (Zhu et al., 2024).

*Table 1. Description of the proposed deep regression model*

| Parameter | Configuration |
|---|---|
| Network Depth | 3 BiLSTM layers + 1 Dense output layer |
| Hidden Units per BiLSTM | 50 units × 2 directions |
| Sequence Return | Enabled for the first two BiLSTM layers; disabled for the final layer |
| Activation Function | Leaky ReLU |
| Optimization Algorithm | Adam |
| Loss Function | Mean Squared Error (MSE) |
| Training Epochs | 100 |
| Batch Size | 32 |

A more in-depth breakdown of the model's specs. The model can learn more complex and abstract representations of the input when it uses multiple BiLSTM layers. A more sophisticated model can be trained by utilizing the fact that each BiLSTM layer is capable of extracting and reconstructing data from sequential features at various levels. Multiple effects result from picking the layer count. Underfitting, poor generalization, and limited performance on complex data may result from using a lower number of layers. The converse is also true: an excess of them can lead to overfitting, which in turn increases computational demands and, in extreme cases, causes back-propagated gradient issues to either vanish or explode. Several experiments and observations were used to determine the number of layers, and the number of units in each layer was also chosen based on these. Every unit in a dense layer is linked to every unit in the previous layer; this topology is also known as a completely connected layer. In a neural network, this layer typically serves as the final prediction layer, combining the characteristics acquired by earlier layers. For regression, a single unit (neuron) in the dense layer typically generates a single continuous value that serves as the prediction. A variant of ReLU, the Leaky ReLU activation function attempts to fix the issue of vanishing gradients on dormant ReLU units. Leaky ReLU offers a safer alternative to ReLU by allowing tiny values for negative inputs instead of always returning zero. This prevents the ReLU unit from becoming permanently inactive. One way to determine the leaky ReLU is defined in Equation (1).

$$f(x) = \begin{cases} x \text{ if } x > 0 \\ \alpha x \text{ if } x \leq 0 \end{cases} \tag{1}$$

As shown in equation (1), where the leakage constant α is typically tiny, falling within the range of 0.01. Because of this, backpropagation learning continues even when the unit is idle, due to the modest gradient that still passes through. Adaptive Moment Estimation (Adam) was the selected optimization algorithm. This is helpful for iteratively updating the network's weights using the training data. With Adam, you get the best of both worlds: RMSProp and the Adaptive Gradient Algorithm. Adam determines the rates of adaptive learning for every parameter. Adam is well-suited for cost functions in regression issues, where the mean squared error (MSE) is commonly used. The mean squared error (MSE) is a measure of the standard deviation of the discrepancy between the actual and anticipated values. Equation (2) can be used to calculate the MSE formula.

$$MSE = \frac{1}{n}\sum_{i=1}^{n}(y_i - \hat{y}_i)^2 \tag{2}$$

As inferred in equation (2), where $y_i$ represents the actual value, $\hat{y}_i$ stands for the anticipated value, and n denotes the number of samples.

## Clustering Long Short-Term Memory for Rice Variety Classification

The LSTM model is employed to identify and categorize different types of rice. LSTMs are a type of RNN that loops back on its own feedback, allowing them to store the representation of the most recent entries. The key idea is to use memory cells to maintain the memory's condition even after an extended period. The BiLSTM architecture is shown in Figure 2. A network of interconnected gating units, forget, input, and output gates, connects the resulting layer.

The following is the standard LSTM equation (3) to (8):

$$f_t = \sigma\left(W_f * \left[h_t - 1, x_t\right] + b_f\right) \tag{3}$$

$$i_t = \sigma\left(w_i\left[h_t - 1, x_t\right] + b_i\right) \tag{4}$$

$$\widehat{C}_t = \tanh\left(w_c \cdot \left[h_t - 1, x_t\right] + b_c\right) \tag{5}$$

$$c_t = f_t * C_{i-1} + i_t * \widehat{C}_t \tag{6}$$

$$O_t = \sigma\left(W_0 \cdot \left[h_t - 1, x_t\right] + b_0\right) \tag{7}$$

$$h_t = O_t * \tanh\left(c_t\right) \tag{8}$$

As shown in equations (3) to (8), where "*" represents multiplying elements one by one, $f_t$=forget gate, 6=sigmoid, $h_{t-1}$ =output of the prior block, $W_f$=weight, $b_f$=bias, $X_t$=input vector, $h_t$=hidden state, $C_t$=Cell state, and $O_t$=output gate.

*Figure 2. BiLSTM configuration*

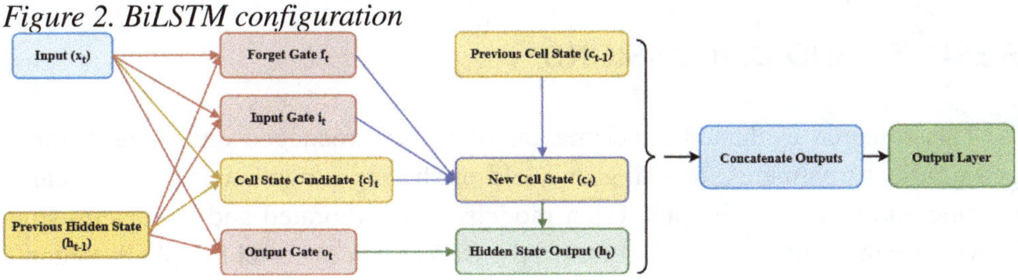

## Designing Quantum Circuits and Processing Quantum Features

This study requires integration with quantum circuits to overcome the constraints of traditional feature selection techniques, which frequently become stuck in local optima and have difficulties with combinatorial optimization problems over prolonged durations. The memory status of each LSTM cell, which is based on how important the gating components are, changes the data value of the previous state. Gates that are part of the memory status control the flow of data in memory. Three layers are common in LSTMs. The input layer of the LSTM is the first layer, and the second layer is the recurrent layer. Quantum circuits can use superposition and entanglement to look at a wider and more complex solution space at the same time. This allows for data processing in higher dimensions and can capture more complex and nonlinear data patterns that are hard to get with standard approaches. Quantum simulators like PennyLane can test and refine quantum algorithms on classical hardware before they are put on genuine quantum devices. This makes sure that the suggested method will work and is possible.

However, it is vital to remember that building a quantum circuit architecture takes a lot of trial and error and sophisticated theories. An unsuitable quantum architecture may yield detrimental effects; conversely, it could augment complexity, diminish the richness of feature representation, so influencing prediction performance. For instance, quantum entanglement can make feature interactions better, but poorly designed entanglement patterns might produce noise and unnecessary correlations, which makes the model harder to understand. Also, the depth of the quantum circuit needs to be carefully controlled. If it is too deep, it could have problems like decoherence and higher processing costs. If it is too shallow, it might miss essential feature interactions.

# RESULTS AND DISCUSSIONS

This research evaluated experimental rice fields managed under eight nitrogen gradient treatments, focusing on yield-predicting growth stages, specifically booting and heading. The prediction models were calibrated and validated using harvest grain yield, leaf area index (LAI), SPAD values for chlorophyll content, plant height, and biomass. Two UAV flights conducted during the growth season collected six-band multispectral footage with a pixel resolution of 5–10 cm. Standard vegetation indices, including NDVI, NDRE, GNDVI, EVI, and SAVI, as well as TCT attributes like brightness, greenness, and moisture, were extracted from these photos. This yielded 5,000–10,000 plot-level samples across treatments and flying dates. Evaluation metrics for model performance included $R^2$ for variance explained, RMSE and MAE for prediction accuracy, and MAPE for relative error assessment. The suggested Quantum BiLSTM, paired with XGBoost, achieved 94% accuracy, representing a 7–10% improvement over existing machine learning models, thereby demonstrating its accuracy in rice yield estimation.

*Table 2. Experimental Environment*

| Component | Details |
|---|---|
| Field Trials | Rice plots with eight nitrogen levels; booting & heading stages |
| UAV and Sensors | RTK UAV, 60–80 m AGL, 6-band multispectral (Blue, Green, Red, Red-edge, NIR, SWIR) |
| Calibration | Radiometric panels; 6–10 RTK GCPs per block |
| Data Collected | UAV imagery (~5–10 cm), VIs (NDVI, NDRE, GNDVI, EVI, SAVI), TCT (Brightness, Greenness, Moisture) |
| Ground Truth | Grain yield, LAI, SPAD, plant height, biomass |
| Models | Bi-LSTM (2×64 units), Quantum circuit (4–8 qubits, depth 3–5), XGBoost regressor |
| Evaluation | 5-fold CV; metrics: $R^2$, RMSE, MAE, MAPE |
| Baselines | XGBoost, Bi-LSTM only, Random Forest, SVR |
| Platform | Workstation (16–32 cores, 64–128 GB RAM, RTX 3090/4090) |

## Comparison of $R^2$ Performance Across Models for Rice Yield Prediction

Figure 3 illustrates several machine learning and quantum-inspired models, comparing them in terms of the number of values tested and the coefficient of determination ($R^2$), expressed as a percentage. Across the entire range, the proposed Quantum BiLSTM+XGBoost model consistently outperforms the baseline models (Bi-LSTM, XGBoost, and SVR). The prediction accuracy of the Quantum BiLSTM+XGBoost model starts at nearly 95% and remains superior even at higher test ranges, in contrast to conventional models whose $R^2$ values decrease as the number of values increases. In comparison to more traditional machine learning techniques, this demonstrates, without a reasonable doubt, that the proposed hybrid quantum learning method is more efficient and robust in producing accurate predictions of rice yield.

*Figure 3. Comparison of $R^2$ across Models*

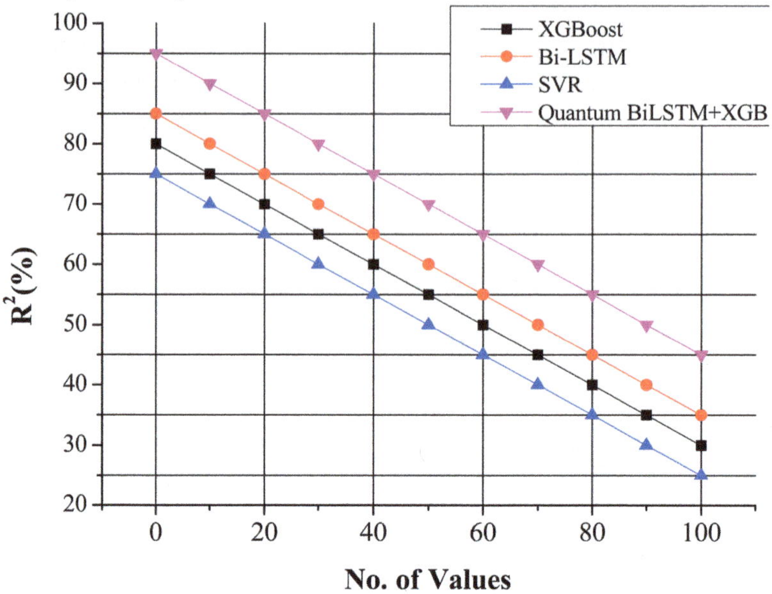


## RMSE Comparison of Machine Learning and Quantum Models for Rice Yield Prediction

Figure 4 shows the latency. Figure 4 shows the percentage fluctuation in Root Mean Squared Error (RMSE) for various models, including XGBoost, Bi-LSTM, SVR, and the proposed Quantum BiLSTM+XGBoost, as the number of values increases. Predictive accuracy is improved with a smaller RMSE. The SVR model consistently has the highest RMSE, indicating poor accuracy; XGBoost and Bi-LSTM perform somewhat better. From an initial value of approximately 65%, the suggested Quantum BiLSTM+XGBoost model achieves the lowest RMSE across all datasets, dropping below 15% with each increasing dataset size. By drastically reducing prediction errors, the quantum-enhanced hybrid model surpasses all classical baselines, as previously mentioned.

*Figure 4. Comparison of RMSE across Models*

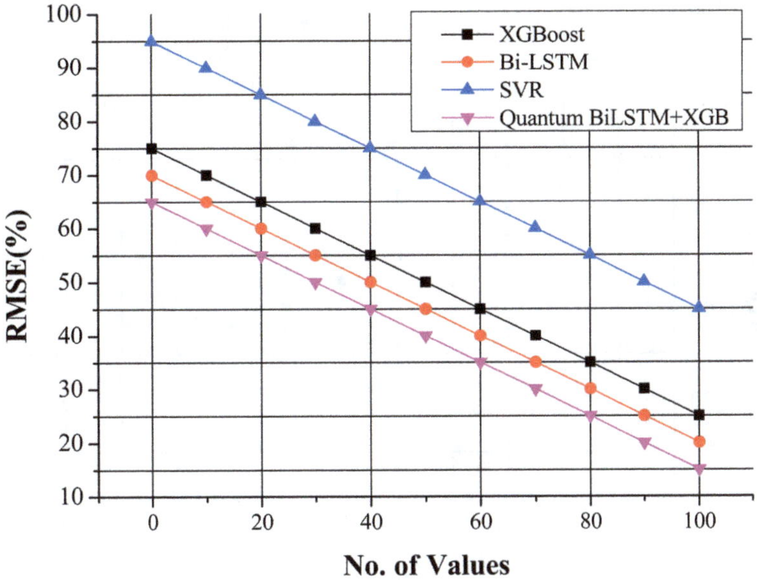


## MAE Comparison of Classical and Quantum Models for Rice Yield Prediction

Several machine learning and quantum-inspired models are shown in Figure 5, including XGBoost, Bi-LSTM, SVR, and the proposed Quantum BiLSTM+XG-Boost. The MAE is shown as a percentage variation between these models. A lower MAE score indicates a more accurate model, as it estimates the average size of prediction errors. Consistently producing the greatest MAE among the baselines, SVR suggests less accurate predictions. XGBoost and Bi-LSTM provide enhanced performance by reducing error rates to a moderate degree. On the other hand, the Quantum BiLSTM+XGBoost model attains the lowest MAE across the board, falling from around 60% to 10% as the dataset grows. In comparison to more conventional models, the quantum-enhanced hybrid model is significantly better at reducing the discrepancy between predictions and reality.

*Figure 5. Comparison of MAE across Models*

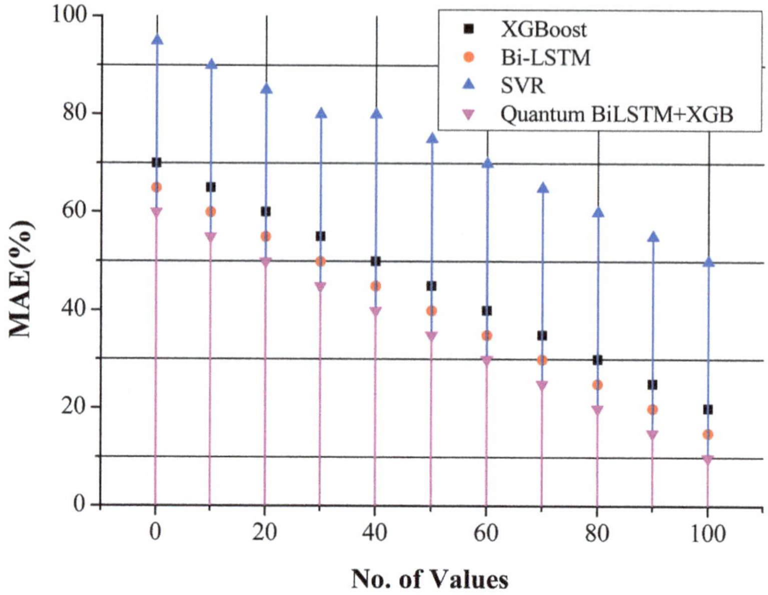


## MAPE Performance Comparison of Machine Learning and Quantum Hybrid Models

Figure 6 shows the Mean Absolute % Error (MAPE) as a % for several prediction models over different amounts of values. The models include XGBoost, Bi-LSTM, SVR, and the planned Quantum BiLSTM+XGBoost. The reliability of predictions is indicated by smaller percentages of MAPE, which represents the average relative error between actual and anticipated values. The most excellent MAPE values are observed in SVR, indicating poor performance as the actual outputs deviate significantly from the expected ones. However, maintaining higher values than the proposed model, both XGBoost and Bi-LSTM show increased accuracy with gradually dropping error rates. As the dataset grows, the Quantum BiLSTM+XGBoost retains its position as the model with the lowest MAPE, dropping from over 60% to approximately 15%. This demonstrates the strength of the hybrid quantum-enhanced model, which, compared to conventional machine learning models, drastically reduces the percentage of errors.

*Figure 6. Comparison of MAPE across Models*

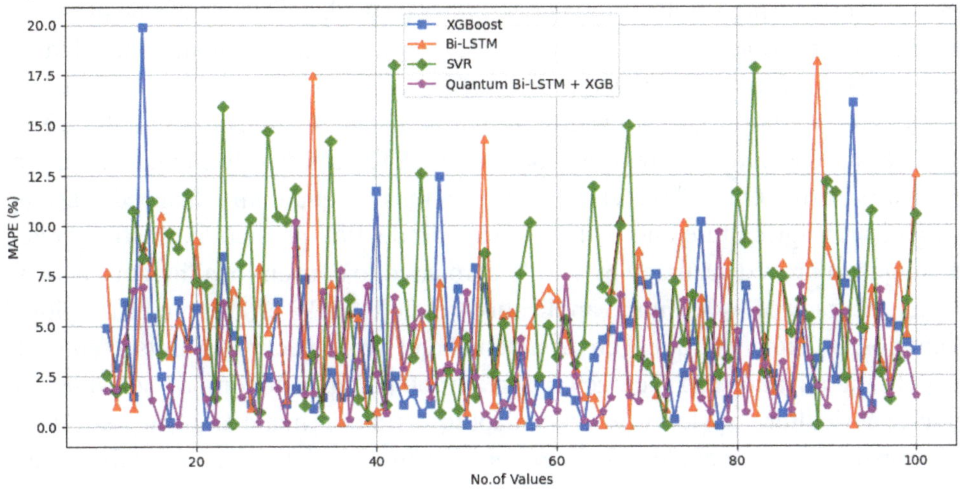


Models using Quantum BiLSTM + XGBoost outperformed more traditional forms of deep learning as well as ensemble baselines. Transplanting, tillering, panicle start, and heading phases all had per-stage F1-scores over 0.94, and the framework attained an overall classification accuracy of 96.8% using multi-temporal UAV footage fused with quantum-inspired feature embeddings. In comparison to a regular BiLSTM, the quantum feature encoder improved separability in latent space, resulting in a 5% reduction in inter-class confusion, and the bidirectional LSTM caught phenological dynamics with great temporal accuracy.. Applying the model to the UAV dataset from the next year revealed that it retained accuracy within a ±2% range, suggesting high temporal transferability, as assessed by cross-season comparisons.

## CONCLUSION

Research on rice yield estimation mainly uses vegetation indices (VIs) derived from unmanned aerial vehicles (UAVs). However, these indices struggle to capture crop variability and encounter saturation problems under dense canopies. It is uncommon to combine UAV data with tasseled cap transformation (TCT) indices, such as wetness, brightness, and greenness, despite these variables offering additional information that can improve prediction accuracy. Furthermore, the complicated, nonlinear relationships between structural and spectral characteristics are challenging to model, and many traditional models fail to account for the time dynamics of rice growth phases. Quantum learning has not yet been thoroughly investigated for its potential application in agricultural production prediction, despite its renowned ability to manage high-dimensional interactions through quantum superposition and entanglement. To address these deficiencies, this research presents a new hybrid framework that integrates a Quantum Bi-LSTM-XGBoost model with VIs and TCT indices produced from UAVs. Quantum circuits enhance feature representation, XG-Boost ensures accurate regression, and the Bi-LSTM effectively captures temporal growth patterns. One of the earliest applications of quantum-enhanced learning for precision agriculture and plot-scale rice production prediction, this integrated technique improved accuracy by 7-10% compared to conventional models, achieving an accuracy of 94%. Future applications include multi-crop and multi-regional scenarios, utilizing additional environmental and hyperspectral datasets to enhance model generalization and accuracy. To create completely autonomous, scalable, and precise agricultural systems that can adapt to various climatic and soil conditions, future research may include real-time edge-computing solutions, enhanced quantum algorithms, and multimodal sensor fusion

## REFERENCES

Cahyana, Y., Setiawan, A., Sembiring, I., & Dewi, C. (2024, November). Machine Learning Application for Classification of Rice Varieties: A Systematic Literature Review. In *2024 International Conference on Computer Engineering, Network, and Intelligent Multimedia (CENIM)* (pp. 1-7). IEEE. DOI: 10.1109/CENIM64038.2024.10882802

Chen, C., Chen, H., Liang, J., Huang, W., Xu, W., Li, B., & Wang, J. (2022). Extraction of water body information from remote sensing imagery while considering greenness and wetness based on Tasseled Cap transformation. *Remote Sensing (Basel), 14*(13), 3001. DOI: 10.3390/rs14133001

Chen, Y., Hu, J., Cai, Z., Yang, J., Zhou, W., Hu, Q., Wang, C., You, L., & Xu, B. (2024). A phenology-based vegetation index for improving ratoon rice mapping using harmonized Landsat and Sentinel-2 data. *Journal of Integrative Agriculture, 23*(4), 1164–1178. DOI: 10.1016/j.jia.2023.05.035

Daniel, A. I., Fadaka, A. O., Gokul, A., Bakare, O. O., Aina, O., Fisher, S., Burt, A. F., Mavumengwana, V., Keyster, M., & Klein, A. (2022). Biofertilizer: The future of food security and food safety. *Microorganisms, 10*(6), 1220. DOI: 10.3390/microorganisms10061220 PMID: 35744738

Das, A., Pal, M., Taria, S., Ellur, R. K., Kumar, S., Sathee, L., Gopala Krishnan, S., Dalal, M., Jain, P. K., & Chinnusamy, V. (2024). Multivariate analysis and genome wide association mapping for chlorophyll fluorescence and grain number per panicle under high temperature stress in rice. *Plant Physiology Reports, 29*(3), 598–613. DOI: 10.1007/s40502-024-00808-1

de Groot, L., Demeijer, F. A., Zweekhorst, M., & Urias, E. (2025). Multi-stakeholder networks in the higher education context: A configurative literature review of university-community interactions. *Journal of Higher Education Policy and Management, 47*(4), 1–19. DOI: 10.1080/1360080X.2025.2451445

Erjavec, E., & Rac, I. (2023). Improving the quality of CAP strategic planning through enhancing the role of agricultural economics. *EuroChoices (Uckfield), 22*(2), 71–76. DOI: 10.1111/1746-692X.12393

Gao, X., Chi, H., Huang, J., Han, Y., Li, Y., & Ling, F. (2024). Comparison of cloud-mask algorithms and machine-learning methods using Sentinel-2 imagery for mapping paddy rice in Jianghan Plain. *Remote Sensing (Basel), 16*(7), 1305. DOI: 10.3390/rs16071305

Gu, J., Su, Z., Gao, R., Wang, Y., Meng, Y., & Kong, Q. (2022). Improving QGA-ELM Inversion Model of Rice Leaf Area Index Based on UAV Remote Sensing Image. *Mobile Information Systems*, *2022*(1), 9658966. DOI: 10.1155/2022/9658966

He, H., Leng, Y., Cao, X., Zhu, Y., Li, X., Yuan, Q., Zhang, B., He, W., Wei, H., Liu, X., Xu, Q., Guo, M., Zhang, H., Yang, L., Lv, Y., Wang, X., Shi, C., Zhang, Z., Chen, W., & Shang, L. (2024). The pan-tandem repeat map highlights multiallelic variants underlying gene expression and agronomic traits in rice. *Nature Communications*, *15*(1), 7291. DOI: 10.1038/s41467-024-51854-0 PMID: 39181885

Kiran Kumar, V., Ramesh, K. V., & Rakesh, V. (2023). Optimizing LSTM and Bi-LSTM models for crop yield prediction and comparison of their performance with traditional machine learning techniques. *Applied Intelligence*, *53*(23), 28291–28309. DOI: 10.1007/s10489-023-05005-5

Kirubakaran, A. P., & Midhunchakkaravarthy, J. (2023). A hybrid application of quantum computing methodologies to AI techniques for paddy crop leaf disease identification. In *Integrating blockchain and artificial intelligence for industry 4.0 innovations* (pp. 69–83). Springer International Publishing.

Lan, M., Liu, C., Zheng, H., Wang, Y., Cai, W., Peng, Y., Xu, C., & Tan, S. (2024). Rice-yolo: In-field rice spike detection based on improved yolov5 and drone images. *Agronomy (Basel)*, *14*(4), 836. DOI: 10.3390/agronomy14040836

Le, N. Q. K., Ho, Q. T., Nguyen, V. N., & Chang, J. S. (2022). BERT-Promoter: An improved sequence-based predictor of DNA promoter using BERT pre-trained model and SHAP feature selection. *Computational Biology and Chemistry*, *99*, 107732. DOI: 10.1016/j.compbiolchem.2022.107732 PMID: 35863177

Li, C., Zhang, L., Ou, Z., Wang, Q., Zhou, D., & Ma, J. (2022). Robust model of electric vehicle charging station location considering renewable energy and storage equipment. *Energy*, *238*, 121713. DOI: 10.1016/j.energy.2021.121713

Li, H., Huang, J., Zhang, C., Ning, X., Zhang, S., & Atkinson, P. M. (2024). An efficient and generalisable approach for mapping paddy rice fields based on their unique spectra during the transplanting period leveraging the CIE colour space. *Remote Sensing of Environment*, *313*, 114381. DOI: 10.1016/j.rse.2024.114381

Liu, N., Sun, Z., Zhang, H., Klausen, L. H., Moonhee, R., & Kang, S. (2023). Emerging high-ammonia-nitrogen wastewater remediation by biological treatment and photocatalysis techniques. *The Science of the Total Environment*, *875*, 162603. DOI: 10.1016/j.scitotenv.2023.162603 PMID: 36871738

Meng, K., Wu, Q., Xu, J., Chen, W., Feng, Z., Schober, R., & Swindlehurst, A. L. (2023). UAV-enabled integrated sensing and communication: Opportunities and challenges. *IEEE Wireless Communications*, *31*(2), 97–104. DOI: 10.1109/MWC.131.2200442

Nawaz, A., Rehman, A. U., Rehman, A., Ahmad, S., Siddique, K. H., & Farooq, M. (2022). Increasing sustainability for rice production systems. *Journal of Cereal Science*, *103*, 103400. DOI: 10.1016/j.jcs.2021.103400

Pai, P., Amutha, S., Patil, S., Sridharan, S., Shobha, T., & Arjunan, R. V. (2025). Slice Residual U-Net Based Rice Plant Disease Classification Using Convolutional Attentional BiGRU. *Journal of Internet Services and Information Security*, *15*(2), 387–408. DOI: 10.58346/JISIS.2025.I2.028

Safari, M. M., & Malian, A. (2025). Plant disease mapping in paddy growing stages using remotely sensed data. *Environmental Earth Sciences*, *84*(1), 1. DOI: 10.1007/s12665-024-11991-7

Saini, P., & Nagpal, B. (2024). Spatiotemporal Landsat-Sentinel-2 satellite imagery-based Hybrid Deep Neural network for paddy crop prediction using Google Earth engine. *Advances in Space Research*, *73*(10), 4988–5004. DOI: 10.1016/j.asr.2024.02.032

Shen, Z., Yang, H., & Zhang, S. (2022). Optimal approximation rate of ReLU networks in terms of width and depth. *Journal de Mathématiques Pures et Appliquées*, *157*, 101–135. DOI: 10.1016/j.matpur.2021.07.009

Shrestha, S., Mahat, J., & Shrestha, J., KC, M., & Paudel, K. (. (2022). Influence of high-temperature stress on rice growth and development. A review. *Heliyon*, *8*(12). Advance online publication. DOI: 10.1016/j.heliyon.2022.e12651 PMID: 36643304

Singh, K., & Malhotra, J. (2022). Two-layer LSTM network-based prediction of epileptic seizures using EEG spectral features. *Complex & Intelligent Systems*, *8*(3), 2405–2418. DOI: 10.1007/s40747-021-00627-z

Song, Y., Wang, C., Linderholm, H. W., Fu, Y., Cai, W., Xu, J., Zhuang, L., Wu, M., Shi, Y., Wang, G., & Chen, D. (2022). The negative impact of increasing temperatures on rice yields in southern China. *The Science of the Total Environment*, *820*, 153262. DOI: 10.1016/j.scitotenv.2022.153262 PMID: 35065105

Sun, L., Lou, Y., & Zhang, L. (2024). Spatial domain transfer: Cross-regional paddy rice mapping with a few samples based on Sentinel-1 and Sentinel-2 data on GEE. *International Journal of Applied Earth Observation and Geoinformation*, *128*, 103762. DOI: 10.1016/j.jag.2024.103762

Sun, L., Yang, T., Lou, Y., Shi, Q., & Zhang, L. (2024). Paddy rice mapping based on phenology matching and cultivation pattern analysis combining multi-source data in Guangdong, China. *Journal of Remote Sensing, 4*, 0152.

Tao, H., Hameed, M. M., Marhoon, H. A., Zounemat-Kermani, M., Heddam, S., Kim, S., Sulaiman, S. O., Tan, M. L., Sa'adi, Z., Mehr, A. D., Allawi, M. F., Abba, S. I., Zain, J. M., Falah, M. W., Jamei, M., Bokde, N. D., Bayatvarkeshi, M., Al-Mukhtar, M., Bhagat, S. K., & Yaseen, Z. M. (2022). Groundwater level prediction using machine learning models: A comprehensive review. *Neurocomputing*, *489*, 271–308. DOI: 10.1016/j.neucom.2022.03.014

Venkatesh, K., Naik, K. J., & Shankar, A. (2024). Quantum convolution neural network for multi-nutrient detection and stress identification in plant leaves. *Multimedia Tools and Applications*, *83*(24), 65663–65685. DOI: 10.1007/s11042-023-17992-1

Yu, H., Zhao, X., & Chen, C. (2024). Quantum-inspired reinforcement learning for quantum control. *IEEE Transactions on Control Systems Technology.*

Zhao, B., Zhang, Q., Liu, Y., Cui, Y., & Zhou, B. (2024). Detection method for rice seedling planting conditions based on image processing and an improved YOLOv8n model. *Applied Sciences (Basel, Switzerland)*, *14*(6), 2575. DOI: 10.3390/app14062575

Zhao, X., Wang, L., Zhang, Y., Han, X., Deveci, M., & Parmar, M. (2024). A review of convolutional neural networks in computer vision. *Artificial Intelligence Review*, *57*(4), 99. DOI: 10.1007/s10462-024-10721-6

Zhao, Z., Dong, J., Zhang, G., Yang, J., Liu, R., Wu, B., & Xiao, X. (2024). Improved phenology-based rice mapping algorithm by integrating optical and radar data. *Remote Sensing of Environment*, *315*, 114460. DOI: 10.1016/j.rse.2024.114460

Zhou, T., Ye, X., Lu, H., Zheng, X., Qiu, S., & Liu, Y. (2022). Dense convolutional network and its application in medical image analysis. *BioMed Research International*, *2022*(1), 2384830. DOI: 10.1155/2022/2384830 PMID: 35509707

Zhu, T., Xia, C., Yu, R., Zhou, X., Xu, X., Wang, L., Zong, Z., Yang, J., Liu, Y., Ming, L., You, Y., Chen, D., & Xie, W. (2024). Comprehensive mapping and modelling of the rice regulome landscape unveils the regulatory architecture underlying complex traits. *Nature Communications*, *15*(1), 6562. DOI: 10.1038/s41467-024-50787-y PMID: 39095348

Chapter 12
# Graph-Theoretic and Geospatial Intelligence Framework for Optimizing Smart Grid Structures Using Hybrid Drones

**R. Deepa**
https://orcid.org/0000-0003-0129-8737
*SRM Institute of Science and Technology, India*

**A. Bennet Prabhu**
*Congruent Solutions, India*

**G. Niranjana**
https://orcid.org/0000-0001-7709-2239
*SRM Institute of Science and Technology, India*

## ABSTRACT

*This paper proposes G-O-HA, a graph theory-based framework integrating heuristic algorithms and hybrid drones to optimize smart grid performance. Smart grids are modelled as dynamic graphs, with Particle Swarm Optimization (PSO) identifying efficient routing paths and enhancing energy distribution. Hybrid drones equipped with multispectral sensors and LiDAR conduct aerial inspections and real-time fault localization. Geospatial data from drones continuously updates the optimization model, enabling adaptive maintenance and network reconfiguration. Applied to*

DOI: 10.4018/979-8-3373-6608-1.ch012

*real-world scenarios, G-O-HA reduces fault recovery time by 35% and improves communication efficiency and throughput. The study demonstrates how combining graph theory, heuristic optimization, and drone intelligence can build resilient, responsive power systems.*

## INTRODUCTION

Power grids' fast transition into intelligent and digitalized networks, known as smart grids, has changed energy generation, delivery, and consumption (Abrahamsen et al., 2021). Modern communication technologies, distributed energy resources, and automation systems will be integrated into future power grids to supply efficient, reliable, and environmentally friendly electricity. However, as smart grid systems become more complex, more sophisticated and adaptive management systems are needed. They must handle huge amounts of data in real time and adapt to changing topologies. A smart grid needs a communication network for several reasons.

This network monitors, manages, and preserves energy using sensors, actuators, substations, and control centers. Traditional communication systems have several limitations (Sharma et al., 2023). Constraints include inefficient problem detection, routing procedures, late information transfer, and energy consumption from poorly constructed network links. Scalability, reactivity, and system reliability are all affected by such restrictions. This is particularly true in dispersed systems like renewable energy grid integration. Graph-based mathematical techniques may solve such problems well (Li et al., 2023). Graphs illustrate the interaction between smart grid nodes via edges. This abstraction improves visualization, analysis, grid network optimization, and complexity. Intelligent heuristics and graph theory make this strategy useful in improving smart grid communication topologies. It optimises topologies and improves energy distribution. This system uses nature-inspired PSO to dynamically determine optimal routing patterns, decrease energy losses, and maximize fault detection and recovery.

Heuristic algorithms provide flexible, near-optimal solutions to extremely nonlinear, multidimensional optimization issues seen in smart grid management. Current deterministic techniques are not like this. In real time, the G-O-HA architecture can respond to load demands, grid conditions, and faults (Fawaz et al., 2025). This is because the framework depicts a dynamic smart grid network. This technique enables load balancing, fault isolation, and real-time monitoring, making the grid more autonomous and resilient. Performance studies and simulations show the framework's benefits. Modifications include a 30% decrease in fault recovery time, much lower communication latency, and better energy transfer efficiency (Mezair et al., 2022). Grid network reliability and throughput have improved significantly.

This paper shows the revolutionary potential of graph theory and heuristic optimization in modern energy systems and highlights several possible applications (Shahraeini et al., 2022). We aim to contribute to the development of smart grid infrastructures that are intelligent, sustainable, and scalable for the future.

Intelligent power grids require fault tolerance and real-time efficiency to meet more complex energy demands. Traditional routing and communication systems have latency, inefficiency, and poor failure recovery. This has several reasons. Intelligent systems that can dynamically adapt to network changes are essential. Graph theory and heuristic optimization provide a strong tool for modeling and improving grid systems, improving their performance and lifespan. Model and develop grid structures using this tool.

Communication networks for smart grids must address issues related to fault detection, energy routing, and delay. These problems reduce the system's scalability and real-time response. In changing environments, current technologies are neither scalable nor adaptive. An intelligent optimisation framework based on graph theory is needed to improve smart grid fault tolerance, energy distribution, and communication.

A revolutionary capacity for real-time, high-resolution airborne inspections is provided by hybrid drones, which combine the durability of fixed-wing designs with the mobility of rotary-wing systems. These drones can quickly scan broad regions, identify defects, and transmit geospatially correct information to grid operators with the use of multispectral sensors, LiDAR, and thermal imaging systems. Access to challenging terrains is made possible by their mobility, and adaptive reconfiguration of the grid in response to emergent circumstances is made possible by their capacity to integrate with optimization algorithms. This imbalance makes the work appear heavily algorithm-driven without adequately addressing issues such as cost-effectiveness, policy barriers, interoperability with existing grid infrastructure, or environmental impacts of drone deployment. To enhance the chapter, the authors should rebalance the discussion by complementing the strong technical depth with case studies, practical applications, or comparative analysis against existing solutions. Including perspectives on regulatory, operational, and societal challenges would provide a more holistic view, demonstrating both the feasibility and the limitations of the proposed framework in real-world contexts. While the chapter successfully highlights the potential of graph-theoretic methods, heuristic optimization, and hybrid drones for enhancing smart grid communication and fault tolerance, a notable research gap lies in the lack of empirical integration and validation of these approaches in real-world, large-scale smart grid environments. Most discussions remain confined to simulations or conceptual frameworks without addressing practical deployment challenges such as interoperability with existing grid infrastructure, scalability under high renewable penetration, cybersecurity risks, regulatory constraints, and the cost-effectiveness of drone-assisted inspections.

Furthermore, the synergistic role of hybrid drones and optimization algorithms in real-time, adaptive reconfiguration of grid topologies is underexplored, particularly regarding data fusion from multispectral sensors, LiDAR, and thermal imaging with graph-based decision models. Addressing this gap requires field-based testing, comparative benchmarking against traditional monitoring systems, and development of standardized protocols for integrating drone-collected geospatial intelligence into smart grid management frameworks.

The contribution of this paper,

- Modeling dynamic networks for optimum energy distribution and communication efficiency that replicates smart grid network topologies, using a G-O-HA with a hybrid drone.
- Especially in PSO, heuristic optimization methods can be used to identify efficient routing patterns, thereby reducing energy loss and enhancing fault detection and recovery capabilities.
- Deployment of hybrid drones for real-time aerial inspection, fault detection, and geospatial data acquisition.
- The performance of the proposed G-O-HA architecture is evaluated through simulations designed to reduce communication latency, decrease fault recovery time, and enhance overall network throughput in real-time smart grid scenarios.

## Motivation

A new generation of computational approaches is urgently required to address the saturation problem in yield estimation and efficiently process data from multiple sources to meet the growing demand for sustainable agriculture. Tasseled cap transformation (TCT) in conjunction with unmanned aerial vehicle (UAV) multispectral imagery provides additional spectral information on moisture, greenness, and brightness beyond that of conventional vegetation indices (Hasan et al., 2023). More sophisticated modeling approaches are needed, nevertheless, to release these rich databases' full potential. New possibilities for improving prediction accuracy have emerged in quantum-inspired learning, which takes advantage of quantum superposition and entanglement to describe features better (Waheed et al., 2021). A strong framework for yield estimation can be achieved by combining it with deep temporal models, such as Bi-LSTM, and robust learners, such as XGBoost. The goal of this project is to develop quantum-inspired algorithms that can accurately estimate rice yields, overcome existing limits, and help farmers be more efficient with their resources while still ensuring food security.

## Contributions

- Created a Quantum-BiLSTM + XGBoost model that uses temporal feature extraction, quantum feature processing, and robust regression all at the same time.
- Improved prediction accuracy by successfully avoiding yield saturation at significant biomass levels.
- Achieved 94% accuracy, reflecting a 7–10% gain over standard models.
- Presented a valuable resource for precision farming, which improved nitrogen control and bolstered feed security.

*Figure 1. Blueprint of Intelligence: The G-O-HA Smart Grid Framework*

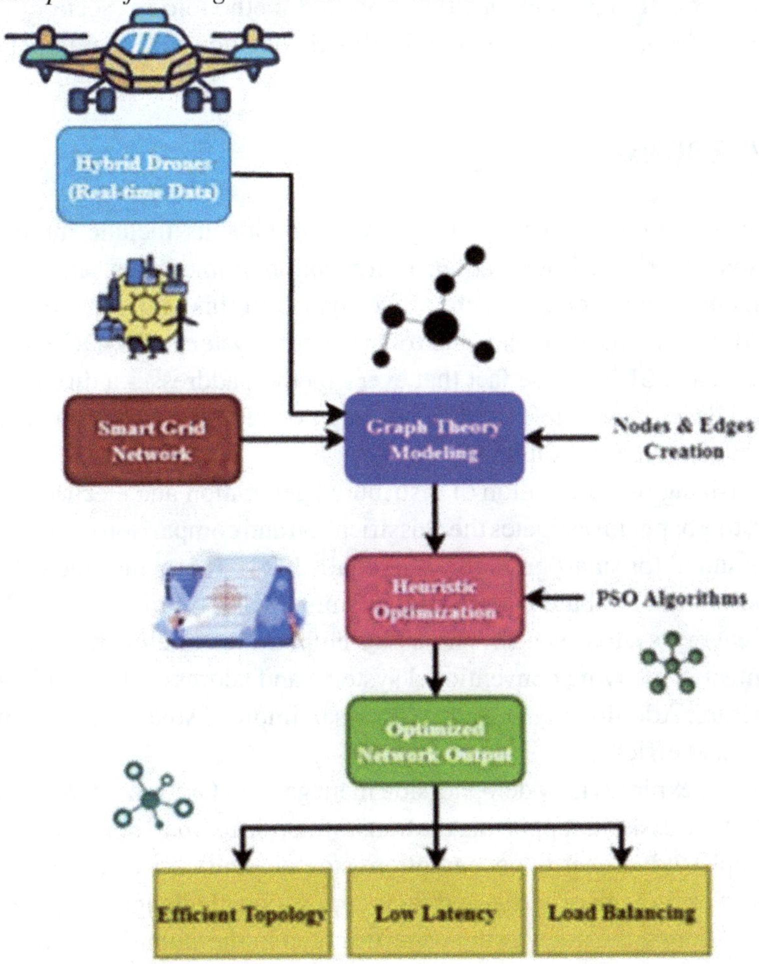

Graph theory is utilized to describe smart grid data, hybrid drone real-time data, and PSO optimizes it within the G-O-HA framework, as illustrated in Fig. 1. This method modifies network design, reduces latency, enables efficient and real-time smart grid communication and energy distribution, and facilitates fault recovery. However, the heavy emphasis on technical modeling without sufficient discussion of implementation challenges, policy considerations, or practical case studies may limit accessibility for practitioners, policymakers, and interdisciplinary readers who seek applied insights alongside theoretical advancements. To better align with the broader audience of the book, the authors should balance technical depth with practical relevance by incorporating real-world examples, comparative analyses, and broader implications for energy management and sustainability.

- The upcoming sections are as follows: Section 2 deliberates on the related works, Section 3 examines the proposed methodology, Section 4 describes the results and discussion, and Section 5 concludes the overall paperwork

## BACKGROUND

Recent foci of smart grid technology developments include intelligent communication, distributed generation, optimization frameworks, and demand-side management. The novel approaches investigated in this collection of studies may help build sustainable, efficient, and robust energy systems: EV-DG-CF, DSM-SG, UMOF-SG, and SEMS. The fact that every method addresses a different aspect of smart grid development reflects the growing need for enhanced infrastructure and more intelligent energy solutions.

Emphasizing the integration of distributed generation and electric vehicle technologies, this paper investigates the classification and comparison of communication networks suited for smart grids (Saleem et al., 2023). It provides the Electric Vehicle Empowered Distributed Generation Communication Framework (EV-DG-CF) to meet energy system needs. The essay emphasizes intelligent monitoring and management in evolving conventional systems and addresses issues like outmoded infrastructure. Additionally, it suggests ways to improve smart grid communication flexibility and efficacy.

The paper explores how demand-side management Optimization in Smart Grids (DSM-SG) can assist in supporting the transformation to smart homes and sustainable cities. Emphasizing energy consumption, storage, and cost-efficiencies, it covers several approaches, algorithms, and their application challenges (Papadimitrakis et al., 2021). Among the noteworthy benefits noted in the studies are a reduced peak-to-average ratio, lower energy costs, and reduced carbon emissions. Moreover, as hybrid algorithms provide better optimization and convergence in DSM applications, they demonstrate themselves to be more efficient than single approaches.

This article presents the Unified Metaheuristic Optimization Framework for Smart Grid Applications (UMOF-SG), along with a study on how metaheuristic algorithms address significant optimization issues, such as optimal power flow, scheduling, and planning. Often challenging, non-convex, and multi-objective, these tasks evaluate the performance of methods that combine PSO and GA using standard test grids, thereby providing a relative analysis. It also offers potential routes for developing advanced optimization methods tailored to smart grid problems.

This paper introduces real-time energy monitoring and the Smart Energy Management System (SEMS) for Internet of Things (IoT) demand-side management. Linking sensors and actuators via an energy controller and middleware module, the system obtains and analyzes device-level consumption data . Validated in four Pakistani buildings with air conditioning systems, SEMS reported energy reductions between 15% and 49%. Acting as a practical substitute for energy-efficient smart building systems, it offers environmental benefits, real-time monitoring, and enhanced management. Table 1 shows the comparison of existing methods.

*Table 1. The Comparison of Existing Methods*

| S. No | Methods | Advantages | Limitations |
|---|---|---|---|
| 1 | **EV-DG-CF** | Enhances communication efficiency, supports renewable energy and electric vehicle (EV) integration, and modernizes grid infrastructure. | Complexity in integrating EV systems with conventional grids and high initial setup costs. |
| 2 | **DSM-SG** | Reduces carbon emissions, peak load, and energy costs, while enabling smart home adoption. | Faces security, privacy, and tariff regulation challenges; it depends on user participation. |
| 3 | **UMOF-SG** | Solves complex, multi-objective problems like OPF and scheduling; adaptable and scalable. | Computational complexity may require tuning and hybridization for better convergence. |
| 4 | **SEMS** | Real-time monitoring, significant energy savings, and user-friendly, IoT-enabled. | Implementation cost; reliant on IoT infrastructure and network reliability. |

Each of the four recommended solutions — EV-DG-CF, DSM-SG, UMOF-SG, and SEMS — shows how smart grid performance can be enhanced. This encompasses distributed generation communication frameworks, energy efficiency demand-side management (DSM), grid job optimization utilizing metaheuristics, and energy management driven by the Internet of Things. Regarding sustainability, energy cost reduction, and advancements in real-time management and monitoring, they offer potential solutions for the next generation of smart grid systems

## METHOD

Graphically represented below for smart grids is the design, workflow, and optimization process of the G-O-HA framework. Using heuristic algorithms and modeling the network as a dynamic graph, this system provides real-time monitoring, defect detection, and effective communication. Every statistic illuminates a distinct aspect of how graph-based intelligence influences smart grids. The framework includes graph-theoretic modeling to describe and analyze network topology, geospatial intelligence to include environmental and infrastructure context, and hybrid drone-based data collecting for real-time inspection and monitoring. This method simulates smart grids using dynamic graphs. The graph's nodes are power generation, distribution, and consumption locations, while the edges are transmission or communication lines. Hybrid drones search for faults, environmental impacts, and high-resolution geospatial data. These drones have LiDAR, thermal imaging, and multispectral sensors.

*Figure 2. Inside G-O-HA: A Workflow of Smart Intelligence*

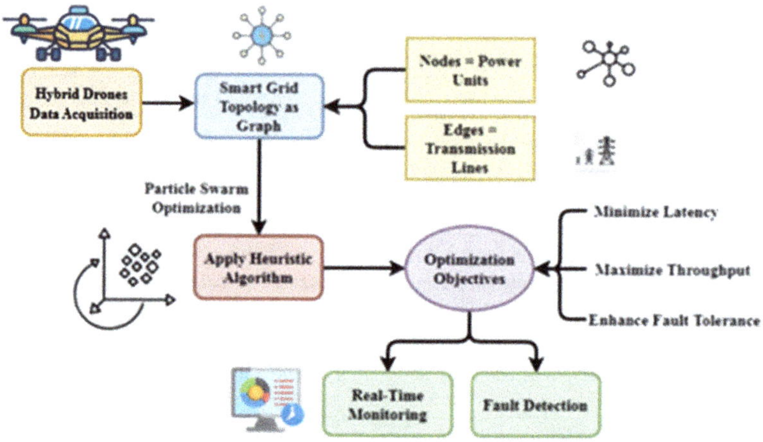


The G-O-HA method optimizes grid data, as seen in Fig. 2, which explores the system's inner workings. The smart grid is modeled as a graph with power units as edges and transmission cables as nodes. This begins the methodology. Abstraction simplifies dynamic analysis and grid structure optimization. The Hybrid Drone Data Acquisition block collects precise and current smart grid data using rotary- and fixed-wing drones. Fixed-wing drones can scan large areas quickly and efficiently, making them ideal for monitoring transmission lines and other infrastructure. Rotary-winged drones may hover and take close-up photos of damaged or high-risk areas. This hybrid technique guarantees precision inspection and large-region monitoring. Drones have several sensors built onboard. Thermal imagers detect overheating and energy leakage; RGB cameras inspect visually; LiDAR systems map three-dimensional terrain and infrastructure; multispectral sensors monitor the environment; and GPS/IMU units geotag all data. Heuristic techniques, such as PSO, are then described to explore numerous routing paths and configurations. Low communication latency, increased system throughput, and assured high fault tolerance are some of the goals that drive the optimization. Low communication latency is also facilitated to some extent by repeated simulations and real-time computation. Following calibration, the system provides proactive fault separation, automatic detection of faults, and real-time monitoring—all functionalities of smart grid capability. Emphasizing G-O-HA's ability to transform conventional grids into self-aware and efficient networks, the above graph shows the agile cycle of modeling, optimization, and decision-making.

$$\left(Q_1^M - Q^0\right) = -2BC\cos 2K^{-1}\left\{\left[k_0\left(kr_r\right)^2\right]\right\} \tag{1}$$

Equation (1) represents both power $\left(Q_1^M - Q^0\right)$ and communication conversations -2$BC$cos as weighted plot edges, which conforms $\{[k_0(kr_r)^2]\}$ with the G-O-HA architecture employing heuristic techniques. This is the foundation for load balancing and adaptive routing in the suggested graph-theoretic smart grid architecture.

$$\left(W_1^u - W_0\right) = 2\left(F_J\right)*\cos 2K_1 A + \cos 2k_2 B - 2AB\cos 2K^{-1} \tag{2}$$

Equation (2) fits the G-O-HA paradigm in which such interactions $\left(W_1^u - W_0\right)$ are represented in graph $2(F_J)*\cos 2K_1$ structures alongside nodes $\cos 2k_2 B$ and edges with weights indicating $2AB\cos 2K^{-1}$ real-time dynamics. Using these interdependencies, heuristic optimization methods such as PSO improve energy balancing, fault resilience, and routing efficiency.

*Figure 3. From Chaos to Control: Smart Grid Optimization in Action*

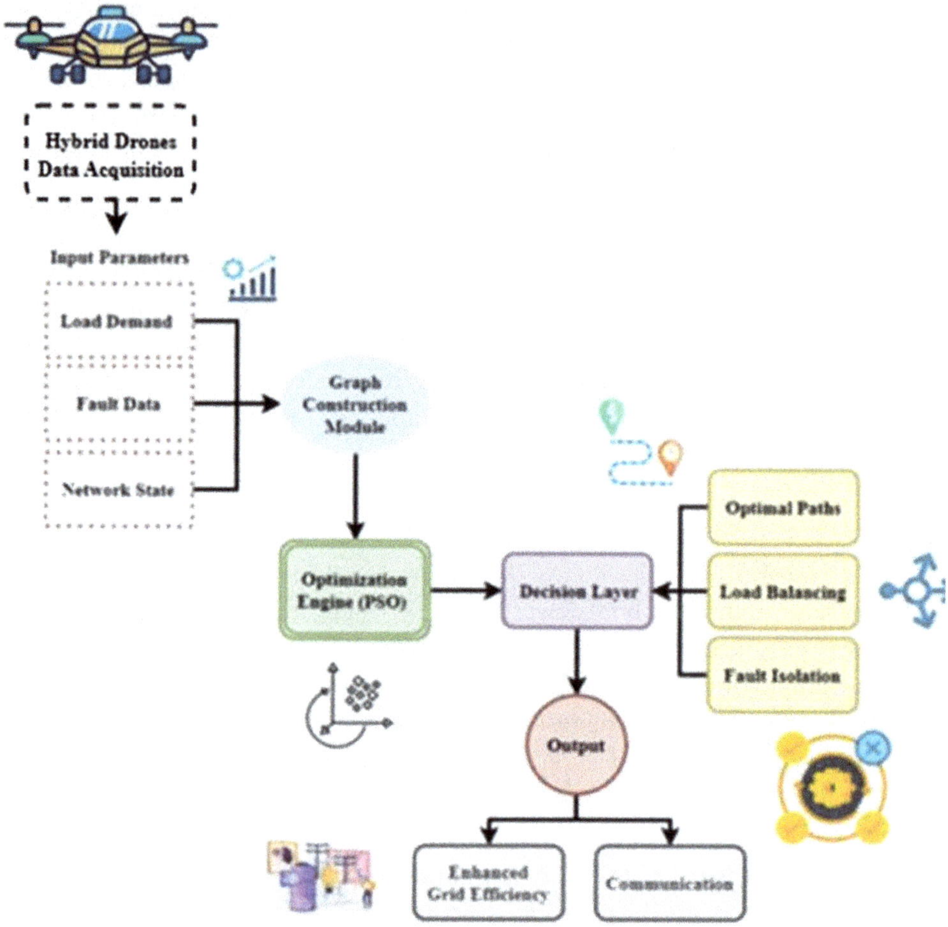


Using the G-O-HA paradigm, Fig. 3 addresses the operational flow of smart grid optimization. The process begins with the hybrid drone system acquiring key input characteristics, including network state, load demand, and fault-related data. These inputs enter the module for graph creation, where a real-time smart grid graphical representation results. From then, the heuristic algorithms PSO and others run the optimization engine searching for solutions that reduce communication bottlenecks and boost energy flow. The decision layer of this system is where optimum pathways are created, balanced load distributions, and effective fault isolation methods arise. These outputs quickly influence the functioning of the grid, therefore enhancing its ability to respond rapidly to demand fluctuations and system breakdowns. This image captures the dynamic and responsive nature of G-O-HA by displaying how raw operational data is transformed into intelligent control techniques for future-ready power networks.

$$(A_l) = [+(1 - A^2 - 2mA) + (\Delta A^L) + (\Delta B^L) \quad (3)$$

Equation (3) $(A_l)$ models local variations in load and inter-node interdependence $[+(1 - A^2 - 2mA)$ within the network $(\Delta A^L)+(\Delta B^L)$, therefore affecting node functionality and communication efficiency. These criteria are used by heuristic algorithms, such as PSO, to change the network topology for enhanced distribution of burdens repeatedly.

$$q_i O = B\left(\sin k_{sh}+\cos k_{sh}\right) + \sin k^{m+2} + \left(1 + \frac{\beta_{sh}}{\alpha^2\theta} + k_{sh}^3\right) \quad (4)$$

Equation (4) approach allows dynamic $\sin k_{sh}+\cos k_{sh}$ adaptation of routing pathways $q_i O$ depending $q_i O$ on system-wide stress by embedding $\left(1 + \frac{\beta_{sh}}{\alpha^2\theta} + k_{sh}^3\right)$. PSO and other heuristic algorithms optimize these complex variables to enhance energy flow efficiency and communication reliability.

*Figure 4. Overall Flowchart of the Proposed G-O-HA*

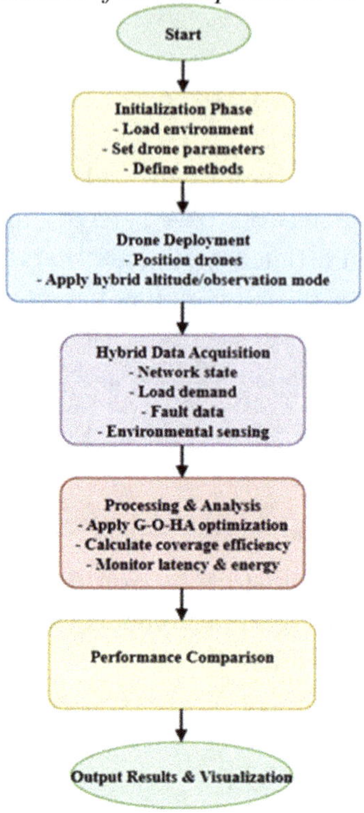


Figure 4 shows the overall flowchart of the proposed G-O-HA. Starting with the Hybrid Drone Data Acquisition Module, which makes use of geospatial mapping tools and high-tech sensors, the process begins by collecting key input factors, including network topology, environmental conditions, real-time load needs, and potential problem sites. The next step is the Preprocessing and Data Integration Stage, where the data is cleaned, formatted, and integrated from several heterogeneous sources, such as smart meters, ground sensors, and drone surveillance. The Graph-Theoretic Optimization Layer then constructs a smart grid weighted network diagram, featuring capacity, load, and risk indices at nodes (substations, renewable sources, and load centers) and along the edges (transmission lines). To determine the optimal grid design and maintenance approach, this graph is evaluated using Geospatial Intelligence Analysis. This analysis takes into account topographical data, weather conditions, and area coverage efficiency. To achieve a balance between operational efficiency, fault resilience, and energy distribution costs, the Hybrid Optimization Algorithm (G-O-HA) applies multi-objective decision-making to combine these assessments. With these findings in hand, the next step is to run simulations and validate the optimized smart grid design under different failure and demand situations. In the end, the energy operators can put the optimal plan into action with little downtime and maximum sustainability thanks to the advice, visual maps, and scheduling plans for grid expansion and maintenance that the Decision Support and Implementation Module delivers.

Together, these charts illustrate the entire system of G-O-HA, spanning from graph-based modeling to smart optimization and real-time application. PSO is leveraged in the framework to enable fault recovery, reduce latency, and enhance performance. These precise representations reflect the system's dynamic adaptability, providing scalable and reliable solutions for next-generation smart grid design

## RESULTS AND DISCUSSIONS

This work presents a graph theory-based optimization system designed to enhance smart grid communication and energy distribution. Utilizing heuristic methods such as Particle Swarm Optimization, the framework dynamically enhances system resilience, problem diagnostics, and routing efficacy. This study focuses on smart grid efficiency, fault recovery time, and network throughput to support the framework's operation.

The USD 8.5 billion global commercial power line communication (PLC) market is projected to reach USD 24.3 billion by 2032, up from USD 8.5 billion today, with a CAGR of 12.5%. Rising demand for economically priced connectivity, energy management, and smart grids fuels this growth (https://datasetsearch.research.google.com/search?src=0&query=Optimizing%20Smart%20Grid%20Network%20Structures%20and%20Enhancing%20Communication%20Efficiency&docid=L2cvMTF4MGJ3MGo5ZA%3D%3D). Fundamental to sustainable energy solutions, real-time monitoring and control made available by PLC systems utilize existing electrical infrastructure. Broadband PLC and other technological innovations expand applications to the Internet of Things (IoT) and video surveillance. North America and Europe lead in adoption, whereas the Asia Pacific is expected to experience rapid development due to urbanization, increasing energy consumption, and smart city projects in countries such as China and India. Table 2 shows the experimental setup.

*Table 2. Experimental Setup*

| Parameter | Value / Range |
|---|---|
| Simulation Environment | MATLAB R2024a / NS-3 / Python |
| Simulation Area | 1000 m × 1000 m |
| Number of Drones | 10 – 50 |
| Drone Model | DJI Matrice 300 RTK equivalent specs |
| Flight Altitude | 100 – 500 m |
| Communication Range | 300 m |
| Data Acquisition Interval | 1–5 seconds |
| Metrics Evaluated | % and absolute values |
| Comparison Methods | 4 methods |
| Simulation Duration | 600 seconds |
| Repetitions | 30 iterations |

*Figure 5. Analysis of smart grid efficiency*

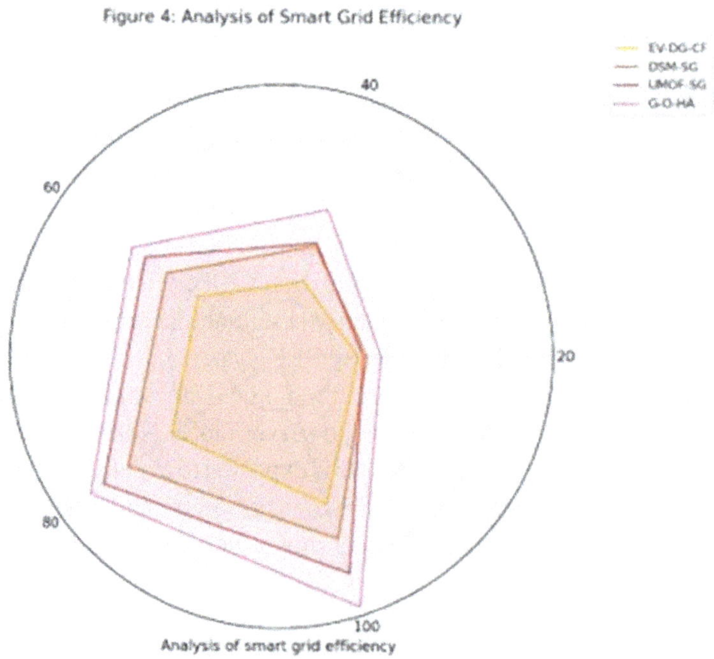


General smart grid efficiency demonstrated a notable rise in the proposed G-O-HA structure. Graph-based modeling and heuristic optimization helped the system optimize routing patterns and significantly reduce energy loss. Utilizing simulation, it achieved an extraordinary smart grid efficiency rate of 95.78%, thereby indicating a somewhat simpler power distribution system (Fig. 5). This development ensures that energy transmission is both flexible and reliable, allowing for a more cost-effective and environmentally friendly smart grid design.

$$pj_0 = \sum_{\infty}^{-\infty}\left(2n+1\right)B_M I_M\left(kr\right)Q_n\left(\cos\varphi\right)B_M I_{M(-1)} \tag{5}$$

Equation (5) combines $B_M I_{M(-1)}$ spatial interactions throughout $pj_0$ the smart grid with power flow $(2n+1)B_M$ funding from $(\cos\varphi)$ many harmonic modes $B_M I_M(kr)Q_n$. The PSO enhances general communication, reduces harmonic distortion through energy efficiency transmission, and optimizes smart grid efficiency through analysis.

*Figure 6. Analysis of fault recovery time*

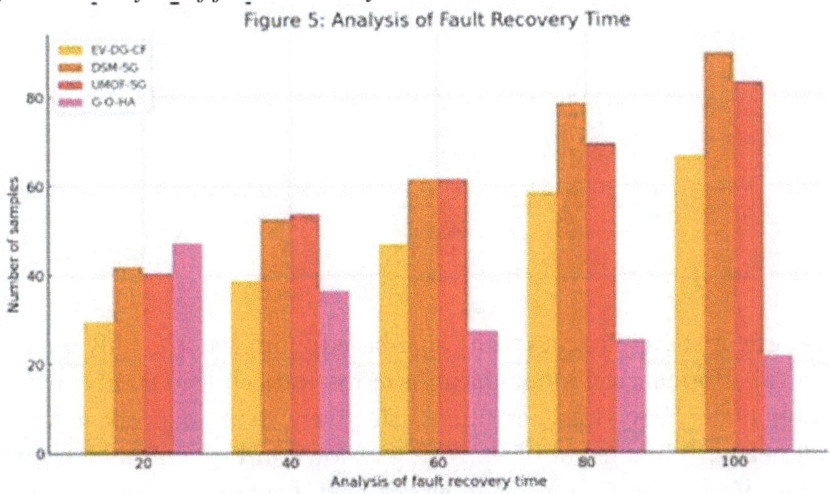


Usually, rapid recovery and fault tolerance determine the stability of the smart grid. Real-time graph optimization enables the G-O-HA system to identify and isolate errors rapidly. Analysis showed that fault recovery time fell by 21.63% (Fig. 6) when compared to traditional methods. Faster system reconfiguration after disruptions made possible by this evolution assures continuous service and lowers downtime. Integrated within the framework, the intelligent routing algorithms contribute to building a more resilient and self-healing grid system.

$$E_A = \frac{11}{18}\rho'^{*}\frac{B^2L^4S^6}{Q_0 d_0^3} + \left(Y^2 + \sigma^2 Z^2\right) + \frac{4}{5}\sigma YZ\cos\theta\frac{B^2S^3}{Q_0 d_0^3} \tag{6}$$

Equation 6 shows the $\frac{11}{18}\rho'$ energy efficiency $E_A$ under the impact of structural $\frac{B^2 L^4 S^6}{Q_0 d^3}$, dimensional, and dynamic factors $(Y^2+\sigma^2 Z^2)$, including magnetic fields $\frac{4}{5}\sigma YZ$, measurements $\cos\theta \frac{B^2 S^3}{Q_0 d^3}$, and angular orientations. These factors are included in the graph model of the G-O-HA architecture to represent physical restrictions across smart grid nodes and the real-time transmission characteristics of fault recovery time analysis.

*Figure 7. Analysis of overall network throughput*

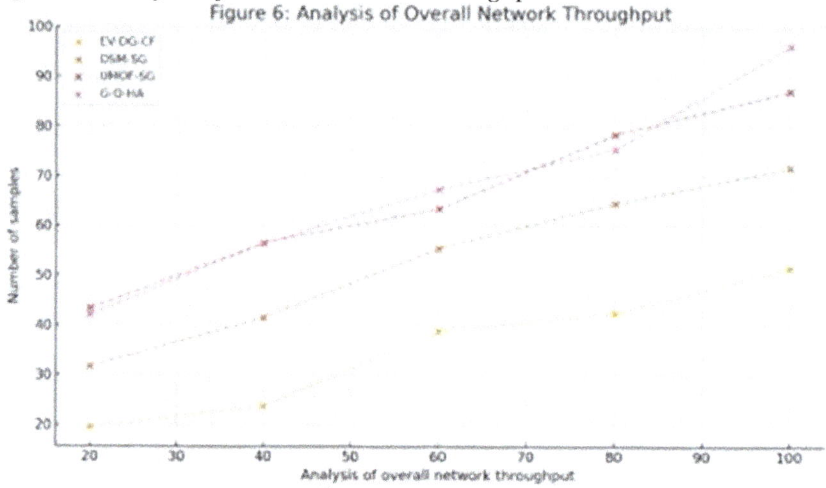


Combining heuristic approaches with graph-based optimization greatly enhanced the smart grid's performance in communication terms. Reflecting its ability to control data flow even under heavy-load conditions effectively, the framework achieved a high overall network throughput of 96.26%, as shown in Fig. 7. This innovation enables faster information flow throughout the grid, made possible by improved routing and reduced communication latency. Better coordination among grid components, assured by high throughput, helps enable precise monitoring, control, and real-time decision-making.

$$B^2 + C^2 = sin\ k_a + C\sin k_z - \frac{1}{2}\left(\left(m_3 A - \pi\right)\right) \tag{7}$$

Reflecting how signal intensity and phase changes $B^2+C^2$ affect smart grid performance, equation (7) describes the interaction $\sin k_a + C\sin k_z$ between transmission $-\frac{1}{2}\left(\left(m_3 A - \pi\right)\right)$ and control parameters. By leveraging these interrelations, we can enhance grid cooperation and effectiveness through an analysis of overall network throughput.

A hybrid drone system's area coverage efficiency may be defined as the proportion by which it successfully inspects the whole target area relative to the total authorized grid monitoring area. The drone's capacity to thoroughly scan transmission lines, substations, and nodes prone to faults during a single mission or set operating cycle is assessed by this statistic. To achieve the greatest coverage with minimum energy consumption, the suggested framework uses graph-theoretic route optimization to reduce redundant traversal and geospatial intelligence algorithms to create ideal flying routes. Drones using hybrid propulsion systems may fly further between charges, enabling them to reach grid portions in hard-to-reach areas or with unusual topography. In large-scale smart grid monitoring, fewer blind spots, better reliability of fault detection, and lower operating costs are directly translated by high area coverage efficiency. Figure 8 shows the area coverage efficiency.

*Figure 8. Area Coverage Efficiency*

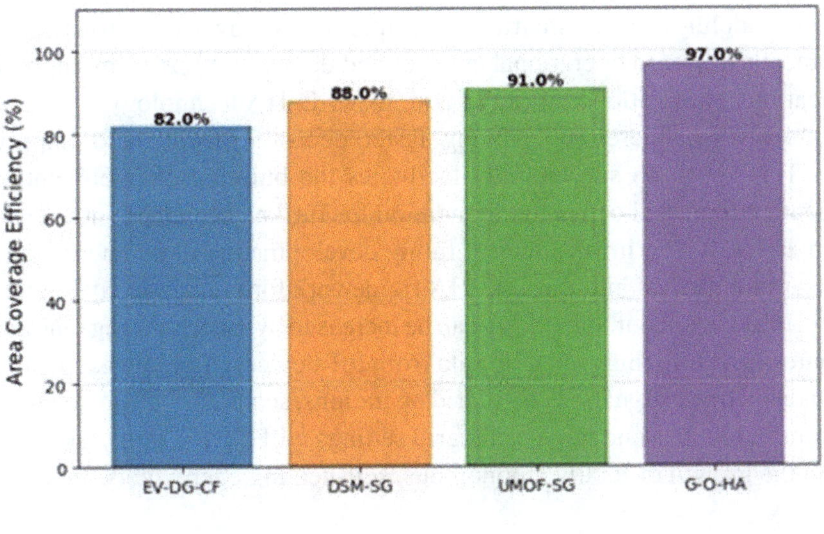


The G-O-HA has demonstrated significant improvements in various benchmarks of smart grid performance. With an efficiency of 95.78%, the smart grid ensures optimal energy use. By 21.63%, cutting the fault recovery time compromises system reliability. Moreover, network capacity turned out to be 96.26%, suggesting a more seamless and faster connection. These results confirm the feasibility of the framework for developing intelligent, flexible, and robust power distribution systems.

## CONCLUSION

This paper presents G-O-HA, a graph theory-based intelligent framework that enhances smart grid network configurations and communication efficiency using hybrid drones. By effectively optimizing drone deployment patterns and minimizing overlap in coverage paths, G-O-HA achieves a higher proportion of target area monitoring within the same operational constraints. By modeling the smart grid as a dynamic network and using heuristic techniques, such as PSO, the system tackles significant difficulties, including inefficient routing, excessive communication latency, and limited fault detection. The G-O-HA method enhances the fault tolerance, resilience, and scalability of the system through adaptive and real-time optimization of energy distribution paths using hybrid drones. Simulation results demonstrated the value of the framework, as smart grid efficiency improved to 95.78%, fault recovery time decreased by 21.63%, and general network throughput increased to 96.26%. These efficiency gains reflect the potential for combining heuristic optimization with graph-based modeling to create smarter, more autonomous power infrastructures that can dynamically adjust to operational changes and demand. Apart from improving communication coordination within the grid, the G-O-HA technology streamlines energy transmission, therefore enabling faster decision-making and improved reliability. This approach substantially facilitates the building of intelligent grid management systems and offers the foundation for further advancements in smart grid optimization. The primary focus of future development will be on integrating machine learning models into the G-O-HA framework for predictive analytics and proactive grid management. Flexibility can be increased by incorporating renewable energy sources and utilizing real-time data from IoT devices. To increase accuracy, speed, and resilience in dynamic smart grid operations, scalability testing on hybrid optimization techniques and large-scale grid settings will also be undertaken. This will enable the growth of totally autonomous, self-healing energy networks

## REFERENCES


Abrahamsen, F. E., Ai, Y., & Cheffena, M. (2021). Communication technologies for smart grid: A comprehensive survey. *Sensors (Basel)*, *21*(23), 8087. DOI: 10.3390/s21238087 PMID: 34884092

Fawaz, A., Mougahrbel, I., Al-Haddad, K., & Kanaan, H. Y. (2025). Graph/game theory-based energy routing methods in the Energy Internet: A review. *IEEE Industrial Electronics Magazine*, *19*(3), 20–39. Advance online publication. DOI: 10.1109/MIE.2025.3552885

Google Dataset Search. (n.d.). *Optimizing smart grid network structures and enhancing communication efficiency* [Dataset]. Google. Retrieved September 21, 2025, from https://datasetsearch.research.google.com/search?src=0&query=Optimizing%20Smart%20Grid%20Network%20Structures%20and%20Enhancing%20Communication%20Efficiency&docid=L2cvMTF4MGJ3MGo5ZA%3D%3D

Hasan, M. K., Habib, A. A., Islam, S., Balfaqih, M., Alfawaz, K. M., & Singh, D. (2023). Smart grid communication networks for electric vehicles empowering distributed energy generation: Constraints, challenges, and recommendations. *Energies*, *16*(3), 1140. DOI: 10.3390/en16031140

Li, Y., Xue, C., Zargari, F., & Li, Y. R. (2023). From graph theory to graph neural networks (GNNs): The opportunities of GNNs in power electronics. *IEEE Access : Practical Innovations, Open Solutions*, *11*, 145067–145084. DOI: 10.1109/ACCESS.2023.3345795

Mezair, T., Djenouri, Y., Belhadi, A., Srivastava, G., & Lin, J. C. W. (2022). A sustainable deep learning framework for fault detection in 6G Industry 4.0 heterogeneous data environments. *Computer Communications*, *187*, 164–171. DOI: 10.1016/j.comcom.2022.02.010

Papadimitrakis, M., Giamarelos, N., Stogiannos, M., Zois, E. N., Livanos, N. I., & Alexandridis, A. (2021). Metaheuristic search in smart grid: A review with emphasis on planning, scheduling and power flow optimization applications. *Renewable & Sustainable Energy Reviews*, *145*, 111072. DOI: 10.1016/j.rser.2021.111072

Saleem, M. U., Shakir, M., Usman, M. R., Bajwa, M. H. T., Shabbir, N., Shams Ghahfarokhi, P., & Daniel, K. (2023). Integrating smart energy management system with internet of things and cloud computing for efficient demand side management in smart grids. *Energies*, *16*(12), 4835. DOI: 10.3390/en16124835

Shahraeini, M., Kotzanikolaou, P., & Nasrolahi, M. (2022). Communication resilience for smart grids based on dependence graphs and eigenspectral analysis. *IEEE Systems Journal*, *16*(4), 6558–6568. DOI: 10.1109/JSYST.2022.3179660

Sharma, T., Sharma, P., & Kumar, N. (2023). Chemical reactive magnetized fluid flow through a vertical channel due to heat source and thermal radiation effects. *Materials Today: Proceedings*, *78*, 481–492. DOI: 10.1016/j.matpr.2022.10.298

Waheed, M. A., & Akogun, O. A. (2021). Quality enhancement of fuel briquette from cornhusk and cassava peel blends for co-firing in coal thermal plant. *International Journal of Energy Research*, *45*(2), 1867–1878. DOI: 10.1002/er.5865

# Compilation of References


Aalto, R., Dunne, T., & Guyot, J. L. (2006). Geomorphic Controls on Andean Denudation Rates. *The Journal of Geology*, *114*(1), 85–99. DOI: 10.1086/498101

Abdi, A. M. (2020). Land cover and land use classification performance of machine learning algorithms in a boreal landscape using Sentinel-2 data. *GIScience & Remote Sensing*, *57*(1), 1–20. 10.1080/15481603.2019.1650447. DOI: 10.1080/15481603.2019.1650447

Abrahamsen, F. E., Ai, Y., & Cheffena, M. (2021). Communication technologies for smart grid: A comprehensive survey. *Sensors (Basel)*, *21*(23), 8087. DOI: 10.3390/s21238087 PMID: 34884092

Abrahart, R. J., & See, L. (2002). Multi-model data fusion for river flow forecasting: An evaluation of six alternative methods based on two contrasting catchments. *Hydrology and Earth System Sciences*, *6*(4), 655–670. 10.5194/hess-6-655-2002. DOI: 10.5194/hess-6-655-2002

Addou, R., Bhiry, N., & Achiban, H. (2025). Dynamics of Wetlands in Ifrane National Park, Morocco: An Approach Using Satellite Imagery and Spectral Indices. *Water (Basel)*, *17*(13), 1869. DOI: 10.3390/w17131869

Adegoke, T. M., Oladoja, O. M., Bashiru, S. O., Mustapha, A. A., Aderupatan, D. E., & Nzei, L. C. (2023). Topp-Leone Inverse Gompertz Distribution: Properties and Different Estimation Techniques and Applications. *Pakistan Journal of Statistics*, *39*(4).

Agatonovic-Kustrin, S., & Beresford, R. (2000). Basic concepts of artificial neural network (ANN) modeling and its application in pharmaceutical research. *Journal of Pharmaceutical and Biomedical Analysis*, *22*(5), 717–727. DOI: 10.1016/S0731-7085(99)00272-1

Agresti, A. (2013). *Categorical Data Analysis* (3rd ed.). Wiley.

Ahmad, S., Wassay, M., & Hussain, M. A. (2025). Cloud Computing: Empowering the Next Generation of Agricultural Research. *Trends in Applied Sciences*. 10.62324/TAPS/2025.060

Ahmed, M. H., El Leithy, B. M., Thompson, J. R., Flower, R. J., Ramdani, M., Ayache, F., & Hassan, S. M. (2009). Application of remote sensing to site characterisation and environmental change analysis of North African coastal lagoons. *Hydrobiologia*, *622*(1), 147–171. DOI: 10.1007/s10750-008-9682-8

Ahmed, Z., Gui, D., Murtaza, G., Liu, Y., & Ali, S. (2023). An Overview of Smart Irrigation Management for Improving Water Productivity Under Climate Change in Drylands. *Agronomy (Basel)*, *13*(8), 2113. Advance online publication. DOI: 10.3390/agronomy13082113

Aitali, R., Snoussi, M., Kolker, A. S., Oujidi, B., & Mhammdi, N. (2022). Effects of Land Use/Land Cover Changes on Carbon Storage in North African Coastal Wetlands. *Journal of Marine Science and Engineering*, *10*(3), 364. DOI: 10.3390/jmse10030364

Akiner, M. E., & Lačnjevac, Č. (2022). Managing Saltwater Intrusion and Agricultural Practices along the Boğaçay River, Turkey: Effects from Excavation and Land Source Pollution. *Journal of Coastal Research*, *38*(3), 567–577. DOI: 10.2112/JCOASTRES-D-21-00101.1

Alaoui, A. M., Choura, M., Maanan, M., Zourarah, B., Robin, M., Conceição, M. F., Andrade, C., Khalid, M., & Carruesco, C. (2010). Metal fluxes to the sediments of the Moulay Bousselham lagoon, Morocco. *Environmental Earth Sciences*, *61*(2), 275–286. DOI: 10.1007/s12665-009-0341-9

Alderman, K., Turner, L. R., & Tong, S. (2012). Floods and human health: A systematic review. *Environment International*, *47*, 37–47. DOI: 10.1016/j.envint.2012.06.003

Alonge, M., & Isreal, O. (2025). GPS and GNSS-based guidance systems for agricultural vehicles.

Alvim, S. J., Guimarães, C. M., Sousa, E. F., Garcia, R. F., & Marciano, C. R. (2022). Application of artificial intelligence for irrigation management: A systematic review. *Engenharia Agrícola*, *42*(spe), e20210159. 10.1590/1809-4430-Eng.Agric.v42nepe20210159/2022. DOI: 10.1590/1809-4430-eng.agric.v42nepe20210159/2022

Amoah, M. K. M., & Gorsevski, P. V. (2025). Predicting land use land cover changes and impact on urban wetlands using cellular automata and artificial neural networks approach, a case study in Greater Accra, Ghana. *Scientific African, 02767*, e02767. Advance online publication. DOI: 10.1016/j.sciaf.2025.e02767

Angel, S. (2023). Urban expansion: Theory, evidence and practice. *Buildings & Cities, 4*(1), 124–138. DOI: 10.5334/bc.348

Anhichem, M., Benbrahim, S., Idrissi, M., Chfiri, R., & Yahyaoui, A. (2021). Environmental state of the Merja Zerga lagoon in Moulay Bousselham, Morocco: Metallic contamination levels and origins in sediments. *Aquaculture, Aquarium, Conservation & Legislation, 14*(2), 774–785.

Arabi, M., Mechkirrou, L., El Malki, M., Alaoui, K., Chaieb, A., Maaroufi, F., & Karmich, S. (2024). Overview of Ecological Dynamics in Morocco – Biodiversity, Water Scarcity, Climate Change, Anthropogenic Pressures, and Energy Resources – Navigating Towards Ecosolutions and Sustainable Development. *E3S Web of Conferences, 527*, 01001. DOI: 10.1051/e3sconf/202452701001

Arrieta, A. B., Díaz-Rodríguez, N., Del Ser, J., Bennetot, A., Tabik, S., Barbado, A., García, S., Gil-López, S., Molina, D., & Benjamins, R. (2020). Explainable Artificial Intelligence (XAI): Concepts, taxonomies, opportunities and challenges toward responsible AI. *Information Fusion, 58*, 82–115. 10.1016/j.inffus.2019.12.012. DOI: 10.1016/j.inffus.2019.12.012

Asawra, K., & Koli, V. K. (2025). Foraging strategies of three co-occurring ibis species (Black-headed, Glossy and Red-naped) in a semi-arid landscape of western India. *Avian Biology Research, 18*(1), 12–20. DOI: 10.1177/17581559251314345

Asefa, T., Kemblowski, M., McKee, M., & Khalil, A. (2006). Multi-time scale stream flow predictions: The support vector machines approach. *Journal of Hydrology (Amsterdam), 318*(1-4), 7–16. 10.1016/j.jhydrol.2005.06.001. DOI: 10.1016/j.jhydrol.2005.06.001

Awoseyi, A. A., Akinsola, J. E. T., Oladoja, O. M., Adeagbo, M., & Adebowale, O. O. (2022). Hybridisation of decision tree algorithm using sequencing predictive model for COVID-19. In *Emerging Technologies for Combatting Pandemics* (pp. 201–221). Auerbach Publications. DOI: 10.1201/9781003324447-10

Ayache, F., Thompson, J. R., Flower, R. J., Boujarra, A., Rouatbi, F., & Makina, H. (2009). Environmental characteristics, landscape history and pressures on three coastal lagoons in the Southern Mediterranean Region: Merja Zerga (Morocco), Ghar El Melh (Tunisia) and Lake Manzala (Egypt). *Hydrobiologia, 622*(1), 15–43. DOI: 10.1007/s10750-008-9676-6

Azghadi, M. R., Olsen, A., Wood, J., Saleh, A., Calvert, B., Granshaw, T., Fillols, E., & Philippa, B. (2024). Precise robotic weed spot-spraying for reduced herbicide usage and improved environmental outcomes—a real-world case study. *arXiv preprint arXiv:2401.13931*. 10.1016/j.compag.2025.110365

Babalola, A. S., Adeogun, A. O., Thabet, H. S., TagEldin, R. A., Oyeniyi, T., Adekunle, O., Izekor, R., Adetunji, O., Olalekan, O., Omotayo, A., Abiodun, O., Adediran, A. D., Adekeye, T., Adegbola, A. O., Isaac, C., Okoko, P. O., & Harwood, J. F. (2025). Geospatial modelling of geographical spread of *Aedes* species, in relation to climatic and topographical factors in Lagos State, Nigeria. *PLoS Neglected Tropical Diseases*, *19*(2), e0012860. DOI: 10.1371/journal.pntd.0012860 PMID: 39913605

Baba, W. M., Chehbouni, A., Ouassanouan, Y., Gascoin, S., Paganini, M., Ottavianelli, G., & Szantoi, Z. (2025). Monitoring water crisis from space across a Mediterranean region. *Scientific Reports*, *15*(1), 23262. DOI: 10.1038/s41598-025-06240-1 PMID: 40603462

Bagja, I., Pangi, P., Mispaki, S., & Rahayu, S. (2025). Utilization of Sentinel-1 imagery for inundation flood mapping in northern part of Java Island. *IOP Conference Series. Earth and Environmental Science*, *1438*(1), 012045. DOI: 10.1088/1755-1315/1438/1/012045

Bahouq, T., Moumane, A., & Touhami, N. (2024). Monitoring Agricultural Dynamics and Its Impact on Water Resources Using GIS and Remote Sensing Data: A Case Study in the Ben Mansour-Mnasra Region, Kenitra, Morocco. In Batchi, M., & Moumane, A. (Eds.), (pp. 183–210). Advances in Geospatial Technologies. IGI Global., DOI: 10.4018/979-8-3693-9651-3.ch007

Bai, S., Kolter, J. Z., & Koltun, V. 2018. "An Empirical Evaluation of Generic Convolutional and Recurrent Networks for Sequence Modeling." *arXiv Preprint arXiv:1803.01271*. https://doi.org//arXiv.1803.01271DOI: 10.48550

Baladia, Y., Ben-Haddad, M., Laadel, N., & Oualid, J. A. (2025). Catastrophic fish mass mortality events in Moroccan freshwater ecosystems: Alarming trends and impacts on biodiversity. *Environmental Monitoring and Assessment*, *197*(3), 290. DOI: 10.1007/s10661-025-13711-5 PMID: 39946026

Barbedo, J. G. A. (2019). Plant disease identification from individual lesions and spots using deep learning. *Biosystems Engineering*, *180*, 96–107. 10.1016/j.biosystemseng.2019.02.002. DOI: 10.1016/j.biosystemseng.2019.02.002

Basir, M. S., Noel, S., Buckmaster, D., & Ashik-E-Rabbani, M. (2024). Enhancing subsurface soil moisture forecasting: A long short-term memory network model using weather data. *Agriculture*, *14*(3), 333. DOI: 10.3390/agriculture14030333

Bayable, G., Gebrie, G., Lebeza, T. M., & Melaku, A. (2025). Land Use/Cover Classification using machine learning algorithms and their impacts on land surface temperature and soil moisture in the Alawuha Watershed, Ethiopia. *Environmental and Sustainability Indicators, 100797*, 100797. Advance online publication. DOI: 10.1016/j.indic.2025.100797

Behzad, M., Asghari, K., & Coppola, E. A.Jr. (2010). Comparative study of SVMs and ANNs in aquifer water level prediction. *Journal of Computing in Civil Engineering, 24*(5), 408–413. 10.1061/(ASCE)CP.1943-5487.0000043. DOI: 10.1061/(ASCE)CP.1943-5487.0000043

Benmokhtar, S., Robin, M., Maanan, M., Boutoumit, S., Badaoui, B., & Bazairi, H. (2023). Monitoring the Spatial and Interannual Dynamic of Zostera noltei. *Wetlands, 43*(5), 43. DOI: 10.1007/s13157-023-01690-7 PMID: 37153812

Bergstra, J., & Bengio, Y. (2012). Random Search for Hyper-Parameter Optimization. *Journal of Machine Learning Research, 13*, 281–305. https://www.jmlr.org/papers/volume13/bergstra12a/bergstra12a.pdf

Beven, K. (2012). Rainfall-Runoff Modelling. DOI: 10.1002/9781119951001

Beven, K., & Kirkby, M. J. (1979). A physically based, variable contributing area model of basin hydrology. *Hydrological Sciences Bulletin, 24*(1), 43–69. DOI: 10.1080/02626667909491834

Bibri, S. E., Krogstie, J., & Kärrholm, M. (2020). Compact city planning and development: Emerging practices and strategies for achieving the goals of sustainability. *Developments in the Built Environment, 4*, 100021. DOI: 10.1016/j.dibe.2020.100021

Birkel, C., & Barahona, A. C. (2019). Rainfall-runoff modeling: a brief overview. *Reference module in earth systems and environmental sciences*. 10.1016/B978-0-12-409548-9.11595-7

Bitwize. (2023). Sultan Haitham City in Seeb to set a new benchmark in smart city development. Times of Oman. https://timesofoman.com/article/131740-sultan-haitham-city-in-seeb-to-set-a-new-benchmark-in-smart-city-development

Blais, M.-A., & Akhloufi, M. A. (2025). Advances in Remote Sensing and Deep Learning in Coastal Boundary Extraction for Erosion Monitoring. *Geomatics, 5*(1). *Scopus., 5*(1), 9. Advance online publication. DOI: 10.3390/geomatics5010009

Boak, E. H., & Turner, I. L. (2005). Shoreline definition and detection : A review. *Journal of Coastal Research, 21*(4), 688–703. DOI: 10.2112/03-0071.1

Bondre, D. A., & Mahagaonkar, S. (2019). Prediction of crop yield and fertilizer recommendation using machine learning algorithms. *International Journal of Engineering Applied Sciences and Technology*, *4*(5), 371–376. https://doi.org/http://www.ijeast.com/. DOI: 10.33564/IJEAST.2019.v04i05.055

Bouazzati, M. E., Batchi, M., Fatah, F., Moumane, A., & Motib, I. (2025). Assessing the Spatio-Temporal Evolution of Lake Aguelmam Azigza Using Remote Sensing and GIS: Middle Atlas – Morocco. In *Remote Sensing and GIS Techniques in Hydrology* (pp. 233–258). IGI Global Scientific Publishing. DOI: 10.4018/979-8-3693-9651-3.ch009

Boudreau, K., Robinson, M., & Farooqi, Z. (2022). IPCC Sixth assessment report: climate change 2021: the physical science basis summary for policymakers. *Canadian Journal of Emergency Management*, *2*(1). Advance online publication. 10.25071/6sw6za31. DOI: 10.25071/6sw6za31

Box George, E., Jenkins Gwilym, M., Reinsel Gregory, C., & Ljung Greta, M. (1976). *Time series analysis: forecasting and control.* Holden Bay., 10.2307/3008255

Boyer, M. (2025). Historicizing The Atlantic Morocco From the Horizon of the New Worlds to the Globalized World: *Afrique(s) en mouvement, n° 9*(3), 32–43. DOI: 10.3917/aem.009.0032

Bradley, A. P. (1997). The use of the area under the ROC curve in the evaluation of machine learning algorithms. *Pattern Recognition*, *30*(7), 1145–1159. DOI: 10.1016/S0031-3203(96)00142-2

Brasington, J., & Richards, K. (1998). Interactions between model predictions, parameters and DTM scales for TopModel. Computers &amp. *Computers & Geosciences*, *24*(4), 299–314. DOI: 10.1016/S0098-3004(97)00081-2

Breiman, L., Friedman, J. H., Olshen, R. A., & Stone, C. J. (2017). Regression Trees. Classification And Regression Trees, 216–265. DOI: 10.1201/9781315139470-8

Breiman, L. (2001). Random Forests. *Machine Learning*, *45*(1), 5–32. DOI: 10.1023/A:1010933404324

Bronson, K. (2019). Looking through a responsible innovation lens at uneven engagements with digital farming. *NJAS Wageningen Journal of Life Sciences*, *90*(1), 100294. 10.1016/j.njas.2019.03.001. DOI: 10.1016/j.njas.2019.03.001

Bucak, T., Trolle, D., Andersen, H. E., Thodsen, H., Erdoğan, Ş., Levi, E. E., Filiz, N., Jeppesen, E., & Beklioğlu, M. (2017). Future water availability in the largest freshwater Mediterranean lake is at great risk as evidenced from simulations with the SWAT model. *The Science of the Total Environment, 581–582*, 413–425. DOI: 10.1016/j.scitotenv.2016.12.149 PMID: 28069301

Bui, D. T., Nguyen, T. T., Hoang, N. D., & Pradhan, B. (2022). Flood susceptibility mapping using deep learning, machine learning, and AHP models in Quang Binh Province, Vietnam. *Environmental Monitoring and Assessment, 194*(5), 349. DOI: 10.1007/s10661-022-10012-0 PMID: 35394223

Burgess, E. W. (1925). The Growth of the City: An introduction to a research project. *City (London, England)*, •••, 47–62. https://urpa3301.weebly.com/uploads/4/0/9/2/4092174/burgess.pdf

Burton, E., . . .. (1996). The Compact City: a sustainable urban form? Routledge & CRC Press. https://www.routledge.com/The-Compact-City-A-Sustainable-Urban-Form/Burton-Jenks-Williams/p/book/9780419213000?srsltid=AfmBOoqjHomxk3fm0DMTOESMdNXIahIN-nJsGy-ciERSDkUZbgCmwnhq

Cahyana, Y., Setiawan, A., Sembiring, I., & Dewi, C. (2024, November). Machine Learning Application for Classification of Rice Varieties: A Systematic Literature Review. In *2024 International Conference on Computer Engineering, Network, and Intelligent Multimedia (CENIM)* (pp. 1-7). IEEE. DOI: 10.1109/CENIM64038.2024.10882802

CESE. (2021). *Le développement durable du littoral marocain : Enjeux et perspectives [Sustainable development of the Moroccan coast : Challenges and prospects]*. https://www.cese.ma

Chafiq, T., Hmamou, M., Ouhammou, I., Azmi, R., & Kumar, M. (2024). Modelling change detection for unveiling urban transitions: Using machine learning algorithms and Sentinel-2 data in Larache City, Morocco. *Modeling Earth Systems and Environment, 10*(2), 1711–1725. DOI: 10.1007/s40808-023-01860-w

Chai, T., & Draxler, R. R. (2014). Root Mean Square Error (RMSE) or Mean Absolute Error (MAE)? – Arguments Against Avoiding RMSE in the Literature. *Geoscientific Model Development, 7*(3), 1247–1250. DOI: 10.5194/gmd-7-1247-2014

Chapi, K., Singh, V. P., Shirzadi, A., Shahabi, H., Bui, D. T., Pham, B. T., & Khosravi, K. (2017). A novel hybrid artificial intelligence approach for flood susceptibility assessment. Environmental modelling &amp. *Environmental Modelling & Software, 95*, 229–245. DOI: 10.1016/j.envsoft.2017.06.012

Chen, S.-J., & Hwang, C.-L. (1992). Fuzzy Multiple Attribute Decision Making Methods. Fuzzy Multiple Attribute Decision Making, 289–486. DOI: 10.1007/978-3-642-46768-4_5

Chen, C., Chen, H., Liang, J., Huang, W., Xu, W., Li, B., & Wang, J. (2022). Extraction of water body information from remote sensing imagery while considering greenness and wetness based on Tasseled Cap transformation. *Remote Sensing (Basel)*, *14*(13), 3001. DOI: 10.3390/rs14133001

Chen, T., & Guestrin, C. 2016. "XGBoost: A Scalable Tree Boosting System." *Proceedings of the 22nd ACM SIGKDD International Conference on Knowledge Discovery and Data Mining*, 785–94. DOI: 10.1145/2939672.2939785

Chen, Y., Hu, J., Cai, Z., Yang, J., Zhou, W., Hu, Q., Wang, C., You, L., & Xu, B. (2024). A phenology-based vegetation index for improving ratoon rice mapping using harmonized Landsat and Sentinel-2 data. *Journal of Integrative Agriculture*, *23*(4), 1164–1178. DOI: 10.1016/j.jia.2023.05.035

Chlingaryan, A., Sukkarieh, S., & Whelan, B. (2018). Machine learning approaches for crop yield prediction and nitrogen status estimation in precision agriculture: A review. *Computers and Electronics in Agriculture*, *151*, 61–69. 10.1016/j.compag.2018.05.012. DOI: 10.1016/j.compag.2018.05.012

Choubin, B., Moradi, E., Golshan, M., Adamowski, J., Sajedi-Hosseini, F., & Mosavi, A. (2019). An ensemble prediction of flood susceptibility using multivariate discriminant analysis, classification and regression trees, and support vector machines. *The Science of the Total Environment*, *651*, 2087–2096. DOI: 10.1016/j.scitotenv.2018.10.064 PMID: 30321730

Chow, V. T., Maidment, D. R., & Mays, L. W. (2010). *Applied Hydrology* (2nd ed.). McGraw-Hill Education.

Chrif El Idrissi, M., Saber, E.-R., & Al Mashoudi, A. (2025). Hydro-climatic stress and ecosystem vulnerability in a highland lake: The case of Dayet Aoua (1935–2022), Middle Atlas, Morocco. *Modeling Earth Systems and Environment*, *11*(5), 312. DOI: 10.1007/s40808-025-02496-8

Chtaibi, N., Magri, N., Belahbib, N., Zidane, L., Douira, A., & Dahmani, J. (2017). Study of Bryophytic Flora in the Ramsar Wetland of Merja Zerga (North-west of Morocco). *International Journal of Environment, Agriculture and Biotechnology*, *2*(3), 1032–1037. DOI: 10.22161/ijeab/2.3.4

Chtioui, T., Hakkou, M., Aangri, A., El Hassani, F., El Mostafa, Z., & Benmohammadi, A. (2024). Assessing coastal flood risk under extreme events and sea level rise in the Casablanca-Mohammedia coastline (Morocco). *Natural Hazards, 120*(12), 10727–10752. DOI: 10.1007/s11069-024-06624-y

Climate Hazards Group. (2014). *CHIRPS Daily: Climate Hazards Group InfraRed Precipitation with Station data (Version 2.0)* [Data set]. University of California, Santa Barbara. https://data.chc.ucsb.edu/products/CHIRPS-2.0/

Cortes, C., & Vapnik, V. (1995). Support-Vector Networks. *Machine Learning, 20*(3), 273–297. DOI: 10.1023/A:1022627411411

Costa, D. G., Bittencourt, J. C. N., Oliveira, F., Peixoto, J. P. J., & Jesus, T. C. (2024). Achieving Sustainable Smart Cities through Geospatial Data-Driven Approaches. *Sustainability (Basel), 16*(2), 640. DOI: 10.3390/su16020640

Costa, J. E. (1987). Hydraulics and basin morphometry of the largest flash floods in the conterminous United States. *Journal of Hydrology (Amsterdam), 93*(3–4), 313–338. DOI: 10.1016/0022-1694(87)90102-8

Cover, T., & Hart, P. (1967). Nearest Neighbor Pattern Classification. *IEEE Transactions on Information Theory, 13*(1), 21–27. DOI: 10.1109/TIT.1967.1053964

Crowell, M., Leatherman, S. P., & Buckley, M. K. (1991). Historical shoreline change : Error analysis and mapping accuracy. *Journal of Coastal Research*, •••, 839–852.

Cutter, S. L., Ash, K. D., & Emrich, C. T. (2014). The geographies of community disaster resilience. *Global Environmental Change, 29*, 65–77. DOI: 10.1016/j.gloenvcha.2014.08.005

Cvetković, V. M., Renner, R., Aleksova, B., & Lukić, T. (2024). Geospatial and temporal patterns of natural and man-made (Technological) disasters (1900–2024): Insights from different socioeconomic and demographic perspectives. *Applied Sciences (Basel, Switzerland), 14*(18), 8129. DOI: 10.3390/app14188129

Dadashpoor, H., Khaleghinia, A., & Shabrang, A. (2024). Explaining the role of land use Changes on land surface temperature in an arid and semi-arid metropolitan area with multi-scale spatial regression analysis. *Environmental Monitoring and Assessment, 196*(2), 124. DOI: 10.1007/s10661-023-12241-2 PMID: 38195837

Daniel, A. I., Fadaka, A. O., Gokul, A., Bakare, O. O., Aina, O., Fisher, S., Burt, A. F., Mavumengwana, V., Keyster, M., & Klein, A. (2022). Biofertilizer: The future of food security and food safety. *Microorganisms, 10*(6), 1220. DOI: 10.3390/microorganisms10061220 PMID: 35744738

Danumah, J. H., Odai, S. N., Saley, B. M., Szarzynski, J., Thiel, M., Kwaku, A., Kouame, F. K., & Akpa, L. Y. (2016). Flood risk assessment and mapping in Abidjan district using multi-criteria analysis model and geoinformation techniques, (Côte d'Ivoire). *Geoenvironmental Disasters*, *3*(1), 10. DOI: 10.1186/s40677-016-0044-y

Darko, A., Chan, A. P. C., Ameyaw, E. E., Owusu, E. K., Pärn, E., & Edwards, D. J. (2018). Review of application of analytic hierarchy process (AHP) in construction. *International Journal of Construction Management*, *19*(5), 436–452. DOI: 10.1080/15623599.2018.1452098

Das, A., Pal, M., Taria, S., Ellur, R. K., Kumar, S., Sathee, L., Gopala Krishnan, S., Dalal, M., Jain, P. K., & Chinnusamy, V. (2024). Multivariate analysis and genome wide association mapping for chlorophyll fluorescence and grain number per panicle under high temperature stress in rice. *Plant Physiology Reports*, *29*(3), 598–613. DOI: 10.1007/s40502-024-00808-1

Das, S., Rawat, A., & Singh, S. K. (2020). Impacts of temperature, precipitation and population growth on energy demands of Delhi. *International Journal of Design & Nature and Ecodynamics*, *15*(4), 539–544. DOI: 10.18280/ijdne.150411

Dawson, C., & Wilby, R. (2001). Hydrological modelling using artificial neural networks. *Progress in Physical Geography*, *25*(1), 80–108. 10.1177/030913330102500104. DOI: 10.1177/030913330102500104

de Groot, L., Demeijer, F. A., Zweekhorst, M., & Urias, E. (2025). Multi-stakeholder networks in the higher education context: A configurative literature review of university-community interactions. *Journal of Higher Education Policy and Management*, *47*(4), 1–19. DOI: 10.1080/1360080X.2025.2451445

De Moel, H., Jongman, B., Kreibich, H., Merz, B., Penning-Rowsell, E., & Ward, P. J. (2015). Flood risk assessments at different spatial scales. *Mitigation and Adaptation Strategies for Global Change*, *20*(6), 865–890. DOI: 10.1007/s11027-015-9654-z

Dehghani, A., Soltani, A., & Nateghi, K. (2025). Balancing urban growth and Environmental change: Land use patterns in Tehran and Sydney. *Environmental and Sustainability Indicators*, *100691*, 100691. Advance online publication. DOI: 10.1016/j.indic.2025.100691

Deka, P. C. (2014). Support vector machine applications in the field of hydrology: A review. *Applied Soft Computing*, *19*, 372–386. 10.1177/030913330102500104. DOI: 10.1016/j.asoc.2014.02.002

DeVantier, B. A., & Feldman, A. D. (1993). Review of GIS applications in hydrologic modeling. *Journal of Water Resources Planning and Management, 119*(2), 246–261. 10.1061/(ASCE)0733-9496(1993)119:2(246). DOI: 10.1061/(ASCE)0733-9496(1993)119:2(246)

Devia, G. K., Ganasri, B. P., & Dwarakish, G. S. (2015). A review on hydrological models. *Aquatic Procedia, 4*, 1001–1007. 10.1016/j.aqpro.2015.02.126. DOI: 10.1016/j.aqpro.2015.02.126

Dhotre, M. T., Niceno, B., Smith, B. L., & Simiano, M. (2009). Large-eddy simulation (LES) of the large scale bubble plume. *Chemical Engineering Science, 64*(11), 2692–2704. DOI: 10.1016/j.ces.2009.02.040

Di Baldassarre, G., Castellarin, A., Montanari, A., & Brath, A. (2009). Probability-weighted hazard maps for comparing different flood risk management strategies: A case study. *Natural Hazards, 50*(3), 479–496. DOI: 10.1007/s11069-009-9355-6

Di Baldassarre, G., & Montanari, A. (2009). Uncertainty in river discharge observations: A quantitative analysis. *Hydrology and Earth System Sciences, 13*(6), 913–921. DOI: 10.5194/hess-13-913-2009

Dilley, M., Chen, R. S., Deichmann, U., Lerner-Lam, A. L., & Arnold, M. (2005). *Natural Disaster Hotspots*. The World Bank., DOI: 10.1596/0-8213-5930-4

Dixit, A. (2003). Floods and vulnerability in Nepal: Need for a paradigm shift. *Natural Hazards, 28*(1), 155–179. DOI: 10.1023/A:1021134218121

Djalante, R. (2018). Review article: A systematic literature review of research trends and authorships on natural hazards, disasters, risk reduction and climate change in Indonesia, 3137–3153. *Natural Hazards and Earth System Sciences, 18*(11), 3137–3153. DOI: 10.5194/nhess-18-3137-2018

Domeneghetti, A., Vorogushyn, S., Castellarin, A., Merz, B., & Brath, A. (2013). Probabilistic flood hazard mapping: effects of uncertain boundary conditions. Hydrology and Earth System Sciences, 17(8), 3127–3140. EEA Report "EEA Signals 2010": stories about people and the environment. (2010). International Journal of Sustainability in Higher Education, 11(3). https://doi.org/DOI: 10.5194/hess-17-3127-2013

Dormann, C. F., Elith, J., Bacher, S., Buchmann, C., Carl, G., Carré, G., Marquéz, J. R. G., Gruber, B., Lafourcade, B., Leitão, P. J., Münkemüller, T., McClean, C., Osborne, P. E., Reineking, B., Schröder, B., Skidmore, A. K., Zurell, D., & Lautenbach, S. (2013). Collinearity: A review of methods to deal with it and a simulation study evaluating their performance. *Ecography*, *36*(1), 27–46. DOI: 10.1111/j.1600-0587.2012.07348.x

Dullo, T., Darkwah, G., Gangrade, S., Morales-Hernández, M., Sharif, M., Kalyanapu, A., Kao, S.-C., Ghafoor, S., & Ashfaq, M. (2021). Assessing climate-change-induced flood risk in the Conasauga River watershed: An application of ensemble hydrodynamic inundation modeling. *Natural Hazards and Earth System Sciences*, *21*(6), 1739–1757. DOI: 10.5194/nhess-21-1739-2021

El Ghizi, S., El Aadel, N., Sadik, M., El Bouch, M., & Hasnaoui, M. (2021). The physicochemical characteristics and the pollution level of Dayet Er-Roumi Lake in Morocco. *E3S Web of Conferences, 234*, 00036. DOI: 10.1051/e3sconf/202123400036

El Ghizi, S., Sadik, M., El Bouch, M., & Hasnaoui, M. (2023). Study of the Ichthyological Populations of Lake Dayet Er-Roumi (Morocco): Use of Indices to Analyze the Ichthyofaunistic Biodiversity. *Egyptian Journal of Aquatic Biology and Fisheries*, *27*(2), 841–851. DOI: 10.21608/ejabf.2023.298312

El Orfi, T., El Ghachi, M., Lebaut, S., & Haidu, I. (2025). Monitoring the water surface dynamics of four Mediterranean mountain lakes in the Middle Atlas (Morocco) using Landsat imagery in a global change context. *Journal of Hydrology. Regional Studies*, *60*, 102516. DOI: 10.1016/j.ejrh.2025.102516

Elabidi, K., Remmal, T., Amraoui, F., Jounaid, H., Amar, M., & Bensalmia, A. (2023). *Master spécialisé : Gestion et Valorisation des Géoressources Projet de Fin d'Etudes L'apport de la Bathymetrie Pour L'etude Hydrogeologique de Dayet Roumi.*

Elith, J., Leathwick, J. R., & Hastie, T. (2008). A working guide to boosted regression trees. Journal of Animal Ecology, 77(4), 802–813. Portico. DOI: 10.1111/j.1365-2656.2008.01390.x

Elmotawakkil, Abdessamad, Adil Moumane, Assia Zahi, Abdelkhalik Sadiki, Jamal Al Karkouri, Mouhcine Batchi, Suraj Kumar Bhagat, Tiyasha Tiyasha, and Nourddine Enneya. 2025. "Artificial Intelligence for Groundwater Recharge Prediction in an Arid Region: Application of Tabular Deep Learning Models in the Feija Basin, Morocco." *Frontiers in Remote Sensing* Volume 6 - 2025. .DOI: 10.3389/frsen.2025.1622360

Elmotawakkil, A., & Enneya, N. (2024). Advanced machine learning for predicting groundwater decline and drought in the Rabat–Salé–Kénitra region, Morocco. *Journal of Hydroinformatics*, *26*(11), 2980–3007. DOI: 10.2166/hydro.2024.328

Elmotawakkil, A., Moumane, A., Sadiki, A., Zahi, A., Al Karkouri, J., Batchi, M., Kumar Bhagat, S., & Enneya, N. (2025). Forecasting short-term rainfall patterns in arid and semi-arid regions using machine learning and deep learning models: A case study from Morocco. *Theoretical and Applied Climatology*, *156*(8), 436. DOI: 10.1007/s00704-025-05677-8

Elmotawakkil, A., Sadiki, A., & Enneya, N. (2024). Predicting Groundwater Level Based on Remote Sensing and Machine Learning: A Case Study in the Rabat-Kénitra Region. *Journal of Hydroinformatics*, *26*(10), 2639–2667. DOI: 10.2166/hydro.2024.494

EM-DAT (Emergency Events Database). (2009) Monitoring hazards & disasters worldwide since 1988. Link: https://www.emdat.be/ last access: 045.09.2025

Endreny, T., & Higgins, M. (2008). Adding Radar Rainfall and Calibration to the TR-20 Watershed Model to Improve Dam Removal Flood Analysis. *Journal of Water Resources Planning and Management*, *134*(3), 314–317. DOI: 10.1061/(ASCE)0733-9496(2008)134:3(314)

England, J. F., Cohn, T. A., Faber, B. A., Stedinger, J. R., Thomas, W. O., Veilleux, A. G., Kiang, J. E., & Mason, R. R. (2018). *Guidelines for determining flood flow frequency — Bulletin 17C*. Techniques and Methods., DOI: 10.3133/tm4B5

England, J. F.Jr. (2011). Flood Frequency and Design Flood Estimation Procedures in the United States: Progress and Challenges. *Australasian Journal of Water Resources*, *15*(1), 33–46. DOI: 10.1080/13241583.2011.11465388

Ennabili, A., Libiad, M., El Haissoufi, M., Khabbach, A., & Bennas, N. (2025). Biodiversity, Change, and Use of Moroccan Wetlands. In Pullaiah, T. (Ed.), *Wetlands of Tropical and Subtropical Asia and Africa* (1st ed., pp. 265–296). Wiley., DOI: 10.1002/9781394235278.ch12

Ennouali, Z., Fannassi, Y., Lahssini, G., Benmohammadi, A., & Masria, A. (2023). Mapping Coastal vulnerability using machine learning algorithms : A case study at North coastline of Sebou estuary, Morocco. *Regional Studies in Marine Science*, *60*, 102829. DOI: 10.1016/j.rsma.2023.102829

Erjavec, E., & Rac, I. (2023). Improving the quality of CAP strategic planning through enhancing the role of agricultural economics. *EuroChoices (Uckfield)*, *22*(2), 71–76. DOI: 10.1111/1746-692X.12393

Erraj Chahid, N., Bouchkara, M., Canu, D. M., El Khalidi, K., & Zourarah, B. (2023). Assessing the Dynamics of the Moulay Bousselham Lagoon: An Integrated DPSIR Approach to Understand Drivers, Impacts, and Human Response. In *IECG 2022* (p. 31). Presented at the IECG 2022, MDPI. DOI: 10.3390/IECG2022-14553

Evenson, R. E., & Gollin, D. (2003). Assessing the impact of the Green Revolution, 1960 to 2000. *Science*, *300*(5620), 758–762. DOI: 10.1126/science.1078710 PMID: 12730592

Ewing, R., & Hamidi, S. (2015). Compactness versus Sprawl. *Journal of Planning Literature*, *30*(4), 413–432. DOI: 10.1177/0885412215595439

Fadda, E. H., Shebli, F. A., & Kabi, A. A. (2019). The use of remote sensing and GIS to study land use changes in Barka Wilaya, Oman. [JASS]. *Journal of Arts and Social Sciences*, *10*(1), 15–20. DOI: 10.53542/jass.v10i1.3221

FAO. 2018. *WaPOR Database Methodology: Level 1. Remote Sensing for Water Productivity Technical Report: Methodology Series*. Rome: FAO. https://openknowledge.fao.org/server/api/core/bitstreams/eb122ea8-99a6-4719-97ff-cadd09a5eced/content

Farhadi, A., & Najafzadeh, M. (2021). Flood susceptibility mapping using Random Forest and remote sensing data in Iran. *Remote Sensing Applications: Society and Environment*, *21*, 100475. DOI: 10.1016/j.rsase.2021.100475

Farooqi, S. M., Kanwal, A., Zaman-ul-Haq, M., Saqib, Z., Akhtar, N., Tariq, A., Abdullah-Al-Wadud, M., Mubbin, M., & Bokhari, S. A. (2025). Integrating Geo-AI with RS & GIS for comprehensive assessments of urban land cover transformations and integrated responses. *Environmental Earth Sciences*, *84*(1), 3. DOI: 10.1007/s12665-024-12005-2

Fauzi, C. (2024). Review Geospatial Artificial Intelligence (Geo-AI): Implementation of Machine Learning on Urban Planning. *Proceedings of the International Conference on Applied Science and Technology on Engineering Science 2023 (iCAST-ES 2023)*. 311-329. https://doi.org/DOI: 10.2991/978-94-6463-364-1_30

Fawaz, A., Mougahrbel, I., Al-Haddad, K., & Kanaan, H. Y. (2025). Graph/game theory-based energy routing methods in the Energy Internet: A review. *IEEE Industrial Electronics Magazine*, *19*(3), 20–39. Advance online publication. DOI: 10.1109/MIE.2025.3552885

Fawcett, T. (2006). An introduction to ROC analysis. *Pattern Recognition Letters*, *27*(8), 861–874. DOI: 10.1016/j.patrec.2005.10.010

Fitri, A., Nurul Hadie, M. S., Agustina, A., Pratiwi, D., Susarman, Pramita, G., & Salah, H. R. A. (2021). Analyses of flood peak discharge in Cimadur River basin, Banten Province, Indonesia. E3S Web of Conferences, 331, 08006. DOI: 10.1051/e3sconf/202133108006

Follum, M. L., Tavakoly, A. A., Niemann, J. D., & Snow, A. D. (2016). AutoRAPID: A Model for Prompt Streamflow Estimation and Flood Inundation Mapping over Regional to Continental Extents. JAWRA Journal of the American Water Resources Association, 53(2), 280–299. Portico. DOI: 10.1111/1752-1688.12476

Forzieri, G., Feyen, L., Russo, S., Vousdoukas, M., Alfieri, L., Outten, S., Migliavacca, M., Bianchi, A., Rojas, R., & Cid, A. (2016). Multi-hazard assessment in Europe under climate change. *Climatic Change*, *137*(1-2), 105–119. DOI: 10.1007/s10584-016-1661-x

Foster, M., Fell, R., & Spannagle, M. (2000). The statistics of embankment dam failures and accidents. *Canadian Geotechnical Journal*, *37*(5), 1000–1024. DOI: 10.1139/t00-030

Fountas, S., Mylonas, N., Malounas, I., Rodias, E., Hellmann Santos, C., & Pekkeriet, E. (2020). Agricultural robotics for field operations. *Sensors (Basel)*, *20*(9), 2672. 10.3390/s20092672. DOI: 10.3390/s20092672 PMID: 32392872

Fox-Kemper, B., Hewitt, H. T., Xiao, C., Aðalgeirsdóttir, G., Drijfhout, S. S., Edwards, T. L., Golledge, N. R., Hemer, M., Kopp, R. E., Krinner, G., Mix, A., Notz, D., Nowicki, S., Nurhati, I. S., Ruiz, L., Sallée, J.-B., Slangen, A. B. A., & Yu, Y. (2021). Ocean, cryosphere, and sea level change. In Masson-Delmotte, V., Zhai, P., Pirani, A., Connors, S. L., Péan, C., Berger, S., Caud, N., Chen, Y., Goldfarb, L., Gomis, M. I., Huang, M., Leitzell, K., Lonnoy, E., Matthews, J. B. R., Maycock, T. K., Waterfield, T., Yelekçi, Ö., Yu, R., & Zhou, B. (Eds.), *Climate Change 2021 : The Physical Science Basis. Contribution of Working Group I to the Sixth Assessment Report of the Intergovernmental Panel on Climate Change* (pp. 1211–1362). Cambridge University Press., DOI: 10.1017/9781009157896.001

Freeze, R. A., & Cherry, J. A. (1979). Groundwater prentice-hall. Englewood Cliffs, NJ, 176, 161-177.

Friedman, J. H. (1991). Multivariate Adaptive Regression Splines. *Annals of Statistics*, *19*(1). Advance online publication. DOI: 10.1214/aos/1176347963

Funk, C., Peterson, P., Landsfeld, M., Pedreros, D., Verdin, J., Shukla, S., Husak, G., Rowland, J., Harrison, L., Hoell, A., & Michaelsen, J. (2015). The climate hazards infrared precipitation with stations—A new environmental record for monitoring extremes. *Scientific Data*, *2*(1), 150066. DOI: 10.1038/sdata.2015.66 PMID: 26646728

Fu, X., He, X., & Ding, L. (2023). Stochastic Flood Simulation Method Combining Flood Intensity and Morphological Indicators. *Sustainability (Basel), 15*(18), 14032. DOI: 10.3390/su151814032

Gaál, L., Szolgay, J., Kohnová, S., Hlavčová, K., Parajka, J., Viglione, A., Merz, R., & Blöschl, G. (2015). Dependence between flood peaks and volumes: A case study on climate and hydrological controls. *Hydrological Sciences Journal, 60*(6), 968–984. DOI: 10.1080/02626667.2014.951361

Gao, X., Chi, H., Huang, J., Han, Y., Li, Y., & Ling, F. (2024). Comparison of cloud-mask algorithms and machine-learning methods using Sentinel-2 imagery for mapping paddy rice in Jianghan Plain. *Remote Sensing (Basel), 16*(7), 1305. DOI: 10.3390/rs16071305

García-Vera, Y. E., Polochè-Arango, A., Mendivelso-Fajardo, C. A., & Gutiérrez-Bernal, F. J. (2024). Hyperspectral image analysis and machine learning techniques for crop disease detection and identification: A review. *Sustainability (Basel), 16*(14), 6064. 10.3390/su16146064. DOI: 10.3390/su16146064

Gebbers, R., & Adamchuk, V. I. (2010). Precision agriculture and food security. *Science, 327*(5967), 828–831. 10.1126/science.1183899. DOI: 10.1126/science.1183899 PMID: 20150492

Generale de la Meteorologie, D. (2024). Etat Du Climat Maroc., https://www.marocmeteo.ma/fr/climat-2024, Retrieved September 12, 2024, from.

Gessler, P. E., Moore, I. D., McKenzie, N. J., & Ryan, P. J. (1995). Soil-landscape modeling and spatial prediction of soil attributes. *International Journal of Geographical Information Systems, 9*(4), 421–432. DOI: 10.1080/02693799508902047

GIEC. (2021). AR6 [*The Physical Science Basis. Chapitre : Sea Level Rise ] [projetions citées par la Banque Mondiale pour le Maroc ] [.* https://climateknowledgeportal.worldbank.org/country/morocco/sea-level-projections]. *Climatic Change, 2021*, •••.

Gilbert, K. M., & Shi, Y. (2024). Quantitatively analysing the driving factors of urban spatial evolution in Lagos (2000-2020). *OAlib, 11*(3), 1–22. DOI: 10.4236/oalib.1111303

Goal 13: Climate action - The Global Goals. (2024, January 23). The Global Goals. https://globalgoals.org/goals/13-climate-action/

Google Dataset Search. (n.d.). *Optimizing smart grid network structures and enhancing communication efficiency* [Dataset]. Google. Retrieved September 21, 2025, from https://datasetsearch.research.google.com/search?src=0&query=Optimizing%20Smart%20Grid%20Network%20Structures%20and%20Enhancing%20Communication%20Efficiency&docid=L2cvMTF4MGJ3MGo5ZA%3D%3D

Grillas, P., Rhazi, L., Lefebvre, G., El Madihi, M., & Poulin, B. (2021). Foreseen impact of climate change on temporary ponds located along a latitudinal gradient in Morocco. *Inland Waters, 11*(4), 492–507. DOI: 10.1080/20442041.2021.1962688

Gu, J., Su, Z., Gao, R., Wang, Y., Meng, Y., & Kong, Q. (2022). Improving QGA-ELM Inversion Model of Rice Leaf Area Index Based on UAV Remote Sensing Image. *Mobile Information Systems, 2022*(1), 9658966. DOI: 10.1155/2022/9658966

Guo, Q., Pu, R., Li, J., & Cheng, J. (2017). A weighted normalized difference water index for water extraction using Landsat imagery. *International Journal of Remote Sensing, 38*(19), 5430–5445. DOI: 10.1080/01431161.2017.1341667

Hakkou, M., & Maanan, M. (2023). *Évaluation multi-décennale des changements du littoral à l ' aide d ' outils géospatiaux et de calculs automatiques sur la côte de Kenitra, au Maroc*. 232-239.

Hakkou, M., Maanan, M., Belrhaba, T., El Khalidi, K., El Ouai, D., & Benmohammadi, A. (2018). Multi-decadal assessment of shoreline changes using geospatial tools and automatic computation in Kenitra coast, Morocco. *Ocean and Coastal Management, 163*, 232–239. DOI: 10.1016/j.ocecoaman.2018.07.003

Hall, J. W., Tarantola, S., Bates, P. D., & Horritt, M. S. (2005). Distributed Sensitivity Analysis of Flood Inundation Model Calibration. *Journal of Hydraulic Engineering (New York, N.Y.), 131*(2), 117–126. DOI: 10.1061/(ASCE)0733-9429(2005)131:2(117)

Hamidifar, H., Nones, M., & Rowinski, P. M. (2024). Flood modeling and fluvial dynamics: A scoping review on the role of sediment transport. *Earth-Science Reviews, 253*, 104775. DOI: 10.1016/j.earscirev.2024.104775

Hancock, G. R. (2005). The use of digital elevation models in the identification and characterization of catchments over different grid scales. *Hydrological Processes, 19*(9), 1727–1749. DOI: 10.1002/hyp.5632

Han, J., Kamber, M., & Pei, J. (2011). *Data Mining: Concepts and Techniques* (3rd ed.). Morgan Kaufmann.

Hao, Z., Singh, V. P., & Xia, Y. (2018). Seasonal drought prediction: Advances, challenges, and future prospects. *Reviews of Geophysics*, *56*(1), 108–141. 10.1002/2016RG000549. DOI: 10.1002/2016RG000549

Hasan, M. K., Habib, A. A., Islam, S., Balfaqih, M., Alfawaz, K. M., & Singh, D. (2023). Smart grid communication networks for electric vehicles empowering distributed energy generation: Constraints, challenges, and recommendations. *Energies*, *16*(3), 1140. DOI: 10.3390/en16031140

He, H., Leng, Y., Cao, X., Zhu, Y., Li, X., Yuan, Q., Zhang, B., He, W., Wei, H., Liu, X., Xu, Q., Guo, M., Zhang, H., Yang, L., Lv, Y., Wang, X., Shi, C., Zhang, Z., Chen, W., & Shang, L. (2024). The pan-tandem repeat map highlights multiallelic variants underlying gene expression and agronomic traits in rice. *Nature Communications*, *15*(1), 7291. DOI: 10.1038/s41467-024-51854-0 PMID: 39181885

He, H., Li, X., & Li, T. (2025). The Sustainable Development of Wetlands and Agriculture: A Literature Review. *Agronomy (Basel)*, *15*(3), 746. DOI: 10.3390/agronomy15030746

Heidecke, C. (2009). *Economic analysis of water use and management in the Middle Drâa Valley in Morocco.* Peter Lang International Academic Publishers. https://books.google.co.ma/books?id=h0l50AEACAAJ

Hengl, T. (2018). Soil texture classes (USDA system) for 6 soil depths (0, 10, 30, 60, 100 and 200 cm) at 250 m (v0.2) [Data set]. Zenodo. DOI: 10.5281/zenodo.2525817

He, Y., Tu, B., Liu, B., Chen, Y., Li, J., & Plaza, A. (2024). Hybrid multi-scale spatial-spectral Transformer for hyperspectral image classification. *IEEE Transactions on Geoscience and Remote Sensing*, *62*, 1–18. Advance online publication. DOI: 10.1109/TGRS.2024.3443662

Hillel, D. (2003). *Introduction to Environmental Soil Physics*. Academic Press.

Himmelstoss, E. A., Henderson, R. E., Kratzmann, M. G., & Farris, A. S. (2021). Digital Shoreline Analysis System (DSAS) version 5.1 user guide. In *Open-File Report* (Nos. 2021-1091). U.S. Geological Survey. DOI: 10.3133/ofr20211091

Hirabayashi, Y., Mahendran, R., Koirala, S., Konoshima, L., Yamazaki, D., Watanabe, S., Kim, H., & Kanae, S. (2013). Global flood risk under climate change. *Nature Climate Change*, *3*(9), 816–821. DOI: 10.1038/nclimate1911

Hochreiter, S., & Schmidhuber, J. (1997). Long Short-Term Memory. *Neural Computation*, *9*(8), 1735–1780. DOI: 10.1162/neco.1997.9.8.1735 PMID: 9377276

Holtz, R. D. (1981). Discussion of "Particle Contacts in Discrete Materials.". *Journal of the Geotechnical Engineering Division, 107*(10), 1437–1440. DOI: 10.1061/AJGEB6.0001199

Horritt, M. S., & Bates, P. D. (2002). Evaluation of 1D and 2D numerical models for predicting river flood inundation. *Journal of Hydrology (Amsterdam), 268*(1–4), 87–99. DOI: 10.1016/S0022-1694(02)00121-X

Hosmer, D. W., Lemeshow, S., & Sturdivant, R. X. (2013). *Applied Logistic Regression* (3rd ed.). Wiley. DOI: 10.1002/9781118548387

Hou, A. Y., Kakar, R. K., Neeck, S., Azarbarzin, A. A., Kummerow, C. D., Kojima, M., Oki, R., Nakamura, K., & Iguchi, T. (2014). The global precipitation measurement mission. *Bulletin of the American Meteorological Society, 95*(5), 701–722. 10.1175/BAMS-D-13-00164.1. DOI: 10.1175/BAMS-D-13-00164.1

IACWD (1982). Water resources data. Annual Report. DOI: 10.3133/891

Immerzeel, W. W., Van Beek, L. P., & Bierkens, M. F. (2010). Climate change will affect the Asian water towers. *science, 328*(5984), 1382-1385. 10.1126/science.1183188

Imran, M., Majeed, M. D., Zaman, M., Shahid, M. A., Zhang, D., Zahra, S. M., Sabir, R. M., Safdar, M., & Maqbool, Z. (2023). Artificial neural networks and regression modeling for water resources management in the upper Indus Basin. *Environmental Sciences Proceedings, 25*(1), 53. 10.3390/ECWS-7-14199

Ingle Smith, D. (1999). Floods: physical processes and human impacts by K. Smith and R. Ward, John Wiley, Chichester, 1998. No. of pages: 382. Earth Surface Processes and Landforms, 24(13), 1261–1261. Portico. https://doi.org/DOI: 10.1002/(sici)1096-9837(199912)24:13<1261:aid-esp22>3.0.co;2-#

Inman, V., & Lyons, M. (2020). Automated inundation mapping over large areas using Landsat data and Google Earth Engine. *Remote Sensing (Basel), 12*(8), 1348. DOI: 10.3390/rs12081348

Intergovernmental Panel on Climate Change (IPCC). (2007). https://www.ipcc.ch/report/ar4/syr/ last access: 05.09.2025

Intergovernmental Panel on Climate Change (IPCC). (2023). Climate Change 2022 – Impacts, Adaptation and Vulnerability. DOI: 10.1017/9781009325844

Ishizaka, A., & Labib, A. (2011). Review of the main developments in the Analytic Hierarchy Process. *Expert Systems with Applications, 38*(11), 14336–14345. DOI: 10.1016/j.eswa.2011.04.143

Iverson, R. M. (1997). The physics of debris flows. Reviews of Geophysics, 35(3), 245–296. Portico. DOI: 10.1029/97RG00426

Iverson, R. M. (2000). Landslide triggering by rain infiltration. Water Resources Research, 36(7), 1897–1910. Portico. DOI: 10.1029/2000WR900090

Jha, K., Doshi, A., Patel, P., & Shah, M. (2019). A comprehensive review on automation in agriculture using artificial intelligence. *Artificial Intelligence in Agriculture, 2*, 1-12. 10.1016/j.aiia.2019.05.004

Jiang, D., Chen, S., Useya, J., Cao, L., & Lu, T. (2022). Crop mapping using the historical crop data layer and deep neural networks: A case study in jilin province, china. *Sensors (Basel), 22*(15), 5853. 10.3390/s22155853. DOI: 10.3390/s22155853 PMID: 35957410

Jonkman, S. N. (2005). Global Perspectives on Loss of Human Life Caused by Floods. *Natural Hazards, 34*(2), 151–175. DOI: 10.1007/s11069-004-8891-3

Jonkman, S. N., & Kelman, I. (2005). An Analysis of the Causes and Circumstances of Flood Disaster Deaths. *Disasters, 29*(1), 75–97. DOI: 10.1111/j.0361-3666.2005.00275.x

Jonkman, S. N., & Penning-Rowsell, E. (2008). Human instability in flood flows 1. *Journal of the American Water Resources Association, 44*(5), 1208–1218. DOI: 10.1111/j.1752-1688.2008.00217.x

Julien, P. Y. (2010). Erosion and Sedimentation. DOI: 10.1017/CBO9780511806049

Junk, W. J. (1999). The flood pulse concept of large rivers: Learning from the tropics. *River Systems, 11*(3), 261–280. DOI: 10.1127/lr/11/1999/261

Jurišić, M., Plaščak, I., Barač, Ž., Radočaj, D., & Zimmer, D. (2021). Sensors and their application in precision agriculture. *Tehnički glasnik, 15*(4), 529-533. 10.31803/tg-20201015132216

Kaczmarek, N., Mahjoubi, I., Naimi, M., Bossenbroek, L., Moumane, A., Znari, M., Luis, M. S.-N. S., Frör, O., & Berger, E. (2025). Nature Conservation in the Draa Basin (Morocco): History, Present Situation, and Future Challenges. *Journal for Nature Conservation, 88*, 127038. DOI: 10.1016/j.jnc.2025.127038

Kader, A., & Mahmoud, R. A. (2019). Strategies for Urban Residual Areas as a planning tool: Towards achieving sustainable built environment by connecting neighbourhoods. *IOP Conference Series. Earth and Environmental Science, 329*(1), 012018. DOI: 10.1088/1755-1315/329/1/012018

Karmaoui, A., & Moumane, A. (2016). Changes in the Environmental Vulnerability of Oasean System (Desert Oasis), Pilot Study in Middle Draa Valley, Morocco. *Expert Opinion on Environmental Biology*, *5*(3). Advance online publication. DOI: 10.4172/2325-9655.1000135

Kaur, A., Bhatt, D. P., & Raja, L. (2024). Developing a hybrid irrigation system for smart agriculture using IoT sensors and machine learning in Sri Ganganagar, Rajasthan. *Journal of Sensors*, *2024*(1), 6676907. 10.1155/2024/6676907. DOI: 10.1155/2024/6676907

Kazakis, N., Giannakis, E., & Verveniotis, G. (2015). Assessment of flood hazard areas at a regional scale using an index-based approach and Analytical Hierarchy Process: Application in Rhodope–Evros region, Greece. *The Science of the Total Environment*, *538*, 555–563. DOI: 10.1016/j.scitotenv.2015.08.055 PMID: 26318691

Kenawy, A. E., Al-Awadhi, T., Abdullah, M., Ostermann, F. O., & Abulibdeh, A. (2025). A Multidecadal Assessment of Drought Intensification in the Middle East and North Africa: The Role of Global Warming and Rainfall Deficit. *Earth Systems and Environment*. Advance online publication. DOI: 10.1007/s41748-025-00576-4

Kesavan, H., Choudhary, S., Khan, S., Samhita, R., & Sharma, A. (2024). AI-based urban resilience planning: Opportunities and challenges. *Journal of Environment and Earth Science*, *6*(2), 200–214. DOI: 10.30564/jees.v6i2.6681

Khaki, S., & Wang, L. (2019). Crop yield prediction using deep neural networks. *Frontiers in Plant Science*, *10*, 621. 10.3389/fpls.2019.00621. DOI: 10.3389/fpls.2019.00621 PMID: 31191564

Khanal, N. R., Shrestha, M., & Ghimire, M. (2007). Preparing for floods: Mapping and assessing hazard in the Ratu watershed, Nepal. *ICIMOD Technical Paper*, *7*, 1–34.

Khatakho, R., Kayastha, P., & Pathak, D. (2021). Multi-hazard risk assessment and mapping in Nepal using geospatial tools and AHP. *Geoscience Frontiers*, *12*(1), 101104. DOI: 10.1016/j.gsf.2020.08.006

Khosravi, K., Pham, B. T., Chapi, K., Shirzadi, A., Shahabi, H., Revhaug, I., Prakash, I., & Tien Bui, D. (2018). A comparative assessment of decision trees algorithms for flash flood susceptibility modeling at Haraz watershed, northern Iran. *The Science of the Total Environment*, *627*, 744–755. DOI: 10.1016/j.scitotenv.2018.01.266

Kidd, C., & Huffman, G. (2011). Global precipitation measurement. *Meteorological Applications*, *18*(3), 334–353. 10.1002/met.284. DOI: 10.1002/met.284

Kiran Kumar, V., Ramesh, K. V., & Rakesh, V. (2023). Optimizing LSTM and Bi-LSTM models for crop yield prediction and comparison of their performance with traditional machine learning techniques. *Applied Intelligence*, *53*(23), 28291–28309. DOI: 10.1007/s10489-023-05005-5

Kirubakaran, A. P., & Midhunchakkaravarthy, J. (2023). A hybrid application of quantum computing methodologies to AI techniques for paddy crop leaf disease identification. In *Integrating blockchain and artificial intelligence for industry 4.0 innovations* (pp. 69–83). Springer International Publishing.

Klerkx, L., Jakku, E., & Labarthe, P. (2019). A review of social science on digital agriculture, smart farming and agriculture 4.0: New contributions and a future research agenda. *NJAS Wageningen Journal of Life Sciences*, *90*(1), 100315. 10.1016/j.njas.2019.100315. DOI: 10.1016/j.njas.2019.100315

Knighton, A. D. (1998). Fluvial Forms and Processes: A New Perspective, Arnold, London,383 p., ill., tabl, pl., 15,5 x 23,5 cm. ISBN 0 340 66313 8. *Géographie Physique et Quaternaire*, *53*(3), 413. DOI: 10.7202/004839ar

Kolluru, V., Mungara, S., Chintakunta, A. N., Kolluru, L., & Telaganeni, C. S. (2024). Geospatial Intelligence Enhancement Using Advanced Data Science and Machine Learning: A Systematic Literature Review. http://dx.doi.org/DOI: 10.5121/csit.2024.141502

Komi, K., Neal, J., Trigg, M. A., & Diekkrüger, B. (2017). Modelling of flood hazard extent in data sparse areas: A case study of the Oti River basin, West Africa. *Journal of Hydrology (Amsterdam)*, *555*, 330–345. DOI: 10.1016/j.jhydrol.2017.10.004

Kotlarski, S., Keuler, K., Christensen, O. B., Colette, A., Déqué, M., Gobiet, A., Goergen, K., Jacob, D., Lüthi, D., van Meijgaard, E., Nikulin, G., Schär, C., Teichmann, C., Vautard, R., Warrach-Sagi, K., & Wulfmeyer, V. (2014). Regional climate modeling on European scales: A joint standard evaluation of the EURO-CORDEX RCM ensemble. *Geoscientific Model Development*, *7*(4), 1297–1333. DOI: 10.5194/gmd-7-1297-2014

Kotsiantis, S. B. (2013). Decision trees: A recent overview. *Artificial Intelligence Review*, *39*(4), 261–283. DOI: 10.1007/s10462-011-9272-4

Kozlowski, T. T. (1997). Responses of woody plants to flooding and salinity. *Tree Physiology*, *17*(7), 490–490. DOI: 10.1093/treephys/17.7.490

Kratzert, F., Klotz, D., Brenner, C., Schulz, K., & Herrnegger, M. (2018). Rainfall–runoff modelling using long short-term memory (LSTM) networks. *Hydrology and Earth System Sciences*, *22*(11), 6005–6022. 10.5194/hess-22-6005-2018. DOI: 10.5194/hess-22-6005-2018

Krishnan, S. R., Nallakaruppan, M., Chengoden, R., Koppu, S., Iyapparaja, M., Sadhasivam, J., & Sethuraman, S. (2022). Smart water resource management using Artificial Intelligence—A review. *Sustainability (Basel)*, *14*(20), 13384. 10.3390/su142013384. DOI: 10.3390/su142013384

Kundzewicz, Z. W., Kanae, S., Seneviratne, S. I., Handmer, J., Nicholls, N., Peduzzi, P., Mechler, R., Bouwer, L. M., Arnell, N., Mach, K., Muir-Wood, R., Brakenridge, G. R., Kron, W., Benito, G., Honda, Y., Takahashi, K., & Sherstyukov, B. (2013). Flood risk and climate change: Global and regional perspectives. *Hydrological Sciences Journal*, *59*(1), 1–28. DOI: 10.1080/02626667.2013.857411

Kundzewicz, Z. W., Mata, L., Arnell, N. W., Döll, P., Jimenez, B., Miller, K., Oki, T., Şen, Z., & Shiklomanov, I. (2008). The implications of projected climate change for freshwater resources and their management. *Hydrological Sciences Journal*, *53*(1), 3–10. 10.1623/hysj.53.1.3. DOI: 10.1623/hysj.53.1.3

Kutner, M. H., Nachtsheim, C., Neter, J., & Li, W. (2005). *Applied Linear Statistical Models* (5th ed.). McGraw-Hill Education.

Kvočka, D., Falconer, R. A., & Bray, M. (2016). Flood hazard assessment for extreme flood events. *Natural Hazards*, *84*(3), 1569–1599. DOI: 10.1007/s11069-016-2501-z

Laaribya, S., & Alaoui, A. (2025). Spatio-Temporal Trends and Climate Change Impacts on Land Cover Dynamics in Talassemtane National Park (Morocco) Using Artificial Intelligence and Google Earth Engine. *Journal of Sustainable Forestry*, *44*(6), 289–318. DOI: 10.1080/10549811.2025.2516072

Laassilia, O. (2013). *Approche multidisciplinaire de suivi de la dynamique marine du littoral de Mohammedia*. Université Cadi Ayyad,Faculté des Sciences et Techniques Marrakech.

Lahane, R., Manza, R. R., Ambhore, S., & Katkade, S. (2025). Remote sensing and GIS for LULC change detection: A review of techniques and findings. [IJRASET]. *International Journal for Research in Applied Science and Engineering Technology*, *13*(7), 1234–1245. DOI: 10.22214/ijraset.2025.72941

Lamb, R., Keef, C., Tawn, J., Laeger, S., Meadowcroft, I., Surendran, S., Dunning, P., & Batstone, C. (2010). A new method to assess the risk of local and widespread flooding on rivers and coasts. Journal of Flood Risk Management, 3(4), 323–336. Portico. DOI: 10.1111/j.1753-318X.2010.01081.x

Lan, M., Liu, C., Zheng, H., Wang, Y., Cai, W., Peng, Y., Xu, C., & Tan, S. (2024). Rice-yolo: In-field rice spike detection based on improved yolov5 and drone images. *Agronomy (Basel)*, *14*(4), 836. DOI: 10.3390/agronomy14040836

Lee, J., & Choi, H. (2018). Comparison of Flood Vulnerability Assessments to Climate Change by Construction Frameworks for a Composite Indicator. *Sustainability (Basel)*, *10*(3), 768. DOI: 10.3390/su10030768

Lee, S., Lee, K.-K., & Yoon, H. (2019). Using artificial neural network models for groundwater level forecasting and assessment of the relative impacts of influencing factors. *Hydrogeology Journal*, *27*(2), 567–579. 10.1007/s10040-018-1866-3. DOI: 10.1007/s10040-018-1866-3

Le, N. Q. K., Ho, Q. T., Nguyen, V. N., & Chang, J. S. (2022). BERT-Promoter: An improved sequence-based predictor of DNA promoter using BERT pre-trained model and SHAP feature selection. *Computational Biology and Chemistry*, *99*, 107732. DOI: 10.1016/j.compbiolchem.2022.107732 PMID: 35863177

Liakos, K. G., Busato, P., Moshou, D., Pearson, S., & Bochtis, D. (2018). Machine learning in agriculture: A review. *Sensors (Basel)*, *18*(8), 2674. 10.3390/s18082674. DOI: 10.3390/s18082674 PMID: 30110960

Liaw, A., & Wiener, M. (2002). Classification and regression by randomForest. *R News*, *2*(3), 18–22.

Li, B., Rodell, M., Kumar, S., Beaudoing, H., Getirana, A., Zaitchik, B. F., de Goncalves, L. G., Cossetin, C., Bhanja, S., Mukherjee, A., Tian, S., Tangdamrongsub, N., Long, D., Nanteza, J., Lee, J., Policelli, F., Goni, I. B., Daira, D., Bila, M., & Bettadpur, S. (2019). Global GRACE data assimilation for groundwater and drought monitoring: Advances and challenges. *Water Resources Research*, *55*(9), 7564–7586. DOI: 10.1029/2018WR024618

Li, C., Zhang, L., Ou, Z., Wang, Q., Zhou, D., & Ma, J. (2022). Robust model of electric vehicle charging station location considering renewable energy and storage equipment. *Energy*, *238*, 121713. DOI: 10.1016/j.energy.2021.121713

Li, H., Huang, J., Zhang, C., Ning, X., Zhang, S., & Atkinson, P. M. (2024). An efficient and generalisable approach for mapping paddy rice fields based on their unique spectra during the transplanting period leveraging the CIE colour space. *Remote Sensing of Environment, 313*, 114381. DOI: 10.1016/j.rse.2024.114381

Li, Q., Li, Z., Shangguan, W., Wang, X., Li, L., & Yu, F. (2022). Improving soil moisture prediction using a novel encoder–decoder model with residual learning. *Computers and Electronics in Agriculture, 195*, 106816. DOI: 10.1016/j.compag.2022.106816

Li, S., Zhu, P., Song, N., Li, C., & Wang, J. (2025). Regional Soil Moisture Estimation Leveraging Multi-Source Data Fusion and Automated Machine Learning. *Remote Sensing (Basel), 17*(5), 837. Advance online publication. DOI: 10.3390/rs17050837

Liu, N., Sun, Z., Zhang, H., Klausen, L. H., Moonhee, R., & Kang, S. (2023). Emerging high-ammonia-nitrogen wastewater remediation by biological treatment and photocatalysis techniques. *The Science of the Total Environment, 875*, 162603. DOI: 10.1016/j.scitotenv.2023.162603 PMID: 36871738

Liu, R., Chen, Y., Wu, J., Gao, L., Barrett, D., Xu, T., Li, L., Huang, C., & Yu, J. (2016). Assessing spatial likelihood of flooding hazard using naïve Bayes and GIS: A case study in Bowen Basin, Australia. *Stochastic Environmental Research and Risk Assessment, 30*(6), 1575–1590. DOI: 10.1007/s00477-015-1198-y

Li, Y., Xue, C., Zargari, F., & Li, Y. R. (2023). From graph theory to graph neural networks (GNNs): The opportunities of GNNs in power electronics. *IEEE Access : Practical Innovations, Open Solutions, 11*, 145067–145084. DOI: 10.1109/ACCESS.2023.3345795

Locatelli, B., Rojas, V., & Salinas, Z. (2008). Impacts of payments for environmental services on local development in northern Costa Rica: A fuzzy multi-criteria analysis. *Forest Policy and Economics, 10*(5), 275–285. DOI: 10.1016/j.forpol.2007.11.007

Loucks, D. P., & Van Beek, E. (2017). *Water resource systems planning and management: An introduction to methods, models, and applications.* Springer. DOI: 10.1007/978-3-319-44234-1

Luijendijk, A., Hagenaars, G., Ranasinghe, R., Baart, F., Donchyts, G., & Aarninkhof, S. (2018). The State of the World's Beaches. *Scientific Reports, 8*(1), 1–11. DOI: 10.1038/s41598-018-24630-6 PMID: 29311619

Lu, J., Hu, J., Zhao, G., Mei, F., & Zhang, C. (2017). An in-field automatic wheat disease diagnosis system. *Computers and Electronics in Agriculture, 142*, 369–379. 10.1016/j.compag.2017.09.012. DOI: 10.1016/j.compag.2017.09.012

Lu, J., Li, J., Fu, H., Tang, X., Liu, Z., Chen, H., Sun, Y., & Ning, X. (2024). Deep learning for multi-source data-driven crop yield prediction in northeast China. *Agriculture, 14*(6), 794. 10.3390/agriculture14060794. DOI: 10.3390/agriculture14060794

Maes, W. H., & Steppe, K. (2019). Perspectives for remote sensing with unmanned aerial vehicles in precision agriculture. *Trends in Plant Science, 24*(2), 152–164. 10.1016/j.tplants.2018.11.007. DOI: 10.1016/j.tplants.2018.11.007 PMID: 30558964

Maier, H. R., Jain, A., Dandy, G. C., & Sudheer, K. P. (2010). Methods used for the development of neural networks for the prediction of water resource variables in river systems: Current status and future directions. *Environmental Modelling & Software, 25*(8), 891–909. 10.1016/j.envsoft.2010.02.003. DOI: 10.1016/j.envsoft.2010.02.003

Maiti, S., & Bhattacharya, A. K. (2009). Shoreline change analysis and its application to prediction : A remote sensing and statistics based approach. *Marine Geology, 257*(1-4), 11–23. DOI: 10.1016/j.margeo.2008.10.006

Makridakis, S. (1993). Accuracy Measures: Theoretical and Practical Concerns. *International Journal of Forecasting, 9*(4), 527–529. DOI: 10.1016/0169-2070(93)90079-3

Malczewski, J. (2006). GIS-based multicriteria decision analysis: A survey of the literature. *International Journal of Geographical Information Science, 20*(7), 703–726. DOI: 10.1080/13658810600661508

Mallett, R., Hagen-Zanker, J., Slater, R., & Duvendack, M. (2012). The benefits and challenges of using systematic reviews in international development research. *Journal of Development Effectiveness, 4*(3), 445–455. DOI: 10.1080/19439342.2012.711342

Maneepong, K., Yamanotera, R., Akiyama, Y., Miyazaki, H., Miyazawa, S., & Akiyama, C. M. (2025). Towards High-Resolution Population Mapping: Leveraging Open Data, Remote Sensing, and AI for Geospatial Analysis in Developing Country Cities—A Case Study of Bangkok. *Remote Sensing (Basel), 17*(7), 1204. DOI: 10.3390/rs17071204

Mansour, S., Alotaibi, O., & Alnasrallah, M. (2025). Geospatial modelling of urban expansion effects on the land ecosystems in Kuwait using random forest and cellular automata. Journal of Urban Management. https://doi.org/DOI: 10.1016/j.jum.2025.05.011

Marks, K., & Bates, P. (2000). Integration of high-resolution topographic data with floodplain flow models. *Hydrological Processes, 14*(11–12), 2109–2122. DOI: 10.1002/1099-1085(20000815/30)14:11/12<2109::AID-HYP58>3.0.CO;2-1

Marshall, J. D., Brauer, M., & Frank, L. D. (2009). Healthy neighborhoods: Walkability and air pollution. *Environmental Health Perspectives, 117*(11), 1752–1759. DOI: 10.1289/ehp.0900595

Martínez-Gomariz, E., Gómez, M., Russo, B., & Djordjević, S. (2016). Stability criteria for flooded vehicles: a state-of-the-art review. Journal of Flood Risk Management, 11(S2). Portico. DOI: 10.1111/jfr3.12262

Massam, B. H. (1988). Multi-Criteria Decision Making (MCDM) techniques in planning. *Progress in Planning, 30*, 1–84. DOI: 10.1016/0305-9006(88)90012-8

Maxwell, A. E., Warner, T. A., & Fang, F. (2018). Implementation of machine-learning classification in remote sensing: An applied review. *International Journal of Remote Sensing, 39*(9), 2784–2817. 10.1080/01431161.2018.1433343. DOI: 10.1080/01431161.2018.1433343

Ma, Y., Wu, H., Wang, L., Huang, B., Ranjan, R., Zomaya, A., & Jie, W. (2015). Remote sensing big data computing: Challenges and opportunities. *Future Generation Computer Systems, 51*, 47–60. 10.1016/j.future.2014.10.029. DOI: 10.1016/j.future.2014.10.029

McFeeters, S. K. (1996). The use of the Normalized Difference Water Index (NDWI) in the delineation of open water features. *International Journal of Remote Sensing, 17*(7), 1425–1432. DOI: 10.1080/01431169608948714

Meinam, M., Lahiri, B., Ghosh, A., Singh, S. K., Bharati, H., Upadhyay, A. D., Pal, P., Sanjenbam, B., & Ngasotter, S. (2025). Aquatic Ecosystem, Fish Diversity and Conservation Strategies of the Fragile Eastern Himalayas: A Comprehensive Review with a Focus on Loktak Lake. *Wetlands, 45*(7), 96. DOI: 10.1007/s13157-025-01978-w

Menard, S. (2002). *Applied Logistic Regression Analysis* (2nd ed.). Sage Publications. DOI: 10.4135/9781412983433

Meng, K., Wu, Q., Xu, J., Chen, W., Feng, Z., Schober, R., & Swindlehurst, A. L. (2023). UAV-enabled integrated sensing and communication: Opportunities and challenges. *IEEE Wireless Communications, 31*(2), 97–104. DOI: 10.1109/MWC.131.2200442

Merz, R., & Blöschl, G. (2003). A process typology of regional floods. Water Resources Research, 39(12). Portico. DOI: 10.1029/2002WR001952

Merz, B., Hall, J., Disse, M., & Schumann, A. (2010). Fluvial flood risk management in a changing world. *Natural Hazards and Earth System Sciences, 10*(3), 509–527. DOI: 10.5194/nhess-10-509-2010

Mezair, T., Djenouri, Y., Belhadi, A., Srivastava, G., & Lin, J. C. W. (2022). A sustainable deep learning framework for fault detection in 6G Industry 4.0 heterogeneous data environments. *Computer Communications*, *187*, 164–171. DOI: 10.1016/j.comcom.2022.02.010

Miljkovic, B., Zizovic, M., Petojevic, A., & Damljanovic, N. (2017). New weighted sum model. *Filomat*, *31*(10), 2991–2998. DOI: 10.2298/FIL1710991M

Millar, R. G., & Quick, M. C. (1998). Stable Width and Depth of Gravel-Bed Rivers with Cohesive Banks. *Journal of Hydraulic Engineering (New York, N.Y.)*, *124*(10), 1005–1013. DOI: 10.1061/(ASCE)0733-9429(1998)124:10(1005)

Miller, T., Mikiciuk, G., Durlik, I., Mikiciuk, M., Łobodzińska, A., & Śnieg, M. (2025). The IoT and AI in Agriculture: The Time Is Now—A Systematic Review of Smart Sensing Technologies. *Sensors (Basel)*, *25*(12), 3583. 10.3390/s25123583. DOI: 10.3390/s25123583 PMID: 40573470

Mir, Y. H., Mir, S., Ganie, M. A., Bhat, J. A., Shah, A. M., Mushtaq, M., & Irshad, I. (2025). Overview of land use and land cover change and its impacts on natural resources. In Ecologically Mediated Development: Promoting Biodiversity Conservation and Food Security (pp. 101-130). Singapore: Springer Nature Singapore. DOI: 10.1007/978-981-96-2413-3_5

Mishra, A. K., & Coulibaly, P. (2009). Developments in hydrometric network design: A review. *Reviews of Geophysics*, *47*(2), 2007RG000243. 10.1029/2007RG000243. DOI: 10.1029/2007RG000243

Montgomery, D. R. (2007). Soil erosion and agricultural sustainability. *Proceedings of the National Academy of Sciences of the United States of America*, *104*(33), 13268–13272. DOI: 10.1073/pnas.0611508104

Moore, I. D., Grayson, R. B., & Ladson, A. R. (1991). Digital terrain modeling: A review of hydrological, geomorphological, and biological applications. *Hydrological Processes*, *5*(1), 3–30. DOI: 10.1002/hyp.3360050103

Mosavi, A., Ozturk, P., & Chau, K. (2018). Flood prediction using machine learning models: Literature review. *Water (Basel)*, *10*(11), 1536. DOI: 10.3390/w10111536

Moumane, A., Al Karkouri, J., & Batchi, M. (2025). Utilizing GIS, remote sensing, and AHP-multi-criteria decision analysis for optimal landfill site selection in Kenitra Province, Morocco: A step towards sustainable development goals. *Discover Environment*, *3*(1), 4. DOI: 10.1007/s44274-025-00183-0

Moumane, A., Al Karkouri, J., & Mouhcine, B. (2024). Detailed JavaScript API for Google Earth Engine to Enhance Monitoring of Surface Water Dynamics in the Drought-Affected Mediterranean Region. In Batchi, M., & Moumane, A. (Eds.), (pp. 105–132). Advances in Geospatial Technologies. IGI Global., DOI: 10.4018/979-8-3693-9651-3.ch004

Moumane, A., Bahouq, T., & Ghazi, A. K. (2025). Integrating GIS, remote sensing, and analytic hierarchical process (AHP) for mapping of groundwater-recharge potential zones in Kenitra Province, Morocco: Toward sustainable development goals achievement. *Euro-Mediterranean Journal for Environmental Integration*, *10*(2), 831–851. DOI: 10.1007/s41207-024-00706-w

Moumane, A., Bahouq, T., Karmaoui, A., Laghfiri, D., Yassine, M., Karkouri, J. A., Batchi, M., Faouzi, M., Boulakhbar, M., & Youssef, A. A. (2025). Lake Iriqui's Remarkable Revival: Field Observations and a Google Earth Engine Analysis of Its Recovery After over Half a Century of Desiccation. *Land (Basel)*, *14*(1), 104. DOI: 10.3390/land14010104

Moumane, A., El Ghazali, F. E., Al Karkouri, J., Delorme, J., Batchi, M., Chafiki, D., & Karmaoui, A. (2021). Monitoring spatiotemporal variation of groundwater level and salinity under land use change using integrated field measurements, GIS, geostatistical, and remote-sensing approach: Case study of the Feija aquifer, Middle Draa watershed, Moroccan Sahara. *Environmental Monitoring and Assessment*, *193*(12), 769. DOI: 10.1007/s10661-021-09581-2 PMID: 34735624

Moumane, A., Elmotawakkil, A., Hasan, M. M., Kranjčić, N., Batchi, M., & Karkouri, J. A. & M. Youssef, Y. (2025). Integrating GIS, Remote Sensing, and Machine Learning to Optimize Sustainable Groundwater Recharge in Arid Mediterranean Landscapes: A Case Study from the Middle Draa Valley, Morocco. *Water (Basel)*, *17*(15), 2336.

Moumane, A., Enajar, A. A., Ghazali, F. E. E., Khouz, A., Karmaoui, A., Al Karkouri, J., & Batchi, M. (2024). *GIS, remote sensing, and analytical hierarchy process (AHP) approach for rainwater harvesting site selection in arid regions: Feija Plain case study, Zagora*. Applied Geomatics., DOI: 10.1007/s12518-024-00585-4

Mukherjee, S., Rizvi, S. S., Biswas, G., Paswan, A. K., Vaiphei, S. P., Warsi, T., & Mitran, T. (2023). Aquatic Eco-systems Under Influence of Climate Change and Anthropogenic Activities. In *Hydrogeochemistry of Aquatic Ecosystems* (pp. 307–331). John Wiley & Sons, Ltd., DOI: 10.1002/9781119870562.ch14

Mulla, D. J. (2013). Twenty five years of remote sensing in precision agriculture: Key advances and remaining knowledge gaps. *Biosystems Engineering*, *114*(4), 358–371. 10.1016/j.biosystemseng.2012.08.009. DOI: 10.1016/j.biosystemseng.2012.08.009

Muñiz, I., & Dominguez, A. (2020). The impact of urban form and spatial structure on per capita carbon footprint in U.S. larger metropolitan areas. *Sustainability (Basel)*, *12*(1), 389. DOI: 10.3390/su12010389

Murase, H., Nagashima, H., Yonezaki, S., Matsukura, R., & Kitakado, T. (2009). Application of a generalized additive model (GAM) to reveal relationships between environmental factors and distributions of pelagic fish and krill: A case study in Sendai Bay, Japan. *ICES Journal of Marine Science*, *66*(6), 1417–1424. DOI: 10.1093/icesjms/fsp105

Mushtaq, M., Ali, H., Raza, A., Maqbool, S., Safdar, M., Ahmed, M., & Sattar, J. (2024). Precision irrigation for sustainable agricultural productivity. In *Emerging Technologies and Marketing Strategies for Sustainable Agriculture* (pp. 184–208). IGI Global Scientific Publishing., DOI: 10.4018/979-8-3693-4864-2.ch010

Mustafa, S., Salahuddin, H., Ashraf, A., Shafique, M., Salahuddin, S., & Akbar, M. (2024). Smart Irrigation System for Wheat in Southern Punjab Using IOT-based Wireless Technology. *Journal of Computing & Biomedical Informatics*, *8*(01). 10.56979/801/2024

Muzammal, H., Zaman, M., Safdar, M., Adnan Shahid, M., Sabir, M. K., Khil, A., Raza, A., Faheem, M., Ahmed, J., & Sattar, J. (2024). Climate change impacts on water resources and implications for agricultural management. In *Transforming agricultural management for a sustainable future: Climate change and machine learning perspectives* (pp. 21–45). Springer., 10.1007/978-3-031-63430-7_2 DOI: 10.1007/978-3-031-63430-7_2

Nagarajan, K., Narwade, R., Pathak, H., Panhalkar, S., Kulkarni, V. S., & Hingmire, A. P. (2022). Review paper for floodplain mapping with applications of HEC-HMS, HEC-RAS, and ArcGIS softwares a remote sensing and GIS approach. [IRJET]. *International Research Journal of Engineering and Technology*, *9*(6), 812–826.

Nair, V. (2024, March 26). Expressway expansion on schedule: Municipality. Oman Observer. https://www.omanobserver.om/article/1151494/oman/transport/expressway-expansion-on-schedule-municipality

Nandi, A., Mandal, A., Wilson, M., & Smith, D. (2016). Flood hazard mapping in Jamaica using principal component analysis and logistic regression. *Environmental Earth Sciences*, *75*(6), 465. DOI: 10.1007/s12665-016-5323-0

NASA. (2024). *NASA Analysis Sees Spike in 2023 Global Sea Level Due to El Niño*. NASA Sea Level Change Portal. https://sealevel.nasa.gov/news/270/nasa-analysis-sees-spike-in-2023-global-sea-level-due-to-el-nino/

Nash, J. E., & Sutcliffe, J. V. (1970). River Flow Forecasting Through Conceptual Models: Part i - a Discussion of Principles. *Journal of Hydrology (Amsterdam)*, *10*(3), 282–290. DOI: 10.1016/0022-1694(70)90255-6

National Agricultural Research Council (NARC). (2020). *Soil Map Data for Nepal*. NARC.

Nawaz, A., Rehman, A. U., Rehman, A., Ahmad, S., Siddique, K. H., & Farooq, M. (2022). Increasing sustainability for rice production systems. *Journal of Cereal Science*, *103*, 103400. DOI: 10.1016/j.jcs.2021.103400

Nayak, P. C., Sudheer, K., Rangan, D., & Ramasastri, K. (2004). A neuro-fuzzy computing technique for modeling hydrological time series. *Journal of Hydrology (Amsterdam)*, *291*(1-2), 52–66. 10.1016/j.jhydrol.2003.12.010. DOI: 10.1016/j.jhydrol.2003.12.010

Neal, J., Schumann, G., Fewtrell, T., Budimir, M., Bates, P., & Mason, D. (2011). Evaluating a new LISFLOOD-FP formulation with data from the summer 2007 floods in Tewkesbury, UK. Journal of Flood Risk Management, 4(2), 88–95. Portico. DOI: 10.1111/j.1753-318X.2011.01093.x

Neumann, B., Vafeidis, A. T., Zimmermann, J., & Nicholls, R. J. (2015). Future Coastal Population Growth and Exposure to Sea-Level Rise and Coastal Flooding—A Global Assessment. *PLoS One*, *10*(3), e0118571. DOI: 10.1371/journal.pone.0118571 PMID: 25760037

Nicklow, J., Reed, P., Savic, D., Dessalegne, T., Harrell, L., Chan-Hilton, A., Karamouz, M., Minsker, B., Ostfeld, A., Singh, A., & Zechman, E. (2010). State of the art for genetic algorithms and beyond in water resources planning and management. *Journal of Water Resources Planning and Management*, *136*(4), 412–432. 10.1061/(ASCE)WR.1943-5452.0000053. DOI: 10.1061/(ASCE)WR.1943-5452.0000053

Nobre, A. D., Cuartas, L. A., Hodnett, M., Rennó, C. D., Rodrigues, G., Silveira, A., Waterloo, M., & Saleska, S. (2011). Height Above the Nearest Drainage – a hydrologically relevant new terrain model. *Journal of Hydrology (Amsterdam)*, *404*(1–2), 13–29. DOI: 10.1016/j.jhydrol.2011.03.051

O'Brien, R. M. (2007). A caution regarding rules of thumb for variance inflation factors. *Quality & Quantity*, *41*(5), 673–690. DOI: 10.1007/s11135-006-9018-6

Obasekore, H., Fanni, M., Ahmed, S. M., Parque, V., & Kang, B.-Y. (2023). Agricultural robot-centered recognition of early-developmental pest stage based on deep learning: A case study on fall armyworm (Spodoptera frugiperda). *Sensors (Basel), 23*(6), 3147. 10.3390/s23063147. DOI: 10.3390/s23063147 PMID: 36991858

Obisesan, K. O., & Oladoja, O. M. (2022). On Normal Process of Diffusion Equation in Monitoring Carbon Monoxide Concentrations in Nigeria. *International Journal of Statistical Distributions and Applications.*, *8*(2), 24–29. DOI: 10.11648/j.ijsd.20220802.11

Odekanle, E. L., Fakinle, B. S., Odejobi, O. J., Akangbe, O. E., Sonibare, J. A., Akeredolu, F. A., & Oladoja, O. M. (2022). COVID-19 induced restriction in developing countries and its impacts on pollution load: A case study of Lagos megacity. *Heliyon*, *8*(8), e10402. DOI: 10.1016/j.heliyon.2022.e10402 PMID: 36065213

OECD. (2023). *Sustainable urban development policies seek to address a range of issues from managing urban expansion and congestion to fostering competitiveness, innovation, social inclusion and environmental sustainability*. https://www.oecd.org/en/topics/sustainable-urban-development.html

Ojha, T., Misra, S., & Raghuwanshi, N. S. (2015). Wireless sensor networks for agriculture: The state-of-the-art in practice and future challenges. *Computers and Electronics in Agriculture*, *118*, 66–84. 10.1016/j.compag.2015.08.011. DOI: 10.1016/j.compag.2015.08.011

Oladoja, O. M., & Adegoke, T. M. (2023). Detection of Cervical Cancer Risk Factors in Venezuela Using the Decision Tree Algorithm. *Reliability: Theory & Applications, 18*(3 74), 492-500.

Oladoja, O. M., Obisesan, K. O., Adebayo, S. A., & Adegoke, T. M. (2023). On the Bayesian modeling of road accidents due to vehicle type in Oyo State, Nigeria. *Reliability: Theory & Applications, 18*(2 73), 223-232.

Oliveira, R., & Loucks, D. P. (1997). Operating rules for multireservoir systems. *Water Resources Research*, *33*(4), 839–852. 10.1029/96WR03745. DOI: 10.1029/96WR03745

Ouarda, T. B. M. J., Girard, C., Cavadias, G. S., & Bobée, B. (2001). Regional flood frequency estimation with canonical correlation analysis. *Journal of Hydrology (Amsterdam)*, *254*(1–4), 157–173. DOI: 10.1016/S0022-1694(01)00488-7

Ougrad, I., Elassassi, Z., Mrabet, A., Mssillou, I., Lim, A., Shahat, A. A., Rezouki, S., & Moubchir, T. (2024). Health Risk Assessment of Trace Elements in Surface Water from Dayat Roumi Lake, Morocco. *Water (Basel), 16*(22), 3231. DOI: 10.3390/w16223231

Oyewole, J. A., & Aro, T. O. (2018). Wind speed pattern in Nigeria (a case study of some coastal and inland areas). *Journal of Applied Science & Environmental Management, 22*(1), 119–123. DOI: 10.4314/jasem.v22i1.22

Ozturk, D., & Batuk, F. (2011). Technique for order preference by similarity to ideal solution (TOPSIS) for spatial decision problems. In Proceedings ISPRS (Vol. 1, No. 4).

Padhiary, M., Saha, D., Kumar, R., Sethi, L. N., & Kumar, A. (2024). Enhancing Precision Agriculture: A Comprehensive Review of Machine Learning and AI Vision Applications in All-Terrain Vehicle for Farm Automation. *Smart Agricultural Technology, 8*, 100483. DOI: 10.1016/j.atech.2024.100483

Pai, P., Amutha, S., Patil, S., Sridharan, S., Shobha, T., & Arjunan, R. V. (2025). Slice Residual U-Net Based Rice Plant Disease Classification Using Convolutional Attentional BiGRU. *Journal of Internet Services and Information Security, 15*(2), 387–408. DOI: 10.58346/JISIS.2025.I2.028

Pandey, P., Kumar, S., Khanna, P., & Pragya, . (2024). Enhancing Agricultural Practices Through IoT-Based Smart Farming Technologies: A Case Study on Banana Crop Yield Optimization. *Asian Journal of Research in Computer Science, 17*(7), 55–66. DOI: 10.9734/ajrcos/2024/v17i7477

Pang, A., Chang, M. W., & Chen, Y. (2022). Evaluation of random forests (RF) for regional and local-scale wheat yield prediction in southeast Australia. *Sensors (Basel), 22*(3), 717. 10.3390/s22030717. DOI: 10.3390/s22030717 PMID: 35161467

Papadimitrakis, M., Giamarelos, N., Stogiannos, M., Zois, E. N., Livanos, N. I., & Alexandridis, A. (2021). Metaheuristic search in smart grid: A review with emphasis on planning, scheduling and power flow optimization applications. *Renewable & Sustainable Energy Reviews, 145*, 111072. DOI: 10.1016/j.rser.2021.111072

Pappenberger, F., Beven, K., Horritt, M., & Blazkova, S. (2005). Uncertainty in the calibration of effective roughness parameters in HEC-RAS using inundation and downstream level observations. *Journal of Hydrology (Amsterdam), 302*(1–4), 46–69. DOI: 10.1016/j.jhydrol.2004.06.036

Parmar, S. (2023). Water Resource Management Using Artificial Intelligence Enabled RS & GIS. Zenodo *(CERN European Organization for Nuclear Research).*

Parmesan, C., Morecroft, M. D., & Trisurat, Y. (2022). [*Impacts, adaptation and vulnerability* GIEC]]. *Climatic Change*, •••, 2022.

Parra-López, C., Ben Abdallah, S., Garcia-Garcia, G., Hassoun, A., Trollman, H., Jagtap, S., Gupta, S., Aït-Kaddour, A., Makmuang, S., & Carmona-Torres, C. (2025). Digital Technologies for Water Use and Management in Agriculture: Recent Applications and Future Outlook. *Agricultural Water Management*, *309*, 109347. DOI: 10.1016/j.agwat.2025.109347

Paudel, Y., & Andersen, T. J. (2013). Monitoring flood and river dynamics in Nepal using remote sensing. *Remote Sensing of Environment*, *136*, 198–210. DOI: 10.1016/j.rse.2013.05.010

Paul, M. J., & Meyer, J. L. (2001). Streams in the Urban Landscape. *Annual Review of Ecology and Systematics*, *32*(1), 333–365. DOI: 10.1146/annurev.ecolsys.32.081501.114040

Peng, C. Y. J., Lee, K. L., & Ingersoll, G. M. (2002). An introduction to logistic regression analysis and reporting. *The Journal of Educational Research*, *96*(1), 3–14. DOI: 10.1080/00220670209598786

Penning-Rowsell, E. C., & Green, C. (2000). New Insights into the Appraisal of Flood-Alleviation Benefits: (1) Flood Damage and Flood Loss Information. Water and Environment Journal, 14(5), 347–353. Portico. DOI: 10.1111/j.1747-6593.2000.tb00272.x

Perera, E. D. P., Hiroe, A., Shrestha, D., Fukami, K., Basnyat, D. B., Gautam, S., Hasegawa, A., Uenoyama, T., & Tanaka, S. (2015). Community-based flood damage assessment approach for lower West Rapti River basin in Nepal under the impact of climate change. *Natural Hazards*, *75*(1), 669–699. DOI: 10.1007/s11069-014-1339-5

Permana, S., Risandi, J., Tito, C., Ramdhan, M., Sufyan, A., Solihuddin, T., Purbani, D., Pryambodo, D. G., Hidayat, R., Rifai, H., & Lukman, K. (2022). Predicting coastal inundation triggered by the oceanic forcing across Jakarta Bay. *IOP Conference Series. Earth and Environmental Science*, *1109*(1), 012005. DOI: 10.1088/1755-1315/1109/1/012005

Pham, Q. B., Kumar, M., Di Nunno, F., Elbeltagi, A., Granata, F., Islam, A. R. M. T., Talukdar, S., Nguyen, X. C., Ahmed, A. N., & Anh, D. T. (2022). Groundwater level prediction using machine learning algorithms in a drought-prone area. *Neural Computing & Applications*, *34*(13), 10751–10773. 10.1007/s00521-022-07009-7. DOI: 10.1007/s00521-022-07009-7

Phillips, S. J., Anderson, R. P., & Schapire, R. E. (2006). Maximum entropy modeling of species geographic distributions. *Ecological Modelling*, *190*(3–4), 231–259. DOI: 10.1016/j.ecolmodel.2005.03.026

Pielke, R. A.Jr, & Downton, M. W. (2000). Precipitation and Damaging Floods: Trends in the United States, 1932–97. *Journal of Climate*, *13*(20), 3625–3637. DOI: 10.1175/1520-0442(2000)013<3625:PADFTI>2.0.CO;2

Pilkey, O. H., & Hume, T. (2001). The shoreline erosion problem : Lessons from the past. *Water and Atmosphere*, *9*(2), 22–23.

Pingali, P. (2007). Agricultural mechanization: adoption patterns and economic impact. *Handbook of agricultural economics, 3*, 2779-2805. 10.1016/S1574-0072(06)03054-4

Plate, E. J. (2002). Flood risk and flood management. *Journal of Hydrology (Amsterdam)*, *267*(1–2), 2–11. DOI: 10.1016/S0022-1694(02)00135-X

Portal, B. I. P. A. D. (2023). Building Information Platform Against Disaster. Government of Nepal. Retrieved from https://bipadportal.gov.np

Pourghasemi, H. R., Moradi, S., & Pradhan, B. (2012). Flood susceptibility mapping using GIS-based support vector machine model. *Journal of Hydrology (Amsterdam)*, *460–461*, 103–113. DOI: 10.1016/j.jhydrol.2012.06.033

Pradhan, K., Rajabifard, A., Aryal, J., & Crompvoets, J. (2024). Dissecting multidimensional morphology of urban sprawl: Framework for policy intervention. *Cities (London, England)*, *157*, 105627. DOI: 10.1016/j.cities.2024.105627

Prasad, A. M., Iverson, L. R., & Liaw, A. (2006). Newer classification and regression tree techniques: Bagging and random forests for ecological prediction. *Ecosystems (New York, N.Y.)*, *9*(2), 181–199. DOI: 10.1007/s10021-005-0054-1

Priestley, C. H. B., & Taylor, R. J. (1972). On the Assessment of Surface Heat Flux and Evaporation Using Large-Scale Parameters. *Monthly Weather Review*, *100*(2), 81–92. DOI: 10.1175/1520-0493(1972)100<0081:OTAOSH>2.3.CO;2

Probst, P., Wright, M., & Boulesteix, A.-L. (2019). Hyperparameters and Tuning Strategies for Random Forest. *Wiley Interdisciplinary Reviews. Data Mining and Knowledge Discovery*, *9*(3), e1301. DOI: 10.1002/widm.1301

Purba, H. J., Azahari, D. H., Dani, F. Z. D. P., Alsyouf, I., Masmoudi, M., & Obaideen, K. (2024). Enhancing rice resilience and sustainability: A bibliometric analysis of innovations for food security and environmental conservation. *BIO Web of Conferences, 119*, 05003. DOI: 10.1051/bioconf/202411905003

Qi, Y., Dou, H., & Wang, Z. (2022). An Adaptive Threshold Selected Method from Remote Sensing Image based on Water Index. *Journal of Physics: Conference Series, 2228*(1), 012001. DOI: 10.1088/1742-6596/2228/1/012001

Quinlan, J. R. (1993). *C4.5: Programs for Machine Learning*. Morgan Kaufmann.

Rad, S. M., Ray, A. K., & Barghi, S. (2022). Water Pollution and Agriculture Pesticide. *Cleanroom Technology, 4*(4), 1088–1102. DOI: 10.3390/cleantechnol4040066

Rahmati, O., Pourghasemi, H. R., & Zeinivand, H. (2019). Flood susceptibility mapping using frequency ratio and weights-of-evidence models in the Golastan Province, Iran. *Geocarto International, 31*(1), 42–70. DOI: 10.1080/10106049.2015.1041559

Rahmati, O., Zeinivand, H., & Besharat, M. (2015). Flood hazard zoning in Yasooj region, Iran, using GIS and multi-criteria decision analysis. *Geomatics, Natural Hazards & Risk, 7*(3), 1000–1017. DOI: 10.1080/19475705.2015.1045043

Rauf, A., & Ghumman, A. R. (2018). Impact assessment of rainfall-runoff simulations on the flow duration curve of the Upper Indus River—A comparison of data-driven and hydrologic models. *Water (Basel), 10*(7), 876. 10.3390/w10070876. DOI: 10.3390/w10070876

Rawls, W. J., Brakensiek, D. L., & Saxtonn, K. E.W. J. RawlsD. L. BrakensiekK. E. Saxtonn. (1982). Estimation of Soil Water Properties. *Transactions of the ASAE. American Society of Agricultural Engineers, 25*(5), 1316–1320. DOI: 10.13031/2013.33720

Rawls, W. J., Pachepsky, Y. A., Ritchie, J. C., Sobecki, T. M., & Bloodworth, H. (2003). Effect of soil organic carbon on soil water retention. *Geoderma, 116*(1–2), 61–76. DOI: 10.1016/S0016-7061(03)00094-6

Reggiani, P., & Weerts, A. H. (2008). A Bayesian approach to decision-making under uncertainty: An application to real-time forecasting in the river Rhine. *Journal of Hydrology (Amsterdam), 356*(1–2), 56–69. DOI: 10.1016/j.jhydrol.2008.03.027

Rehman, S. U., Bhatti, A., Kraus, S., & Ferreira, J. J. M. (2021). The role of environmental management control systems for ecological sustainability and sustainable performance. *Management Decision, 59*(9), 2217–2237. DOI: 10.1108/MD-06-2020-0800

Robin, M. (2002). *Télédétection et modélisation du trait de côte et de sa cinématique. Le Littoral: Regards, Pratiques et Savoirs: études offertes à F.* Verger, Presses de l'Ecole Nationale Supérieure.

Rodell, M., Velicogna, I., & Famiglietti, J. S. (2009). Satellite-based estimates of groundwater depletion in India. *Nature, 460*(7258), 999–1002. 10.1038/nature08238. DOI: 10.1038/nature08238 PMID: 19675570

Rodriguez-Galiano, V. F., Ghimire, B., Rogan, J., Chica-Olmo, M., & Rigol-Sanchez, J. P. (2012). An assessment of the effectiveness of a random forest classifier for land-cover classification. *ISPRS Journal of Photogrammetry and Remote Sensing, 67*, 93–104. DOI: 10.1016/j.isprsjprs.2011.11.002

Rong, Y., Zhang, T., Zheng, Y., Hu, C., Peng, L., & Feng, P. (2020). Three-dimensional urban flood inundation simulation based on digital aerial photogrammetry. *Journal of Hydrology (Amsterdam), 584*, 124308. DOI: 10.1016/j.jhydrol.2019.124308

Rouse, J. W., Haas, R. H., Schell, J. A., & Deering, D. W. (1974). Monitoring vegetation systems in the Great Plains with ERTS. *NASA Special Publication, 351*, 309.

Saaty, T. L. (1980). *The Analytic Hierarchy Process*. McGraw-Hill.

Saaty, T. L. (2008). Decision making with the analytic hierarchy process. *International Journal of Services Sciences, 1*(1), 83. DOI: 10.1504/IJSSCI.2008.017590

Sabir, R. M., Mehmood, K., Sarwar, A., Safdar, M., Muhammad, N. E., Gul, N., Athar, F., Majeed, M. D., Sattar, J., & Khan, Z. (2024). Remote sensing and precision agriculture: a sustainable future. In *Transforming agricultural management for a sustainable future: climate change and machine learning perspectives* (pp. 75–103). Springer., 10.1007/978-3-031-63430-7_4 DOI: 10.1007/978-3-031-63430-7_4

Sadki, M., Chahlaoui, A., Belkhiri, A., Maliki, A., Saidi, A., & Khaffou, M. (2024). Ornithological characterisation of the natural habitats's Lake Dayet Er Roumi (Khemisset, Morocco). *IOP Conference Series. Earth and Environmental Science, 1398*(1), 012003. DOI: 10.1088/1755-1315/1398/1/012003

Sadki, M., Chahlaoui, A., Maliki, A., Khaffou, M., & Lahrouz, S. (2022). Mapping the natural habitats of a continental wetland: Dayet Er Roumi lake (Khémisset, Morocco). *IOP Conference Series. Earth and Environmental Science, 1090*(1), 012030. DOI: 10.1088/1755-1315/1090/1/012030

Safari, M. M., & Malian, A. (2025). Plant disease mapping in paddy growing stages using remotely sensed data. *Environmental Earth Sciences, 84*(1), 1. DOI: 10.1007/s12665-024-11991-7

Safdar, M., Shahid, M. A., Sarwar, A., Rasul, F., Majeed, M. D., & Sabir, R. M. (2023). Crop water stress detection using remote sensing techniques. *Environmental Sciences Proceedings, 25*(1), 20. 10.3390/ECWS-7-14198

Sahbani, S., Béjaoui, B., Benabdallah, S., Toujani, R., Fathalli, A., Zaaboub, N., Aouissi, J., Kassouk, Z., Hamdi, N., Mbarek, N. B., Missaoui, H., Basti, L., Blake, R., & Norouzi, H. (2022). Systematic review of a RAMSAR wetland and UNESCO biosphere reserve in a climate change hotspot (Ichkeul Lake, Tunisia). *Journal of Sea Research, 190*, 102288. DOI: 10.1016/j.seares.2022.102288

Saini, P., & Nagpal, B. (2024). Spatiotemporal Landsat-Sentinel-2 satellite imagery-based Hybrid Deep Neural network for paddy crop prediction using Google Earth engine. *Advances in Space Research, 73*(10), 4988–5004. DOI: 10.1016/j.asr.2024.02.032

Saleem, M. F., Raza, A., Sabir, R. M., Safdar, M., Faheem, M., Al Ansari, M. S., & Hussain, S. (2024). Applications of sensors in precision agriculture for a sustainable future. In *Agriculture and Aquaculture Applications of Biosensors and Bioelectronics* (pp. 109–137). IGI Global Scientific Publishing., DOI: 10.4018/979-8-3693-2069-3.ch006

Saleem, M. U., Shakir, M., Usman, M. R., Bajwa, M. H. T., Shabbir, N., Shams Ghahfarokhi, P., & Daniel, K. (2023). Integrating smart energy management system with internet of things and cloud computing for efficient demand side management in smart grids. *Energies, 16*(12), 4835. DOI: 10.3390/en16124835

Sanyal, J., & Lu, X. X. (2004). Application of remote sensing in flood management with special reference to monsoon Asia: A review. *Natural Hazards, 33*(2), 283–301. DOI: 10.1023/B:NHAZ.0000037035.65105.95

Sarker, I. H., Kayes, A., Badsha, S., Alqahtani, H., Watters, P., & Ng, A. (2020). Cybersecurity data science: An overview from machine learning perspective. *Journal of Big Data, 7*(1), 41. 10.1186/s40537-020-00318-5. DOI: 10.1186/s40537-020-00318-5

Sauer, V. B., Thomas, W. O.Jr, Stricker, V. A., & Wilson, K. V. (1983). *Flood characteristics of urban watersheds in the United States (No. 2207)*. USGPO.

Schulz, O., Busche, H., & Benbouziane, A. (2008). Decadal precipitation variances and reservoir inflow in the semi-arid Upper Drâa basin (South-Eastern Morocco). In *Climatic changes and water resources in the Middle East and North Africa* (pp. 165–178). Springer Berlin Heidelberg.

Schumann, G., Di Baldassarre, G., & Bates, P. D. (2009). The Utility of Spaceborne Radar to Render Flood Inundation Maps Based on Multialgorithm Ensembles. *IEEE Transactions on Geoscience and Remote Sensing, 47*(8), 2801–2807. DOI: 10.1109/TGRS.2009.2017937

Scott, K. M., Macias, J. L., Naranjo, J. A., Rodriguez, S., & McGeehin, J. P. (2001). *Catastrophic debris flows transformed from landslides in volcanic terrains : mobility, hazard assessment and mitigation strategies*. Professional Paper., DOI: 10.3133/pp1630

See, L., & Abrahart, R. J. (2001). Multi-model data fusion for hydrological forecasting. *Computers & Geosciences*, *27*(8), 987–994. 10.1016/S0098-3004(00)00136-9. DOI: 10.1016/S0098-3004(00)00136-9

Seel, N., Verma, L. P., & Gupta, A. (2025). IoT-Based Smart Water Resource Management. In *Unleashing the Power of Blockchain and IoT for Water Informatics* (pp. 147–160). Springer., 10.1007/978-981-96-3886-4_9 DOI: 10.1007/978-981-96-3886-4_9

Senoo, E. E. K., Anggraini, L., Kumi, J. A., Karolina, L. B., Akansah, E., Sulyman, H. A., Mendonça, I., & Aritsugi, M. (2024). IoT solutions with artificial intelligence technologies for precision agriculture: Definitions, applications, challenges, and opportunities. *Electronics (Basel)*, *13*(10), 1894. 10.3390/electronics13101894. DOI: 10.3390/electronics13101894

Senthil Kumar, A. R., Goyal, M. K., Ojha, C. S. P., Singh, R. D., & Swamee, P. K. (2013). Application of artificial neural network, fuzzy logic and decision tree algorithms for modelling of streamflow at Kasol in India. *Water Science and Technology*, *68*(12), 2521–2526. DOI: 10.2166/wst.2013.491

Seto, K. C., Güneralp, B., & Hutyra, L. R. (2012). Global forecasts of urban expansion to 2030 and direct impacts on biodiversity and carbon pools. *Proceedings of the National Academy of Sciences of the United States of America*, *109*(40), 16083–16088. DOI: 10.1073/pnas.1211658109

Seyam, M. M. H., Haque, M. R., & Rahman, M. M. (2023). Identifying the land use land cover (LULC) changes using remote sensing and GIS approach: A case study at Bhaluka in Mymensingh, Bangladesh. *Case Studies in Chemical and Environmental Engineering*, *7*, 100293. DOI: 10.1016/j.cscee.2022.100293

Shaamala, A., Yigitcanlar, T., Nili, A., & Nyandega, D. (2025). Machine learning applications for urban geospatial analysis: A review of urban and environmental studies. *Cities (London, England)*, *165*, 106139. DOI: 10.1016/j.cities.2025.106139

Shahraeini, M., Kotzanikolaou, P., & Nasrolahi, M. (2022). Communication resilience for smart grids based on dependence graphs and eigenspectral analysis. *IEEE Systems Journal*, *16*(4), 6558–6568. DOI: 10.1109/JSYST.2022.3179660

Shamshiri, R., Kalantari, F., Ting, K., Thorp, K. R., Hameed, I. A., Weltzien, C., Ahmad, D., & Shad, Z. M. (2018). Advances in greenhouse automation and controlled environment agriculture: A transition to plant factories and urban agriculture. *International Journal of Agricultural and Biological Engineering, 11*(1), 1–22. https://doi.org/https://dx.doi.org/10.25165/j.ijabe.20181101.3210. DOI: 10.25165/j.ijabe.20181101.3210

Sharifi, A., & Mahdipour, H. (2024). Utilising geospatial artificial intelligence for remote sensing applications. *Environmental Earth Sciences, 83*(9), 274. DOI: 10.1007/s12665-024-11584-4

Sharifi, E., Pradhan, B., & Pourghasemi, H. R. (2019). GIS-based flood susceptibility mapping using multi-criteria decision analysis. *Catena, 173*, 501–512. DOI: 10.1016/j.catena.2018.10.030

Sharma, S., Sharma, U., Sharma, S., Sharma, S., Likhita, J., Thakur, S., & Sharma, N. (2025). Strategies for Sustainable Land Management Through Climate-Resilient Horticulture: A Systematic Review. *Land Degradation & Development*, ldr.70082. DOI: 10.1002/ldr.70082

Sharma, T., Sharma, P., & Kumar, N. (2023). Chemical reactive magnetized fluid flow through a vertical channel due to heat source and thermal radiation effects. *Materials Today: Proceedings, 78*, 481–492. DOI: 10.1016/j.matpr.2022.10.298

Shaw, H. F. (1994). M. J. Wilson, Ed. Clay Mineralogy: Spectroscopic and Chemical Determinative Methods. Hall, London. 1994. 367 pp. ISBN 0-412-533804. Mineralogical Magazine, 58(393), 701–701. DOI: 10.1180/minmag.1994.058.393.30

Shen, Z., Yang, H., & Zhang, S. (2022). Optimal approximation rate of ReLU networks in terms of width and depth. *Journal de Mathématiques Pures et Appliquées, 157*, 101–135. DOI: 10.1016/j.matpur.2021.07.009

Shiklomanov, I. A., & Rodda, J. C. (2003). *World water resources at the beginning of the twenty-first century*. Cambridge University Press.

Shrestha, M., Acharya, S. C., & Shrestha, P. K. (2020). Climate change impact on water resources in Nepal: A review. *Journal of Water and Climate Change, 11*(2), 281–301. DOI: 10.2166/wcc.2019.080

Shrestha, S., Mahat, J., & Shrestha, J., KC, M., & Paudel, K. (. (2022). Influence of high-temperature stress on rice growth and development. A review. *Heliyon, 8*(12). Advance online publication. DOI: 10.1016/j.heliyon.2022.e12651 PMID: 36643304

Shrivastava, R., & Sharma, A. (2011). Smart growth: A modern urban principle. *Architecture Research, 1*(1), 8–11. DOI: 10.5923/j.arch.20110101.02

Shustikova, I., Domeneghetti, A., Neal, J. C., Bates, P., & Castellarin, A. (2019). Comparing 2D capabilities of HEC-RAS and LISFLOOD-FP on complex topography. *Hydrological Sciences Journal, 64*(14), 1769–1782. DOI: 10.1080/02626667.2019.1671982

Siddique, I. (2024). Machine learning empowered geographic information systems: Advancing Spatial analysis and decision making. https://doi.org/.DOI: 10.30574/wjarr.2024.22.1.1200

Sigafoos, R. S. (1964). *Botanical evidence of floods and floodplain deposition.* Professional Paper., DOI: 10.3133/pp485A

Simonovic, S. P. (2012). *Managing water resources: methods and tools for a systems approach.* Routledge. DOI: 10.4324/9781849771917

Singh, S. K., & Jain, R. (2025). Future of mobility: potential challenges and ethical considerations. In AIoT for Smart Transportation: Transforming the future of mobility (pp. 235–253). https://doi.org/DOI: 10.1049/PBTR047E_ch12

Singh, A., Ganapathysubramanian, B., Singh, A. K., & Sarkar, S. (2016). Machine learning for high-throughput stress phenotyping in plants. *Trends in Plant Science, 21*(2), 110–124. 10.1016/j.tplants.2015.10.015. DOI: 10.1016/j.tplants.2015.10.015 PMID: 26651918

Singha, C., Rana, V. K., Pham, Q. B., Nguyen, D. C., & Łupikasza, E. (2024). Integrating machine learning and geospatial data analysis for comprehensive flood hazard assessment. *Environmental Science and Pollution Research International, 31*(35), 48497–48522. DOI: 10.1007/s11356-024-34286-7 PMID: 39030454

Singh, J., Sandhu, S. S., Singh, D., & Hadda, M. S. (2017). Soil Management to Optimize Water in Rice-Wheat Cropping. In Lichtfouse, E. (Ed.), *Sustainable Agriculture Reviews* (pp. 253–279). Springer International Publishing., DOI: 10.1007/978-3-319-48006-0_8

Singh, K., & Malhotra, J. (2022). Two-layer LSTM network-based prediction of epileptic seizures using EEG spectral features. *Complex & Intelligent Systems, 8*(3), 2405–2418. DOI: 10.1007/s40747-021-00627-z

Singh, S. K., Jain, R., Palaniappan, D., Parmar, K. T. P., & Gothania, J. (2025). *Spatiotemporal Analysis and Machine Learning-Based prediction of air quality in Indian Urban.* Environmental Research and Technology., DOI: 10.35208/ert.1587308

Singh, V. P. (1997). Effect of spatial and temporal variability in rainfall and watershed characteristics on stream flow hydrograph. *Hydrological Processes*, *11*(12), 1649–1669. DOI: 10.1002/(SICI)1099-1085(19971015)11:12<1649::AID-HYP495>3.0.CO;2-1

Singh, V. P., & Woolhiser, D. A. (2002). Mathematical modeling of watershed hydrology. *Journal of Hydrologic Engineering*, *7*(4), 270–292. 10.1061/(ASCE)1084-0699(2002)7:4(270). DOI: 10.1061/(ASCE)1084-0699(2002)7:4(270)

Sishodia, R. P., Ray, R. L., & Singh, S. K. (2020). Applications of remote sensing in precision agriculture: A review. *Remote Sensing (Basel)*, *12*(19), 3136. 10.3390/rs12193136. DOI: 10.3390/rs12193136

Sit, M., Demiray, B. Z., Xiang, Z., Ewing, G. J., Sermet, Y., & Demir, I. (2020). A comprehensive review of deep learning applications in hydrology and water resources. *Water Science and Technology*, *82*(12), 2635–2670. 10.2166/wst.2020.369. DOI: 10.2166/wst.2020.369 PMID: 33341760

Sivapalan, M., Takeuchi, K., Franks, S., Gupta, V., Karambiri, H., Lakshmi, V., Liang, X., McDonnell, J., Mendiondo, E. M., O'connell, P., Oki, T., Pomeroy, J. W., Schertzer, D., Uhlenbrook, S., & Zehe, E. (2003). IAHS Decade on Predictions in Ungauged Basins (PUB), 2003–2012: Shaping an exciting future for the hydrological sciences. *Hydrological Sciences Journal*, *48*(6), 857–880. 10.1623/hysj.48.6.857.51421. DOI: 10.1623/hysj.48.6.857.51421

Smith, D. P., & Bedient, P. B. (1981). Preliminary model of an urban floodplain under changing land use. *Journal of Hydrology (Amsterdam)*, *51*(1–4), 179–185. DOI: 10.1016/0022-1694(81)90127-X

Snoek, J., Larochelle, H., & Adams, R. P. 2012. "Practical Bayesian Optimization of Machine Learning Algorithms." *Advances in Neural Information Processing Systems (NeurIPS)* 25: 2951–59 https://papers.nips.cc/paper_files/paper/2012/hash/05311655a15b75fab86956663e1819cd-Abstract.html

Snoussi, M. (2020). *Diagnostic des zones côtières de la Région de Rabat-Salé-Kenitra*. 1-70.

Song, Y., Wang, C., Linderholm, H. W., Fu, Y., Cai, W., Xu, J., Zhuang, L., Wu, M., Shi, Y., Wang, G., & Chen, D. (2022). The negative impact of increasing temperatures on rice yields in southern China. *The Science of the Total Environment*, *820*, 153262. DOI: 10.1016/j.scitotenv.2022.153262 PMID: 35065105

Sørensen, R., Zinko, U., & Seibert, J. (2006). On the calculation of the topographic wetness index: Evaluation of different methods based on field observations. *Hydrology and Earth System Sciences, 10*(1), 101–112. DOI: 10.5194/hess-10-101-2006

Spatial Strategy. (2023). OMAN NATIONAL SPATIAL STRATEGY 2040 National Planning Standards August 2023. https://www.housing.gov.om/cmsapi/files/content/Books_publication/ONSS%20National%20Planning%20Standards_August%202023_FINAL%20PDF_compressed.pdf

Spencer, K. (2024). As floodwaters rise, toxic contaminants released from old landfills pose more of a hazard to nature and to us. DOI: 10.64628/AB.vajrrfwsd

Stedinger, J. R., & Griffis, V. W. (2008). Flood Frequency Analysis in the United States: Time to Update. *Journal of Hydrologic Engineering, 13*(4), 199–204. DOI: 10.1061/(ASCE)1084-0699(2008)13:4(199)

Stefanidis, S., & Stathis, D. (2013). Assessment of flood hazard based on natural and anthropogenic factors using AHP. *Natural Hazards, 68*, 569–585. DOI: 10.1007/s11069-013-0639-5

Stuckey, M. H., & Reed, L. A. (2000). *Techniques for estimating magnitude and frequency of peak flows for Pennsylvania streams (No. 4189).* US Department of the Interior, US Geological Survey.

Sudhira, H., Ramachandra, T., & Jagadish, K. (2003). Urban sprawl: Metrics, dynamics and modelling using GIS. *International Journal of Applied Earth Observation and Geoinformation, 5*(1), 29–39. DOI: 10.1016/j.jag.2003.08.002

Sukanya, S., & Joseph, S. (2023). Climate change impacts on water resources: An overview. *Visualization techniques for climate change with machine learning and artificial intelligence*, 55-76. 10.1016/B978-0-323-99714-0.00008-X

Sun, L., Yang, T., Lou, Y., Shi, Q., & Zhang, L. (2024). Paddy rice mapping based on phenology matching and cultivation pattern analysis combining multi-source data in Guangdong, China. *Journal of Remote Sensing, 4*, 0152.

Sun, L., Lou, Y., & Zhang, L. (2024). Spatial domain transfer: Cross-regional paddy rice mapping with a few samples based on Sentinel-1 and Sentinel-2 data on GEE. *International Journal of Applied Earth Observation and Geoinformation, 128*, 103762. DOI: 10.1016/j.jag.2024.103762

Tafraouti, M. (2025, janvier 9). *The Erosion of the Mediterranean Coast of Morocco : Reality and Implications | Earth Journalism Network.* https://earthjournalism.net/stories/the-erosion-of-the-mediterranean-coast-of-morocco-reality-and-implications

Tao, H., Hameed, M. M., Marhoon, H. A., Zounemat-Kermani, M., Heddam, S., Kim, S., Sulaiman, S. O., Tan, M. L., Sa'adi, Z., Mehr, A. D., Allawi, M. F., Abba, S. I., Zain, J. M., Falah, M. W., Jamei, M., Bokde, N. D., Bayatvarkeshi, M., Al-Mukhtar, M., Bhagat, S. K., & Yaseen, Z. M. (2022). Groundwater level prediction using machine learning models: A comprehensive review. *Neurocomputing*, *489*, 271–308. DOI: 10.1016/j.neucom.2022.03.014

Tehrany, M. S., Lee, M. J., Pradhan, B., Jebur, M. N., & Lee, S. (2014). Flood susceptibility mapping using integrated bivariate and multivariate statistical models. *Environmental Earth Sciences*, *72*(10), 4001–4015. DOI: 10.1007/s12665-014-3289-3

Tehrany, M. S., Pradhan, B., & Jebur, M. N. (2014). Flood susceptibility mapping using a novel ensemble weights-of-evidence and support vector machine models in GIS. *Journal of Hydrology (Amsterdam)*, *512*, 332–343. DOI: 10.1016/j.jhydrol.2014.03.008

Teng, J., Jakeman, A. J., Vaze, J., Croke, B. F. W., Dutta, D., & Kim, S. (2017). Flood inundation modelling: A review of methods, recent advances and uncertainty analysis. Environmental Modelling &amp. *Environmental Modelling & Software*, *90*, 201–216. DOI: 10.1016/j.envsoft.2017.01.006

Thieler, E. R., Himmelstoss, E. A., Zichichi, J. L., & Ergul, A. (2009). *The digital, Shoreline Analysis System (DSAS) version 4.0 An ArcGIS extension for calculating shoreline change. (Ver. 4.3.4730, April 2012). US. Geological Survey Open-File Report 2008-1278, accessed June 2012. En ligne :* Https://pubs.er.usgs.gov/publicat. https://doi.org/DOI: 10.3133

Thior, M., ASy, M. A., Cisse, I., Dieye, E. H. B., Sane, T., Ba, B. D., Solly, B., & Descroix, L. (2021). A cartographic approach of the evolution of the coastline of the Casamance estuary. *Mappemonde, 131*, 0-22. DOI: 10.4000/mappemonde.5939

Thompson, J. R., Flower, R. J., Ramdani, M., Ayache, F., Ahmed, M. H., Rasmussen, E. K., & Petersen, O. S. (2009). Hydrological characteristics of three North African coastal lagoons: Insights from the MELMARINA project. *Hydrobiologia*, *622*(1), 45–84. DOI: 10.1007/s10750-008-9680-x

Tien Bui, D., Hoang, N.-D., Martínez-Álvarez, F., Ngo, P.-T. T., Hoa, P. V., Pham, T. D., Samui, P., & Costache, R. (2020). A novel deep learning neural network approach for predicting flash flood susceptibility: A case study at a high frequency tropical storm area. *The Science of the Total Environment*, *701*, 134413. DOI: 10.1016/j.scitotenv.2019.134413

Tockner, K., & Stanford, J. A. (2002). Riverine flood plains: Present state and future trends. *Environmental Conservation, 29*(3), 308–330. DOI: 10.1017/S037689290200022X

Tripoli, M., & Schmidhuber, J. (2020). Emerging opportunities for the application of blockchain in the agri-food industry.

Tsouros, D. C., Bibi, S., & Sarigiannidis, P. G. (2019). A review on UAV-based applications for precision agriculture. *Information (Basel), 10*(11), 349. 10.3390/info10110349. DOI: 10.3390/info10110349

Tucker, C. J. (1979). Red and photographic infrared linear combinations for monitoring vegetation. *Remote Sensing of Environment, 8*(2), 127–150. DOI: 10.1016/0034-4257(79)90013-0

UICN. (2010). *Blue Planet : Coastal Zones. Union Internationale pour la Conservation de la Nature*. https://www.iucn.org/resources/report/blue-planet-coastal-zones

Ullah, S., & Zhang, J. (2020). GIS-based flood hazard mapping using relative frequency ratio method: A case study from eastern Hindu Kush, Pakistan. *Journal of Flood Risk Management, 13*(2), e12587. DOI: 10.1111/jfr3.12587 PMID: 32210424

UNDRR. (2021) The human cost of disasters: an overview of the last 20 years (2000-2019) Link: https://www.undrr.org/publication/human-cost-disasters-overview-last-20-years-2000-2019?utm_source=chatgpt.com Last access: 05.09.2025

UN-Habitat. (2023). *Goal 11: Sustainable Cities and Communities*. https://unhabitat.org/about-us/sustainable-development-goals

van der Zaag, P., & Savenije, H. G. (2004). *Principles of integrated water resources management*. Unesco-IHE Delft.

Venkatesh, K., Naik, K. J., & Shankar, A. (2024). Quantum convolution neural network for multi-nutrient detection and stress identification in plant leaves. *Multimedia Tools and Applications, 83*(24), 65663–65685. DOI: 10.1007/s11042-023-17992-1

Waheed, M. A., & Akogun, O. A. (2021). Quality enhancement of fuel briquette from cornhusk and cassava peel blends for co-firing in coal thermal plant. *International Journal of Energy Research, 45*(2), 1867–1878. DOI: 10.1002/er.5865

Walling, D. E., & Fang, D. (2003). Recent trends in the suspended sediment loads of the world's rivers. *Global and Planetary Change, 39*(1–2), 111–126. DOI: 10.1016/S0921-8181(03)00020-1

Wan, Z., Hook, S., & Hulley, G. (2021). *MODIS/Terra Land Surface Temperature/ Emissivity 8-Day L3 Global 1km SIN Grid V061* [Data set]. NASA Land Processes Distributed Active Archive Center. Date Accessed: 2025-10-09DOI: 10.5067/ MODIS/MOD11A2.061

Wang, J., Liu, Q., & Zhao, P. (2023). Multi-Source Data Integration for Soil Moisture Estimation Using Machine Learning. *Remote Sensing*, *15*(5), 1123. DOI: 10.3390/ rs15174214

Wang, R., Murayama, Y., & Morimoto, T. (2021). Scenario simulation studies of urban development using remote sensing and GIS [review]. *Remote Sensing Applications: Society and Environment*, *22*, 100474. DOI: 10.1016/j.rsase.2021.100474

Wang, Y., Shi, L., Hu, Y., Hu, X., Song, W., & Wang, L. (2024). A comprehensive study of deep learning for soil moisture prediction. *Hydrology and Earth System Sciences*, *28*(4), 917–943. DOI: 10.5194/hess-28-917-2024

Wang, Z., Li, D., Jiang, R., & Okumura, M. (2025). *Continuous sign language recognition with multi-scale spatial-temporal feature enhancement*. IEEE., DOI: 10.1109/ACCESS.2025.3526330

Ward, A. D., & Trimble, S. W. (2003). *Environmental hydrology*. Crc Press. DOI: 10.1201/b13148

Ward, P. J., Blauhut, V., Bloemendaal, N., Daniell, J. E., de Ruiter, M. C., Duncan, M. J., Emberson, R., Jenkins, S. F., Kirschbaum, D., Kunz, M., Mohr, S., Muis, S., Riddell, G. A., Schäfer, A., Stanley, T., Veldkamp, T. I. E., & Winsemius, H. C. (2020). Review article: Natural hazard risk assessments at the global scale. *Natural Hazards and Earth System Sciences*, *20*(4), 1069–1096. DOI: 10.5194/nhess-20-1069-2020

Werner, A. D., & Simmons, C. T. (2009). Impact of Sea-Level Rise on Sea Water Intrusion in Coastal Aquifers. Groundwater, 47(2), 197–204. Portico. DOI: 10.1111/j.1745-6584.2008.00535.x

Werner, M. G. F. (2001). Impact of grid size in GIS based flood extent mapping using a 1D flow model. *Physics and Chemistry of the Earth. Part B: Hydrology, Oceans andAtmosphere*, *26*(7–8), 517–522. DOI: 10.1016/S1464-1909(01)00043-0

Werner, M. G. F. (2004). A comparison of flood extent modelling approaches through constraining uncertainties on gauge data. *Hydrology and Earth System Sciences*, *8*(6), 1141–1152. DOI: 10.5194/hess-8-1141-2004

Whitehead, P. G., Wilby, R. L., Battarbee, R. W., Kernan, M., & Wade, A. J. (2009). A review of the potential impacts of climate change on surface water quality. *Hydrological Sciences Journal, 54*(1), 101–123. 10.1623/hysj.54.1.101. DOI: 10.1623/hysj.54.1.101

Willmott, C. J. (1981). On the Validation of Models. *Physical Geography*, *2*(2), 184–194. DOI: 10.1080/02723646.1981.10642213

Wing, O. E. J., Bates, P. D., Smith, A. M., Sampson, C. C., Johnson, K. A., Fargione, J., & Morefield, P. (2018). Estimates of present and future flood risk in the conterminous United States. *Environmental Research Letters, 13*(3), 034023. DOI: 10.1088/1748-9326/aaac65

Winter, B., Schneeberger, K., Förster, K., & Vorogushyn, S. (2020). Event generation for probabilistic flood risk modelling: Multi-site peak flow dependence model vs. weather-generator-based approach. *Natural Hazards and Earth System Sciences, 20*(6), 1689–1703. DOI: 10.5194/nhess-20-1689-2020

World Bank. (2024a). *Climate Change impacts on the Blue Economy in Morocco : Prospects for jobs in Coastal Tourism. Maghreb Technical Note*(12).

World Bank. (2024b). *Disappearing coasts in the Maghreb : Coastal erosion and its costs*. World Bank. https://www.worldbank.org/en/country/morocco/publication/disappearing-coasts-in-the-maghreb-coastal-erosion-and-its-costs

Wu, Y., & Jiang, Q. (2024). Flood risk assessment using Random Forest: A case study of Wuhan City, China. *Water (Basel), 16*(2), 321. DOI: 10.3390/w16020321

Xie, M. (2006). Integrated water resources management (IWRM)–introduction to principles and practices. Africa Regional Workshop on IWRM, Nairobi, Oct

Xu, H. (2006). Modification of normalised difference water index (NDWI) to enhance open water features in remotely sensed imagery. *International Journal of Remote Sensing*, *27*(14), 3025–3033. DOI: 10.1080/01431160600589179

Yilmaz, K. K., Gupta, H. V., & Wagener, T. (2008). A process-based diagnostic approach to model evaluation: Application to the NWS distributed hydrologic model. Water Resources Research, 44(9). Portico. DOI: 10.1029/2007WR006716

Yu, H., Zhao, X., & Chen, C. (2024). Quantum-inspired reinforcement learning for quantum control. *IEEE Transactions on Control Systems Technology*.

Yu, Y., Zhang, Y., Cheng, Z., Song, Z., & Tang, C. (2024). Multi-scale spatial pyramid attention mechanism for image recognition: An effective approach. *Engineering Applications of Artificial Intelligence, 133*, 108261. DOI: 10.1016/j.engappai.2024.108261

Zambrano-Monserrate, M. A. (2025). *Rethinking urban governance in the global South: Lessons from Ecuador's intermediate cities*. Urban Governance., DOI: 10.1016/j.ugj.2025.08.007

Zevenbergen, L. W., & Thorne, C. R. (1987). Quantitative analysis of land surface topography. *Earth Surface Processes and Landforms*, *12*(1), 47–56. DOI: 10.1002/esp.3290120107

Zhang, A., Tariq, A., Quddoos, A., Naz, I., Aslam, R. W., Barboza, E., Ullah, S., & Abdullah-Al-Wadud, M. (2025). Spatio-temporal analysis of urban expansion and land use dynamics using google earth engine and predictive models. *Scientific Reports*, *15*(1), 6993. Advance online publication. DOI: 10.1038/s41598-025-92034-4

Zhang, B., Chen, W., Li, M. Z., Guo, X., Zheng, Z., & Yang, R. (2024). MGAtt-LSTM: A multi-Scale spatial correlation prediction model of PM2.5 concentration based on multi-graph attention. *Environmental Modelling & Software*, *179*, 106095. DOI: 10.1016/j.envsoft.2024.106095

Zhang, D., Xu, H., Wang, T., & He, Q. (2019). Flood susceptibility mapping using machine learning models based on remote sensing big data. *Remote Sensing*, *11*(19), 2283. DOI: 10.3390/rs11192283

Zhang, K., Hu, J., & Xu, L. (2023). Ensemble Machine Learning Approaches for Daily Surface Soil Moisture Estimation. *Geoscientific Model Development*, *16*, 5825–5843.

Zhang, L., Dawes, W. R., & Walker, G. R. (2008). Response of mean annual evapotranspiration to vegetation changes at catchment scale. *Water Resources Research*, *37*(3), 701–708. DOI: 10.1029/2000WR900325

Zhang, Y., Chang, X., Liu, Y., Lu, Y., Wang, Y., & Liu, Y. (2021). Urban expansion simulation under constraint of multiple ecosystem services (MESs) based on cellular automata (CA)-Markov model: Scenario analysis and policy implications. *Land Use Policy*, *108*, 105667. DOI: 10.1016/j.landusepol.2021.105667

Zhang, Y., Shen, J., He, L., Feng, J., Chi, L., & Wang, X. (2024). Challenge to Lake Ecosystems: Changes in Thermal Structure Triggered by Climate Change. *Water (Basel)*, *16*(6), 888. DOI: 10.3390/w16060888

Zhao, B., Zhang, Q., Liu, Y., Cui, Y., & Zhou, B. (2024). Detection method for rice seedling planting conditions based on image processing and an improved YOLOv8n model. *Applied Sciences (Basel, Switzerland)*, *14*(6), 2575. DOI: 10.3390/app14062575

Zhao, J., Jin, J., Guo, Q., Liu, L., Chen, Y., & Pan, M. (2014). Dynamic risk assessment model for flood disaster on a projection pursuit cluster and its application. *Stochastic Environmental Research and Risk Assessment, 28*(8), 2175–2183. DOI: 10.1007/s00477-014-0881-8

Zhao, X., Wang, L., Zhang, Y., Han, X., Deveci, M., & Parmar, M. (2024). A review of convolutional neural networks in computer vision. *Artificial Intelligence Review, 57*(4), 99. DOI: 10.1007/s10462-024-10721-6

Zhao, Z., Dong, J., Zhang, G., Yang, J., Liu, R., Wu, B., & Xiao, X. (2024). Improved phenology-based rice mapping algorithm by integrating optical and radar data. *Remote Sensing of Environment, 315*, 114460. DOI: 10.1016/j.rse.2024.114460

Zheng, B., & Agresti, A. (2000). Summarizing the predictive power of a generalized linear model. *Statistics in Medicine, 19*(13), 1771–1781. DOI: 10.1002/1097-0258(20000715)19:13<1771::AID-SIM485>3.0.CO;2-P

Zhou, T., Ye, X., Lu, H., Zheng, X., Qiu, S., & Liu, Y. (2022). Dense convolutional network and its application in medical image analysis. *BioMed Research International, 2022*(1), 2384830. DOI: 10.1155/2022/2384830 PMID: 35509707

Zhou, Y., Ma, Z., & Wang, L. (2002). Chaotic dynamics of the flood series in the Huaihe River Basin for the last 500 years. *Journal of Hydrology (Amsterdam), 258*(1–4), 100–110. DOI: 10.1016/S0022-1694(01)00561-3

Zhu, T., Xia, C., Yu, R., Zhou, X., Xu, X., Wang, L., Zong, Z., Yang, J., Liu, Y., Ming, L., You, Y., Chen, D., & Xie, W. (2024). Comprehensive mapping and modelling of the rice regulome landscape unveils the regulatory architecture underlying complex traits. *Nature Communications, 15*(1), 6562. DOI: 10.1038/s41467-024-50787-y PMID: 39095348

Zhu, Z., Qiu, S., & Ye, S. (2022). Remote sensing of land change: A multifaceted perspective. *Remote Sensing of Environment, 282*, 113266. DOI: 10.1016/j.rse.2022.113266

# About the Contributors

**Sathiyaraj A.** is an Associate Professor, Department of Information Technology under School of Computing at Sathyabama Institute of Science and Technology, Chennai, Tamil Nadu. He received his B.Tech. Degree in Department of at Information Technology, Adhiparasakthi Engineering College, Melmaruvathur, Affiliated to Anna University, Chennai, India, at 2007. M.E. degree in Department of Computer Science and Engineering, Anna University, Chennai, India, at 2012 and Ph.D in Department of Computer Science and Engineering at SSSUTMS, MP, India, at 2021. He has published a research article in IEEE and International Conference, National Conference Proceedings, Scopus and SCI Journal and Patent. He has organized various events in Department, His area of research work is Networking Routing, MANET, Sensor Networking, Machine Learning and IoT.

**Latifat Abimbola** is well organized, self motivated and confident mathematician with ability to impart knowledge coupled with good computing skills. She is a passionate advocate of ethical practice in scientific research. She had a BSc, MSc, MPhil and PhD in Mathematics from the University of Ibadan, Nigeria. She is currently a Senior Lecturer and acting Head of the Department of Mathematics and Statistics, Abiola Ajimobi Technical University, Ibadan, Nigeria. She had publications in Local and International journals. Her areas of expertise is Stochastic Analysis.

**Muhammed Akiner** graduated from the Civil Engineering Department of Eastern Mediterranean University in 2002. The author completed his MS study on Structural Mechanics at the same university in 2004. He held another MS degree in Environmental Engineering from SUNY at Buffalo in 2007. The author obtained his Ph.D. from the Civil Engineering Department of Bogazici University. He published several scientific publications on engineering, mainly in the field of environmental engineering. His research interests include environmental engineering, water resources, ecology, civil engineering, construction & building technology.

**Hamza Anjum** is a Research Associate at Department of Computer Science Ghulam Ishaq khan Institute of Engineering Sciences and Technology, Khyber Pakhtunkhwa, Pakistan.

**Hafiz Muhammad Awais** is a Research Associate at Agricultural Remote Sensing Lab (ARSL), University of Agriculture, Faisalabad, Pakistan

**Mehmet Ali Çelik** is an Assoc. Prof. of Physical Geography. He focuses on the interaction between climate variables, nature and the environment. His multidisciplinary research interests include remote sensing and GIS studies involving climate, natural resource management, and disasters. Dr. Çelik uses multispectral and hyperspectral remote sensing and GIS to monitor changes to ecosystems and analyse the contribution of natural and anthropogenic factors to these changes.

**R. Deepa** is an Assistant Professor in the Department of Computing Technology with over 15 years of experience in teaching, research, and student mentorship. Her research focuses on Artificial Intelligence and Machine Learning, with applications in data analytics, geospatial technologies, and intelligent systems. She has guided numerous student projects, published scholarly works, and actively contributes to curriculum development and academic workshops.

**C Geetha** is an Associate Professor, Department of Information Technology under School of Computing at Sathyabama Institute of Science and Technology, Chennai, Tamil Nadu. She received her B.Tech. degree in Department of at Information Technology, Manonmaniam Sundaranar University, India, at 2004. M.Tech. degree in Department of Information Technology at Sathyabama University, India, at 2009 and Ph.D in Department of Computer Science and Engineering at Bharath Institute of Higher Education and Research, Chennai, Tamil Nadu, India, at 2023. She has published a research article in IEEE and International Conference, National Conference Proceedings, Scopus and Web Of Sciecne Journal and Patent. She has organized various events in Department, Her area of research work is machine learning and data analytics

**Jaishri Gothania** is an academician, author, researcher, and data scientist with over 9 years of experience. She holds a Ph.D. in Computer Science Engineering, an M.Tech in Computer Science, a B.Tech in Computer Science and Engineering, and a PGD in Education. She has published 1 book, 2 book chapters, and 12 research articles; presented 6 conference papers; and holds multiple patents. Her work bridges data science and innovation, contributing significantly to academia, research, and technological advancement.

**Niranjana Govindan** is a distinguished academic and researcher serving as a Professor and the Head of the Department of Computing Technologies at SRM Institute of Science and Technology (SRMIST), Kattankulathur Campus, Tamil Nadu, India With over two decades of experience in teaching and research, Dr. Niranjana has made significant contributions to fields such as Machine Learning, Deep Learning, Artificial Intelligence, Image Processing, Internet of Things (IoT), and Networking. Her research interests focus on real-world applications, including medical imaging, smart healthcare systems, and secure communications

**J Jayanthi** completed M.E Computer Science Engineering from R.M.K Engineering College and worked has Junior Research Fellow (JRF) for this ISRO Funded Project in Vellore Institue of Technology during the year 2019 - 2022.I'm Currently working as Principal Software Engineer in Innova Solutions, Chennai. My area of Interest include Digital Image Processing, Artificial Intelligence, Machine Learning.

**L.K Shoba** Received her B.E degree in Electronics and Communication Engineering from Noorul Islam college of Engineering, Kumaracoil, Manonmanium sundaranar university, India in 2003. She has received her M.Tech. degree in Information Technology from Madras Institute of Technology, Chennai, Anna University, India in 2009. She holds a Ph.D degree in Information and communication Engineering from Anna University, India in 2023. She is in teaching profession since June 2009. Currently she is working as Assistant Professor in the Department of Computing Technologies, SRM Institute of Science and Technology, kattankulathur, Her research area of interest is Artificial Intelligence and Machine Learning.

**Kashif Mehmood** is a Lecturer at Department of Irrigation and Drainage University of Agriculture, Faisalabad, Pakistan.

**Muhammad Sajid Mehmood** is an Associate Professor, School of Tourism and Planning, Pingdingshan University, Pingdingshan 467000, China.

**Raghav Mehra** received his B-Tech. degree in electronics and communication engineering from Guru Gobind Singh Indraprastha University, Delhi, India, in 2007. Since then, he has been with the Space Applications Center, Indian Space Research Organization, Ahmedabad 380015, India, and has been involved in the data processing of microwave payloads, namely, synthetic aperture radars (SARs), scatterometers, and radar altimeters. His research interests include SAR data calibration, geocoding, and SAR value-added product generation.

**Rajat Mishra** is an agricultural engineer and researcher specializing in soil and water conservation engineering, watershed hydrology, and climate change adaptation. He is currently serving as a Teaching Associate in the Department of Soil Conservation and Water Management at Chandra Shekhar Azad University of Agriculture and Technology, Kanpur, Uttar Pradesh. Born and raised in Uttar Pradesh, Dr. Mishra pursued his passion for agricultural sciences through a strong academic journey. He earned his B.Tech in Agricultural Engineering from Mahatma Gandhi Chitrakoot Gramoday Vishwavidyalaya, Satna (2016), followed by an M.Tech in Agricultural Engineering (Soil and Water Conservation Engineering) from SHUATS, Prayagraj (2018). He further advanced his expertise by completing his Ph.D. in Soil and Water Conservation Engineering at G.B. Pant University of Agriculture and Technology, Pantnagar, in 2024, with a thesis focused on Runoff-Sediment-Nutrient Dynamics in the Gadarwara Watershed under Climate Change Scenarios and Best Management Practices. Throughout his career, Dr. Mishra has worked with esteemed institutions such as the Punjab Agricultural University, Ludhiana, the National Institute of Hydrology, Roorkee, and the Indian Institute of Technology, Roorkee, contributing to multiple national-level projects on watershed management, hydrology, and climate resilience in agriculture. He has authored numerous research papers, articles, and book chapters in reputed journals and publications, with key contributions in areas such as irrigation scheduling, evapotranspiration modeling, drought quantification, and sustainable crop management. His work has been recognized with Best Paper Awards (2021, 2023), as well as the Best Working Model Award (2023) in innovative digital agriculture. Beyond research, Dr. Mishra has participated in specialized workshops and trainings across India, gaining expertise in remote sensing, GIS, hydrological modeling, and machine learning applications in agricultural water management. He is proficient in multiple technical tools, including ArcGIS, QGIS, SWAT, HEC-RAS, SPHY, and CROPWAT, and is expanding his skills in Python, R, and Google Earth Engine. A recipient of the ICAR-SRF fellowship (AIR-23, 2018), Dr. Mishra continues to focus on advancing sustainable agricultural practices, with particular interest in climate-smart farming, irrigation efficiency, and watershed-scale conservation strategies. Through his academic, professional, and research contributions, Dr. Rajat Mishra is making a significant impact in the field of agricultural engineering, working toward a future where water and soil resources are managed sustainably for food security and environmental resilience.

**Hafiz Muhammad Mohsin Raza** is a Visiting Lecturer, Department of Agricultural Sciences Allama Iqbal Open University, Islamabad, Pakistan.

**Nalain E Muhammad** is a Visiting Lecturer at Department of Irrigation and Drainage University of Agriculture, Faisalabad, Pakistan.

**Mutaha Munir** is a Research Scholar, Institute of Botany, University of the Punjab Lahore,54000, Punjab Pakistan.

**Aisha Nazir** is an Associate Professor, Institute of Botany, University of the Punjab, Lahore.

**Elizabeth Oladejo** has a BSc (Hons) in Statistics (First Class Grade) from Abiola Ajimobi Technical University, Nigeria. She is skilled in R and Python Programming. Her area of expertise is Environmental Data Science and Computational Statistics

**Oladapo Oladoja** is a well-organised, self-motivated and confident statistician with the ability to impart knowledge, coupled with good computing skills in how to successfully assess and analyse data, as well as proffer recommendations to analyse data, with an in-depth knowledge of the R Programming Language. He had a Professional Diploma in Statistics, BSc in Statistics (First Class Hons), MSc in Environmental Statistics and PhD in Environmental Statistics (first to be awarded by the University of Ibadan). He is currently a Lecturer of Statistics at Abiola Ajimobi Technical University, Ibadan, Nigeria (Formerly First Technical University) through teaching, research, supervising and motivating students to attain their full potential. He is a member of IBS-GNi, TIES, ISBA, AMMNet and IMS. He has several publications in various journals. His area of interest is Bayesian Environmetrics, Generative AI, SpatioTemporal Modelling (Emission of Greenhouse Gases and Pollutants of all forms), Dilution Theory, Sports Modelling, Computational Statistics, Operations Research, Global Logistics and Supply Chain Management.

**A. Bennet Prabhu** is currently working with Congruent Solutions Ltd., where he contributes his expertise in advanced computing technologies and innovative problem-solving. Alongside his professional career, he has demonstrated a strong passion for research and development, publishing papers in reputed international journals. His research interests span across modern computational techniques and their practical applications in industry, with a focus on leveraging technology for efficient and scalable solutions. With both academic and industry contributions, he continues to bridge the gap between theoretical advancements and real-world implementations.

**Ramya R.** completed her B.E and M.E degree in Computer science and engineering. She had 14 years of teaching experience in various Engineering colleges. She did PhD in computer science and engineering and published many articles and book chapters in her domain. She specialises in edge computing, cloud computing, Machine learning, artificial intelligence, Quality of services and resource management.

**Amina Rashid** is a Lecturer at Department of Agronomy University of Agriculture, Faisalabad.

**Fahd Rasul** is an Associate Professor at Department of Agronomy University of Agriculture, Faisalabad.

**Sudhakar Reddy** completed his UG,PG and Ph.D in reputed institutes of Andhra pradesh, He has published many papers, he has more experience in teaching field.

**Pushpalatha S.** is Professor, Department of Computer Science and Engineering, Saveetha School of Engineering, Saveetha Institute of Medical and Technical Sciences, Chennai, Tamil Nadu, India. Specializes in Security and Machine Learning with 18 years of experience.

**Padmini S** completed her UG in Adhiparasakthi engineering college at 2007 and M.E in Anna university affiliated college, She published research papers and conferences.

**Naeem Saddique** is an Assistant Professor at Department of Irrigation and Drainage, University of Agriculture, Faisalabad 38000, Pakistan.

**Muhammad Safdar** is a Research Assistant at Agricultural Remote Sensing Lab (ARSL), University of Agriculture Faisalabad Pakistan.

**Santhi V.** has received her Ph.D. in Computer Science and Engineering from VIT University, Vellore, India. She has pursued her M.Tech. in Computer Science and Engineering from Pondicherry University, Puducherry. She has received her B.E. in Computer Science and Engineering from Bharathidasan University, Trichy, India. Currently she is working as Associate Professor in the School of Computing Science and Engineering, VIT University, Vellore, India. She has authored many national and international journal papers and one book. She is currently in the process of editing two books. Also, she has published many chapters in different books published by International publishers. She is senior member of IEEE and she is holding membership in many professional bodies like CSI, ISTE, IACSIT, IEEE and IAENG. Her areas of research include Image Processing, Digital Signal Processing, Digital Watermarking, Data Compression and Computational Intelligence.

**Muhammad Adnan Shahid** is a Professor at Agricultural Remote Sensing Lab (ARSL), University of Agriculture, Faisalabad 38000, Pakistan.

**Imran Shauket** is a Lecturer at Department of Structures and Environmental Engineering University of Agriculture, Faisalabad 38000, Pakistan.

**Sitesh Kumar Singh** is an accomplished academic and Chartered Engineer with a Ph.D. in Civil Engineering and over nine years of diverse experience in education, research, consultancy, and educational administration. Currently serving as an Assistant Professor in the Department of Civil and Environmental Engineering at the National University of Science and Technology, Muscat, Sultanate of Oman. Dr. Singh holds multiple advanced degrees including M.Tech, B.Tech, PGDIPR, PGDUPDL and PGDE. His career encompasses curriculum development, research supervision, quality assurance, and strategic academic planning. He has supervised over 40 postgraduate and 2 doctoral students, contributing to academic leadership through workshops, conferences, and editorial roles, and has published over 5 books, 10 book chapters, 58 research articles & conference presentations, and 7 patents.

**R Swathy** (Swathy Ramasamy) received B.E. Information Technology from Periyar Maniammai College of Technology for Women affiliated to Bharathidasan University, Thanjavur, Tamil Nadu in 2004 and M.Tech. Information Technology from Sathyabama University, Chennai, Tamil Nadu in 2006. She holds a Ph.D. in Information and Communication from Anna University, where she developed a deep passion for research in the domain of Cloud Computing. Her current research interests include Cloud Computing, Data Analytics and IoT and she is dedicated to pushing the boundaries of knowledge in these areas. In addition to her research, Swathy has over 17 years of teaching experience in a reputed university, and she is currently working as Assistant Professor in SRM University in the Department of Networking and Communications.

# Index

## A

Analytic Hierarchy Process 23, 30, 64, 86, 89
Anthropogenic 1, 5, 35, 36, 37, 54, 56, 57, 61, 89, 116, 230, 243, 245, 266
Artificial intelligence 1, 23, 87, 119, 120, 121, 122, 124, 126, 127, 128, 135, 136, 138, 139, 140, 141, 142, 144, 147, 148, 149, 150, 151, 155, 160, 161, 163, 164, 165, 167, 168, 169, 172, 173, 175, 197, 201, 202, 203, 205, 213, 222, 223, 224, 225, 226, 227, 248, 272, 281, 298, 300

## C

CHIRPS 67, 196
Climate change 1, 21, 26, 27, 35, 36, 56, 60, 61, 62, 64, 65, 85, 88, 89, 102, 116, 120, 121, 122, 144, 148, 149, 150, 151, 153, 155, 169, 170, 171, 172, 173, 174, 196, 202, 203, 205, 214, 222, 223, 228, 229, 230, 244, 246, 247, 248, 249, 256, 258, 263, 264, 282
Coastal Erosion 231, 232, 244, 246, 249
Coastline 114, 206, 229, 230, 231, 232, 233, 234, 235, 236, 237, 243, 244, 245, 246, 247, 249
Compact Cities 251, 256

## E

Ecosystem 7, 36, 53, 54, 57, 59, 60, 92, 98, 100, 101, 104, 113, 139, 149, 177, 219, 230, 251, 262, 263, 264, 266, 268, 276
Environmental Monitoring 59, 85, 118, 198, 201, 203, 204, 214, 215, 221, 222, 224, 226
Environmental Sustainability 251, 256, 264, 267, 268, 269, 275

## F

Flood Hazard 2, 10, 19, 24, 27, 30, 65, 70, 73, 84, 86, 87, 88, 89, 226
Flood mapping 4, 22, 84, 163, 164
Flood Risk Assessment 2, 63, 65, 71, 83, 85, 89, 164
Frequency Ratio 63, 64, 88, 89

## G

Geo-Spatial Technologies 252, 253, 254, 259, 268, 271
GIS 1, 2, 4, 5, 18, 19, 28, 30, 32, 59, 84, 87, 88, 89, 97, 98, 116, 118, 163, 164, 165, 168, 170, 172, 198, 202, 203, 205, 206, 208, 209, 214, 224, 227, 229, 232, 236, 237, 251, 252, 253, 254, 256, 259, 260, 261, 262, 265, 267, 268, 270, 271, 274, 275, 276
GIS and Remote Sensing 116, 262
GNSS 120, 124, 127, 141

## H

Hybrid method 65, 281
Hydrological Modeling 10, 58, 150, 153, 154, 164, 165, 182

## L

Landsat 8 35, 40, 41, 69, 94, 96, 97, 98
Linear Regression 13, 201, 212, 213, 215, 227
Logistic Regression 63, 64, 86, 87, 88, 259

## M

Machine Learning 11, 12, 14, 17, 18, 19, 20, 23, 60, 64, 71, 82, 83, 84, 85, 88, 90, 117, 121, 122, 124, 127, 129, 134, 135, 136, 141, 142, 143, 144, 145, 148, 150, 153, 160, 161, 162, 163, 164, 168, 169, 171, 172, 173, 175, 176, 177, 178, 179, 180, 182, 185, 188, 194, 196, 197, 198, 199, 200, 201, 202, 203, 204, 205, 207, 210,

211, 212, 213, 214, 215, 222, 223, 224, 225, 226, 227, 232, 247, 260, 261, 272, 273, 275, 279, 280, 281, 283, 286, 291, 292, 293, 294, 297, 298, 300, 318
Mean Absolute Error 196, 201, 212, 213

## N

Nepal 63, 64, 65, 66, 85, 86, 87, 88, 89

## P

Precision Agriculture and Artificial Intelligence 119, 122
Precision Irrigation 121, 143, 176, 177, 179, 180, 193, 195

## Q

Quantum-Inspired Algorithms 277, 280, 304

## R

Random Forest 12, 63, 64, 86, 89, 135, 176, 177, 178, 182, 183, 186, 188, 192, 193, 195, 199, 201, 212, 213, 215, 216, 227, 260, 274, 284, 291
Remote Sensing 1, 2, 18, 19, 26, 28, 30, 35, 36, 41, 58, 59, 60, 61, 62, 86, 87, 88, 89, 90, 96, 97, 98, 112, 113, 114, 116, 117, 118, 119, 120, 121, 123, 130, 131, 135, 143, 144, 145, 147, 148, 150, 151, 153, 154, 162, 163, 164, 165, 168, 169, 171, 173, 175, 176, 177, 178, 179, 180, 193, 194, 197, 198, 199, 201, 202, 203, 204, 205, 218, 221, 222, 223, 224, 225, 227, 232, 233, 245, 247, 248, 251, 252, 253, 254, 256, 259, 260, 261, 262, 265, 267, 268, 270, 271, 274, 275, 276, 280, 282, 286, 297, 298, 300
Root Mean Square Error 178, 185, 188, 192, 196, 212, 213

## S

Smart Agriculture 137, 142, 175, 177, 180, 195
Smart Growth 251, 269, 270, 271, 275
Soil Moisture Prediction 175, 176, 177, 178, 180, 188, 192, 194, 198, 200
Sustainable Farming 119, 120, 122, 128, 140

## U

Urban Planning 203, 204, 205, 217, 224, 254, 258, 260, 261, 262, 263, 264, 268, 269, 271
Urban Sprawl 203, 251, 252, 253, 254, 255, 256, 257, 258, 259, 260, 261, 262, 263, 264, 265, 266, 267, 268, 269, 270, 271, 275

## W

Water Resource Management 58, 98, 142, 148, 149, 150, 151, 164, 168, 172, 173, 176, 177, 178
Water Resources 8, 24, 25, 26, 28, 33, 36, 73, 89, 90, 92, 94, 104, 116, 147, 148, 149, 151, 153, 155, 156, 160, 161, 163, 165, 167, 168, 170, 172, 173, 174, 198, 199, 252, 262, 263
West Rapti River 65, 88
Wind Rose Analysis 209, 218
Wind Speed 125, 201, 204, 207, 208, 210, 211, 212, 213, 215, 216, 217, 219, 220, 225, 228

www.ingramcontent.com/pod-product-compliance
Lightning Source LLC
LaVergne TN
LVHW080906300926
849162LV00003B/577